VOLUME EIGHTY TWO

ADVANCES IN BOTANICAL RESEARCH

How plants communicate with their biotic environment

ADVANCES IN BOTANICAL RESEARCH

Series Editors

Jean-Pierre Jacquot
Professor, Membre de L'Institut Universitaire de France, Unité Mixte de Recherche INRA, UHP 1136 "Interaction Arbres Microorganismes", Université de Lorraine, Faculté des Sciences, Vandoeuvre, France

Pierre Gadal
Honorary Professor, Université Paris-Sud XI, Institut Biologie des Plantes, Orsay, France

VOLUME EIGHTY TWO

ADVANCES IN BOTANICAL RESEARCH

How plants communicate with their biotic environment

Volume Editor

GUILLAUME BECARD

Laboratoire de Recherche en Sciences Végétales (LRSV), UMR 5546 CNRS/Université Paul Sabatier, Castanet-Tolosan, France

ELSEVIER

ACADEMIC PRESS

An imprint of Elsevier

Academic Press is an imprint of Elsevier
125 London Wall, London EC2Y 5AS, United Kingdom
The Boulevard, Langford Lane, Kidlington, Oxford OX5 1GB, United Kingdom
50 Hampshire Street, 5th Floor, Cambridge, MA 02139, United States
525 B Street, Suite 1800, San Diego, CA 92101-4495, United States

First edition 2017

Notices

Knowledge and best practice in this field are constantly changing. As new research and experience broaden our understanding, changes in research methods, professional practices, or medical treatment may become necessary.

Practitioners and researchers must always rely on their own experience and knowledge in evaluating and using any information, methods, compounds, or experiments described herein. In using such information or methods they should be mindful of their own safety and the safety of others, including parties for whom they have a professional responsibility.

To the fullest extent of the law, neither the Publisher nor the authors, contributors, or editors, assume any liability for any injury and/or damage to persons or property as a matter of products liability, negligence or otherwise, or from any use or operation of any methods, products, instructions, or ideas contained in the material herein.

ISBN: 978-0-12-801431-8
ISSN: 0065-2296

Publisher: Zoe Kruze
Acquisition Editor: Kristen Shankland
Editorial Project Manager: Thomas Van Der Ploeg
Production Project Manager: Magesh Kumar Mahalingam
Designer: Vicky Pearson Esser

Typeset by TNQ Books and Journals

CONTENTS

CONTRIBUTORS

G. Arimura
Tokyo University of Science, Tokyo, Japan

M. Barret
IRHS, INRA Site d'Angers Beaucouzé, Beaucouzé, France

T.J. Baum
Iowa State University, Ames, IA, United States

J.D. Blande
University of Eastern Finland, Kuopio, Finland

G. Brun
Université de Nantes, Nantes, France

G. Chiapusio
Université de Bourgogne Franche-Comté, UMR CNRS 6249, Montbéliard, France

P. Delavault
Université de Nantes, Nantes, France

T. Fort
BIOGECO, INRA, University of Bordeaux, Pessac, France

E. Gaulin
Laboratoire de Recherche en Sciences Végétales, Université de Toulouse, CNRS, UPS, France

R. Geurts
Wageningen University, Wageningen, The Netherlands

L. Gilbert
James Hutton Institute, Aberdeen, United Kingdom

T. Hewezi
University of Tennessee, Knoxville, TN, United States

R. Holmer
Wageningen University, Wageningen, The Netherlands

D. Johnson
University of Aberdeen, Aberdeen, United Kingdom

W. Kohlen
Wageningen University, Wageningen, The Netherlands

S. Latif
Charles Sturt University, Wagga Wagga, NSW, Australia

P. Lemanceau
Agroécologie, AgroSup Dijon, INRA, University of Bourgogne Franche-Comté, Dijon, France

K. Lunau
Heinrich-Heine-University, Düsseldorf, Germany

S. Mazurier
Agroécologie, AgroSup Dijon, INRA, University of Bourgogne Franche-Comté, Dijon, France

S. Mondy
Agroécologie, AgroSup Dijon, INRA, University of Bourgogne Franche-Comté, Dijon, France

G. Montiel
Université de Nantes, Nantes, France

I.S. Pearse
Illinois Natural History Survey, Champaign, IL, United States

B. Pivato
Agroécologie, AgroSup Dijon, INRA, University of Bourgogne Franche-Comté, Dijon, France

J.-B. Pouvreau
Université de Nantes, Nantes, France

L. Rutten
Wageningen University, Wageningen, The Netherlands

C.-M. Ryu
Korea Research Institute of Bioscience and Biotechnology, Daejeon, South Korea; University of Science and Technology, Daejeon, South Korea

R. Sharifi
Razi University, Kermanshah, Iran

P. Simier
Université de Nantes, Nantes, France

S. Thoiron
Université de Nantes, Nantes, France

C. Vacher
BIOGECO, INRA, University of Bordeaux, Pessac, France

R. van Velzen
Wageningen University, Wageningen, The Netherlands

P. Wester
Heinrich-Heine-University, Düsseldorf, Germany

L.A. Weston
Charles Sturt University, Wagga Wagga, NSW, Australia

PREFACE

As photosynthetic organisms that convert solar energy into chemical energy, plants are a primary source of organic compounds of all sorts and the object of a great covetousness for a myriad of biotrophic organisms. They represent a source of food for countless herbivorous, pollinating, parasitic, pathogenic and symbiotic organisms. These organisms are insects, nematodes, plants, bacteria, fungi and oomycetes to only cite the prominent ones. With no eyes, ears, nose and motility, plants would be most vulnerable if they had not developed sophisticated ways to perceive, recognize and manage the presence of all these swarming biotrophic organisms surrounding them. Upon all these biotic stimuli, plants know how to find the proper responses consisting in modifying their metabolism to produce repellent compounds and triggering their immune system against foes, or consisting in unrolling the red carpet out to attract and accommodate beneficial friends. This book provides our most recent knowledge on how plants communicate with their biotic environment, not only to safeguard their integrity but also to develop win–win cooperation. It shows that the chemical identity of the molecular signals, the pathways of their perception and transduction, the plant responses as well as the importance of these signalling processes from an ecological and evolutionary perspective are increasingly being unveiled. The book includes twelve chapters organized in three main parts that describe how plants communicate with other plants, with microbes and with animals.

Plant–plant signalling involves specific secondary metabolites, including volatile organic chemicals (VOCs), that are produced by plants under biotic stress. Then eavesdropping neighbouring plants can adjust their defences and increase their fitness after perceiving these VOCs (Chapter 1). Other secondary metabolites or allelochemicals can also be constitutively released by some plants and have a negative effect on neighbouring plants (Chapter 2), thus playing a role in plant competition and vegetation dynamics. In more intimate plant–plant interactions, like in root parasitic interactions, the plant parasite must recognize very specific host-released molecules to complete its life cycle (Chapter 3). Plants interconnected by the mycelium of symbiotic fungi, the common mycorrhizal network, can also exchange information. The fungus can transport warning signals produced by plants under biotic stress to neighbouring plants before they are themselves attacked (Chapter 4).

Plants also communicate with all kinds of different microbes. Some of them, mainly composed of bacteria, fungi and oomycetes, represent plant microbiota and live on the surface of seeds, roots and leaves. Their community is shaped by various trophic interactions and exchange of specific molecules (Chapter 5). Others are rhizobacteria living in the rhizosphere. They establish a close relationship with plant roots and play important roles for promoting plant growth and activating plant defences (Chapter 6). Less frequent but harmful microbes can also invade plant tissues and cause severe diseases. In a complex arms race, plants have learned to combat these pathogens while these latter have developed specific proteic weapons (effectors) to circumvent plant defence mechanisms (Chapter 7). To improve their nutrient acquisition, plants also establish symbiotic relationship with highly beneficial microbes such as nitrogen-fixing bacteria and mycorrhizal fungi. Intimate signalling pathways take place here. These pathways show striking commonalities as a result of a long history of plant–microbe coevolution (Chapter 8).

The interactions between flowering plants and their animal pollinators such as flies, bees, beetles, birds, bats and others are based on tricky communication involving visual, olfactory, gustatory or tactile signals that can be honest but also mimicked and deceptive (Chapters 9 and 10). Constitutively or after induction, plants also emit volatile chemicals which provide cues to repulse or attract herbivores. Plants can also perceive specific insect pheromones and modify their defence response (Chapter 11). Combinations of volatile, nonvolatile and chemotactic compounds originating from roots may also constitute long- and short-distance attractants to parasitic nematodes. Inversely nematodes produce specific chemicals and effectors that modify plant development and impede plant defence.

By describing the multiple and subtle ways by which plants interact with their biotic environment, this book addresses a fundamental aspect of plant biology and ecology. It has been conceived to target researchers specialized in one field of plant communication and interested to learn about other fields of plant communication. It is also for nonspecialist biologists and ecologists interested in plant biology such as biology/ecology teachers or students. Its originality is that the contributors, who are renowned scientists, belong to distinct scientific communities having different expertises in ecology, evolutionary biology and molecular biology. It addresses a topical subject as the natural biological processes described represent basic knowledge to develop low input sustainable agriculture.

Guillaume Bécard

PART ONE

Plant-Plant Communication

CHAPTER ONE

From the Lab Bench to the Forest: Ecology and Defence Mechanisms of Volatile-Mediated 'Talking Trees'

G. Arimura*[,1], I.S. Pearse[§]
*Tokyo University of Science, Tokyo, Japan
[§]Illinois Natural History Survey, Champaign, IL, United States
[1]Corresponding author: E-mail: garimura@rs.tus.ac.jp

Contents

Abstract

Transfer of information is important for almost all biotic interactions, but has received less attention in plants. Volatile organic chemicals (VOCs) emitted from floral tissues and green tissues of plants affect how plants interact with each other (informally termed 'plant communication'), with mutualists such as pollinators and with enemies such as herbivores. In this chapter, we give an overview of VOC-based plant communication, in which plants that eavesdrop on VOC signals emitted from neighbouring herbivore-damaged plants increase their defence. While historically controversial, recent research has begun to establish the details of these interactions in various plant systems, including the identity of the VOC signal, the mechanism of VOC perception and the transduction pathways that link VOC perception to a defensive response. At the same time, ecological studies have begun to establish the evolutionary drivers of plant communication. Key insights from these studies are that these responses are kin-specific, memorable and mediated by known internal defence signals and epigenetic

Advances in Botanical Research, Volume 82
ISSN 0065-2296
http://dx.doi.org/10.1016/bs.abr.2016.08.001

regulations in plant cells. Moving forward, studies that connect our increasing understanding of the mechanisms of plant–plant communication with their ecological consequences will help determine the importance of this type of defensive induction as well as the targets of selection within the plant–plant communication apparatus.

1. INTRODUCTION

Plants release volatile organic chemicals (VOCs, Fig. 1) into the atmosphere for many purposes, including the attraction of mutualist animals, resisting environmental stress and directly controlling plant pests. One class of plant-emitted VOCs is released only upon attack by herbivores. The adaptive value of emitting these VOCs has focused on interactions between the plant, its herbivores and the herbivore's natural enemies. Substantial work has demonstrated that plants benefit from VOC emission post herbivore attack for several reasons. Herbivore-induced VOCs can directly repel herbivores such as ovipositing butterflies and host-seeking aphids (Unsicker, Kunert, & Gershenzon, 2009). VOCs can attract natural enemies of herbivores, including predators and parasitic wasps (Arimura, Matsui, & Takabayashi, 2009). Moreover, several plant species are able to eavesdrop on

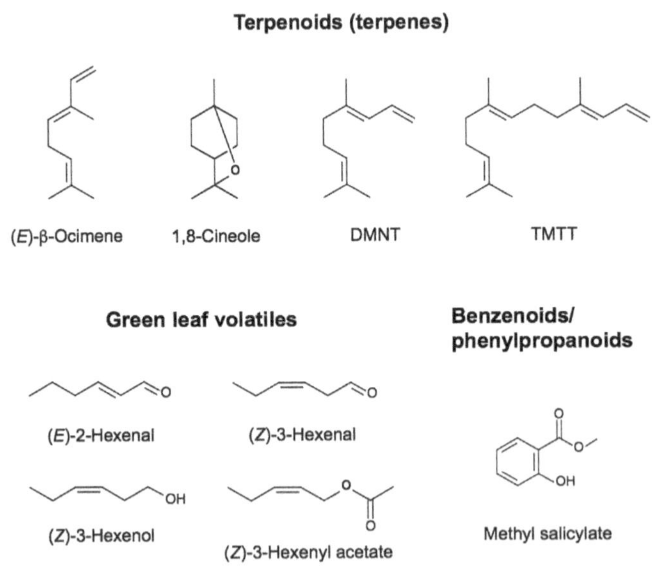

Figure 1 Representative volatile organic chemicals serving as signalling cues for plant–plant communications. *DMNT*, (*E*)-4,8-dimethyl-1,3,7-nonatriene; *TMTT*, (*E*,*E*)-4,8,12-trimethyltrideca-1,3,7,11-tetraene.

such VOC cues emitted from infested plants to adjust their defences to fit their current or future risk of herbivory and to increase their fitness as a result. For instance, a classic study reported that undamaged poplar and sugar maple trees can respond to VOC infochemicals sent from their damaged neighbours, as indicated by the fact that undamaged trees in the neighbourhood of conspecific damaged trees obtain defensive properties following herbivore attack of their neighbours (Baldwin & Schultz, 1983). Those plant behaviours have been dubbed plant–plant communication or between-plant communication (though these terms deviate from strict behavioural definitions of communication and signals, see later) and are popularly called 'talking trees', as told in fairy tales.

Plant–plant signals have had a rocky history in the scientific literature, but have become an accepted component of plant biology in recent years. This acceptance is due to extensive laboratory studies geared at understanding the mechanism of release and perception of volatile signals by plants as well as field studies that have sought to understand the evolutionary drivers of these plant behaviours. Here, we summarize the physiological laboratory studies and the ecological field studies that have provided convincing evidence of the mechanisms and consequences of plant–plant signals. We further suggest future directions for research that combines a better understanding of plant–plant signalling mechanisms with field studies to uncover the role of plant–plant signals for plant–herbivore interactions.

2. PLANTS BIOSYNTHESIZE AND EMIT VOLATILE ORGANIC CHEMICALS IN RESPONSE TO HERBIVORY: HOW DO PLANT–PLANT SIGNALS EMERGE?

Plant–plant signals are one of the roles of the diverse suite of VOC metabolites that plants emit. Hundreds of millions of tons of VOCs, consisting mostly of monoterpenes and isoprene, are expelled into the atmosphere annually by terrestrial plants (Guenther et al., 1995). In contrast to VOCs emitted constitutively from plants, higher plants induce the emission of VOCs in response to biotic stresses such as herbivore damage. Here we outline the physiological pathways that mediate VOC production and release.

Recent advances in metabolome analysis, in combination of biologic assays, allow the characterization of herbivore-induced plant volatiles (HIPVs). HIPVs consist mainly of terpenes, green leaf volatiles (GLVs) and phenylpropanoids (Fig. 1). Among HIPVs, terpenes contain the highest proportion of metabolite diversity. The plant-derived terpenes are originally derived from the mevalonate pathway and the 2-*C*-methyl-D-erythritol 4-phosphate pathway. The structural diversity of volatile terpenes is predominantly achieved with catalysis caused by enzymes in the terpene synthase (TPS) family. The diversity of the large family of TPSs is believed to originate from repeated duplication and the functional divergence of an ancestral TPS that is involved in plant primary metabolism, for example, hormone biosynthesis (Trapp & Croteau, 2001).

Volatile benzenoids and phenylpropanoids, which contain benzene ring(s), are derived from the shikimate pathway and can be categorized into different subclasses depending on their basic structures (C6—C1: methyl salicylate, benzaldehyde; C6—C2: 2-phenylethanol, phenylacetaldehyde; C6—C3: eugenol, chavicol). The structural diversity of those compounds is predominantly achieved via side-chain diversity caused by the actions of methyltransferases, acetyltransferases and the other species-specific modification enzymes (Arimura, Matsui, Koeduka, & Holopainen, 2016). Accordingly, benzenoids and phenylpropanoids are often specific to family, genus or species of plants.

GLVs are formed from lipids through an oxygenation reaction. In comparison to terpenes, benzenoids and phenylpropanoids, GLVs lack structural diversity. (*Z*)-3-Hexenal, (*E*)-2-hexenal (*Z*)-3-hexen-1-ol and (*Z*)-3-hexen-1-yl acetate are present as the major GLVs in terrestrial plants, whereas there are minor products derived from modifications of the above basic structures of GLVs.

VOC production is elicited directly by damage and also by recognition of particular herbivores. From the standpoint of animals and plants that perceive VOCs emitted by a plant, this means that different volatile metabolites might have different meanings as information. Whether in response to a particular herbivore or as a generalized damage response, the signalling molecule jasmonate (JA) and repressor proteins of the jasmonate ZIM domain (JAZ) are typically involved in the induction of most HIPVs (Okada, Abe, & Arimura, 2015). When the plant responds to a particular herbivore, recognition of that herbivore is thought to be achieved via elicitor molecules from herbivore saliva (e.g., the fatty acid—amino acid conjugate volicitin). Such elicitors can positively regulate the signalling pathway involved in JA and its derivatives (Schmelz, Engelberth, Alborn, Tumlinson, & Teal, 2009). JAs can synergistically crosstalk with other phytohormones (e.g., ethylene and salicylic acid) to play a pivotal role in regulating the biosynthesis of herbivore-induced plant defence responses, including the biosynthesis of HIPVs (Okada et al., 2015). The details have not been elucidated but some other, unknown elements, may also be concomitantly involved in the orchestration of HIPV blends, because a few members of master signalling molecules such as JAs are not sufficient to explain how plants induce specific blends of HIPVs that depend on the herbivore species mounting the attack. For instance, lima bean plants emit different blends of HIPVs when attacked by beet armyworm (*Spodoptera exigua*), common armyworm (*Mythimna separata*), or two-spotted spider mite (*Tetranychus urticae*) (Ozawa, Arimura, Takabayashi, Shimoda, & Nishioka, 2000). Making such herbivore-specific blends allows damaged plants to attract the specialist natural enemy of each herbivore, thereby leading to specific tritrophic interactions consisting, for instance, of lima bean plants – *T. urticae* – *Phytoseiulus persimilis* the specialist predators of the spider mites.

HIPVs are also transmitted systemically from damaged parts to undamaged parts within an individual plant. Systemic induction allows leaves distant from a current herbivore attack to elicit defence response before the aggregation of the pests from damaged area. One hypothesis as to the origin of plant–plant signals is that these systems were originally used as within-plant signals in plant species with poor vascular connections. Such a hypothesis could indicate an ancient origin of such signalling mechanisms, as ancient plant lineages lack efficient vascular connections. Within-plant signalling via HIPVs was first described in hybrid poplar (*Populus deltoides* x *nigra*) damaged by gypsy moth larvae (*Lymantria dispar* L.) (Frost et al., 2007). Subsequently, it has been reported that HIPVs, released from lima bean leaves damaged by the herbivorous leaf beetles, *Cerotoma ruficornis* and *Gynandrobrotica guerreroensis* (Chrysomelidae), prime the secretion of extrafloral nectar that attracts predatory arthropods (e.g., ants) in undamaged parts (Heil & Silva Bueno, 2007), and HIPVs released from sagebrush and willow induce defence in clonal stands that lack vascular connection (Karban & Shiojiri, 2009; Pearse, Hughes, Shiojiri, Ishizaki, & Karban, 2013).

Since both within- and between-plant communications via VOCs occur only at short distances (~60 cm) (Heil & Adame-Alvarez, 2010; Karban, Shiojiri, Huntzinger, & McCall, 2006), within-plant communication may be the most important consequence of HIPV perception by a plant, except in cases where neighbouring plants are very close together. In summary, HIPV emission has likely evolved as a means to strengthen mutualisms between plants and natural enemies of herbivores, and to achieve within-plant signalling without relying on vascular connections. By chance, the neighbour plants consequently eavesdropped on HIPV cues, thus initiating behaviour of between-plant communications (Fig. 2).

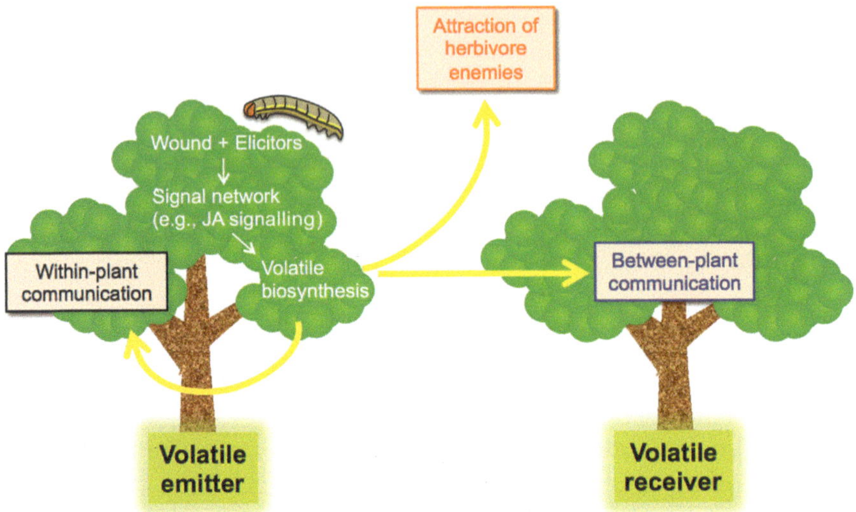

Figure 2 Ecological communication networks via herbivore-induced plant volatiles (HIPVs). HIPVs enable to promote an array of communications with herbivore enemies, uninfested parts of leaves within the individual plants and neighbouring plants. *JA*, jasmonate.

3. LAB-BASED ADVANCES IN PLANT—PLANT COMMUNICATION

Plant—plant communication has become an accepted phenomenon for scientists, but there are many questions left unanswered. In particular, only little is known about the mechanisms for VOC reception. A partial mechanism for VOC reception has been explored in tomato plants exposed to (*Z*)-3-hexenol emitted from *Spodoptera litura*-infested tomato plants. The receiver plants accumulated the volatile compound in its glycosylated form ((Z)-3-hexenyl-vicianoside) in the leaf cells (Sugimoto et al., 2014). It remains unclear whether the alcoholic GLV is chemically modified by the plant simply as a means of detoxification, whether the GLV product is stored and used directly by the plant or whether chemical modification is part of the mechanism by which VOC molecules elicit a defensive response in these plants. In fact, nothing is known about both reception and transport systems of VOCs, except for fragments of information such as the fact that VOCs from peppermint essential oil decrease plant root plasma membrane potential, thus increasing membrane depolarization (Maffei, Camusso, & Sacco, 2001). Changes in the bioelectric potential of cells imply changes in the flux of ions across the plasma membrane. Once VOCs cross the lipophilic plasma

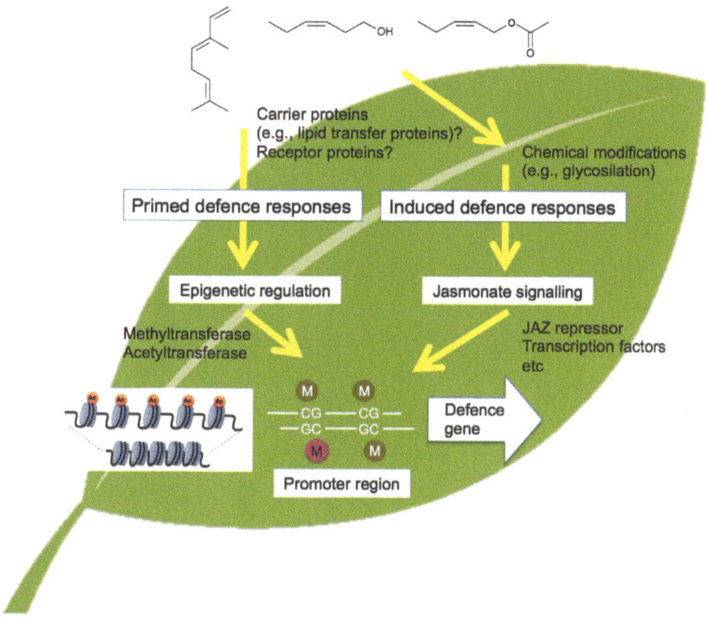

Figure 3 Possible mechanisms underlying volatile organic chemical (VOC) reception in eavesdropping plants. When the VOC signal is strong enough, plants may trigger induced plant defence responses. Otherwise, when the VOC signal is not sufficient to induce plant defence responses, plants are primed to respond more strongly to future damage such as with greater activation of defence genes when subsequently challenged by pest attack. The possible mechanisms underlying induced or primed defence responses are illustrated.

membrane, they must traverse the hydrophilic cell wall, and then influence membrane potential. Small carrier proteins, such as extracellular lipid transfer proteins, may be required to prevent repartitioning of VOCs into the plasma membrane (Widhalm, Jaini, Morgan, & Dudareva, 2015) (Fig. 3).

In some cases, receiver plants do not induce defence responses immediately after exposed to HIPVs, but respond more strongly and more rapidly than nonreceiver plants upon damage by herbivores. This ecological phenomenon is referred to as 'priming'. Priming may provide a realistic function of plant–plant communications in the fitness of plants in natural habitats, because HIPVs can be emitted by damaged plants for only the limited period of days following attack by herbivores (Arimura, Huber, & Bohlmann, 2004). However, receiver plants cannot be aware of how much later herbivores will arrive and how long they invest in a potentially costly defence. Priming of defences may be regarded as a rudimentary system of plant 'memory' in which reception of a volatile cue may affect defensive responses much later in the future. The function of priming systems is an area of very active research. Recently, it was found that maize receiver plants exposed to HIPVs elicited by plants damaged by the herbivore *M. separata* had 'memory' of the HIPV signal for at least 5 days after exposure (Ali, Sugimoto, Ramadan, & Arimura, 2013). In some cases, 'long-term memory' of herbivore attack may be caused by epigenetic regulation of genes. For example, in the promoter region of a defence gene (Bowman-Birk type trypsin inhibitor), nonmethylated cytosine residues were observed in the genome of plants exposed to HIPVs more frequently than in the genome of plants exposed to the volatiles of uninfested plants. Epigenetic regulation of genes, according to DNA methylation and histone posttranslational modifications, can often alter the switch of gene transcriptional mode, i.e., activation or inactivation (Jaskiewicz, Conrath, & Peterhänsel, 2011). These systems may underlie plants' 'memory' of volatile signals via plant–plant communications (Fig. 3). The length of memory of an attack that could be caused by epigenetic modifications is consistent with patterns observed in the field. For example, it has been shown that herbivore resistance of sagebrush (*Artemisia tridentata*) plants increased for about 2 months after exposure to VOCs (Karban & Shiojiri, 2009; Karban, Shiojiri, & Ishizaki, 2010).

4. THE EVOLUTION OF PLANT–PLANT SIGNALS: HAS PLANT–PLANT SIGNALLING SELECTED FOR EMISSION AND PERCEPTION OF PARTICULAR VOLATILE ORGANIC CHEMICALS?

For decades, the adaptive benefits of plant–plant signals have been debated. This debate is closely connected to how behavioural ecologists define signals (Karban, 2015). For a trait to evolve as a signal, it must benefit both the organism that emits the signal as well as the organism that receives it (Pearse & Mescher, 2016; Scott-Phillips, 2008). For receiver plants, it is likely that the perception of a plant–plant signal provides a forecast of future risk of herbivory, and risk assessment can allow plants to optimally defend against attack (Orrock et al., 2015).

It is less clear what benefit a plant might gain from emitting a signal that is perceived by its neighbour. Research on plant–plant signals has taken two paths on this subject. First, research has explored the possibility that plant–plant signals may have evolved to coordinate defences within a single plant or between close relatives, as described above. In this case, the plant–plant signal does benefit the emitter either directly by a coordination of its own defences or indirectly via the inclusive fitness benefits of protecting its close relatives (Frost et al., 2007; Karban, Shiojiri, Ishizaki, Wetzel, & Evans, 2013). Second, it is possible that there is no selective benefit for a plant to emit VOCs that are perceived by its neighbours, such that the neighbours simply 'eavesdrop' on VOCs that are emitted for other purposes. In this case, from a strict behavioural perspective, VOCs emitted and perceived by neighbouring plants are rarely 'signals', because they do not benefit the emitter, though we continue to use this term since it has been widely used in referring to this process (Heil & Karban, 2010).

What evidence exists that plant–plant signalling might have evolved as a means of signalling within a single plant or between close relatives? In poplar, suppression of VOC signals decreased the defensive response in leaves that had little vascular connection (Frost et al., 2007), suggesting that VOCs can be a potent means of self-signalling. Additionally, VOC signals passed between different individuals of 'self' genotypes of a plant have been shown to induce defence to a greater degree than VOCs from nonself in sagebrush (Karban & Shiojiri, 2009) and willow (Pearse et al., 2013). In sagebrush, VOCs from a closely related plant induced defence to a greater degree than VOCs from an unrelated plant (Karban et al., 2013), probably because close relatives tend to emit similar VOC compositions (Karban, Wetzel, et al., 2014). On the contrary, self-signalling was not more effective at inducing defence than nonself signals in three annual forbs (Pearse et al., 2012). Interestingly, while evidence from a few systems has supported the notion that self- or close-relative signals are more effective at inducing defence than nonself signals, this is not actually a requirement for the evolution of plant–plant signalling via coordination of defences within a plant because a volatile signal could be effective both within a plant as well as between individuals. These studies actually support a more complicated scenario in which it may be costly to provide or receive a signal between nonself and unrelated individuals. Why this would be is still unclear, though competition between neighbouring plants or aspects of associational resistance may be reasons to hide or ignore signals from neighbours.

5. THE ECOLOGICAL CONSEQUENCES OF PLANT–PLANT SIGNALS: DO PLANT–PLANT SIGNALS MATTER COMPARED TO OTHER TYPES OF DEFENCE?

From a broader ecological standpoint, it would be interesting to know how much plant–plant signals actually contribute to plant defence. In this case, we might care less about the evolutionary pathway that has resulted in a plant—plant signal, but focus more on the extent of the consequences of that signal. Plants possess a huge arsenal of defences against herbivores. Many of those defences are expressed constitutively or change in a predictable way with the development of a plant tissue. Others are induced following damage to the plant or following feeding by a particular herbivore (Felton & Tumlinson, 2008; Karban & Baldwin, 1997). How important are defences induced by plant—plant signals compared to defences with other patterns of expression?

In general, plants tend to coordinate their defences in a conservative way; plants often invest in defence even in situations with little risk of herbivory (Orrock et al., 2015). Do plants really gain that much by inducing defences in anticipation of damage by their herbivores? And if so, why not invest in these defences all the time? Induction due to plant—plant signals should only benefit a receiver plant under a relatively narrow set of assumptions about the costs of defences and the way in which herbivore pressure changes over space (i.e., between plants) and over time (Fig. 4). For example, if there are few costs of defence, then a receiver plant would be better off investing in defence all the time. Likewise, if herbivores are extremely mobile, such that damage at time T does not predict the risk of damage at time T+1, then there would be little use in inducing defence at a later time based on

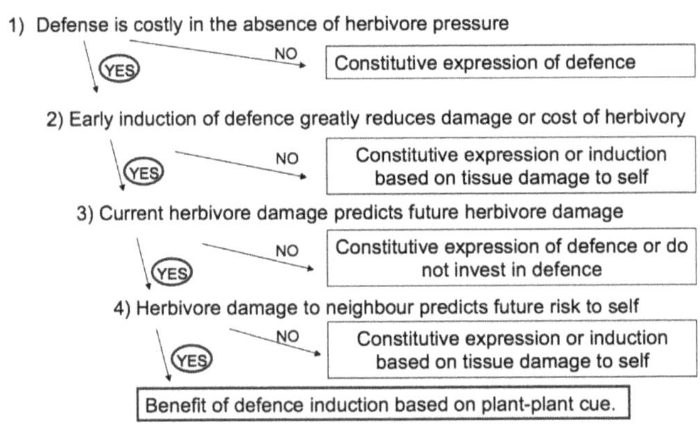

Figure 4 A schematic of the contexts in which perception of a damage induced plant—plant signal could benefit the receiver plant by providing information that allows for early induction or priming of a defence. If each of the ecological requirements is met (a 'YES' to each of the statements), selection might favour a plant—plant signal induction of a defence. In all other cases, selection should favour constitutive expression of the defence, no expression of the defence, or induction of the defence based on damage to the receiver plant itself.

damage currently happening to a neighbour. Finally, if herbivores are unlikely to move between neighbouring plants, there is little reason for a receiver plant to induce defences based on damage to a neighbour, because that cue will not be informative about herbivory risk. Nevertheless, the majority of studies that have tested for induction by plant—plant signals have found evidence supporting plant—plant signals (Karban, Yang, et al., 2014). Plants appear to have evolved numerous strategies to anticipate future damage by herbivores, and plant—plant signals may be only ones of these. For example, plants recognize insect eggs and induce localized defences even before the plant-feeding larva emerges (Hilker, Kobs, Varama, & Schrank, 2002). One reason why plants might invest heavily in anticipating herbivory is that the host choice by the herbivore often occurs long before the appearance of damage. For example, for insects that do not readily move between plants, a key point in plant defence might be to avoid oviposition. Moreover, early stages of many herbivories are more susceptible to plant defence than late stages (Zalucki, Clarke, & Malcolm, 2002), so early elicited defences might be particularly important.

Controlled laboratory studies and highly replicate field studies can observe significant effects even if the magnitude of those effects is not ecologically relevant (Scheiner & Gurevitch, 2001). Most studies of plant—plant signals fall into these categories (controlled lab study or highly replicated field study), so what evidence exists that the effects of plant—plant signals are of large enough magnitude to be ecologically relevant? In a recent meta-analysis, the magnitude of the effect of a plant—plant signal varied dramatically among studies (Karban, Yang, et al., 2014). Some systems have shown consistent evidence supporting a role of plant—plant signalling in inducing plant defence (i.e., sagebrush Karban et al., 2013), whereas others have shown a tendency toward induced susceptibility, which would probably be detrimental to the plant (Pearse et al., 2012). The magnitude of the defensive effect of plant—plant signals is often on par with the effects of other factors that are thought to be important aspects of plant defence. Using the well-studied system of willows as an example, plant—plant signals reduced herbivore damage to willows by 13—41% (Pearse et al., 2013) and relative growth rate (RGR) of a caterpillar by 16—51% (Hughes, Pearse, Grof-Tisza, & Karban, 2015). The magnitude of this response is similar to that of other important determinants of plant—herbivore interactions on willows such as CO_2 addition [14% decrease in RGR of a herbivorous beetle] (Veteli, Kuokkanen, Julkunen-Tiitto, Roininen, & Tahvanainen, 2002), temperature [25% increase in RGR (Veteli et al., 2002)], nitrogen environment and the presence of avian predators [0–83% reduction in herbivore abundance (Sipura, 1999)]. The role of plant–plant signals appears to depend on the plant–herbivore system in question. However, when plant–plant signals reduce herbivore damage, they can do so with a magnitude that is ecologically important.

Plant–plant signals can affect herbivore–plant interactions, but do they have broader consequences for food webs surrounding those interactions? Support for broader consequences of plant–plant signals is currently sparse; however, a few lines of evidence suggest that herbivore-induced plant–plant signals may affect predators as well as plants and herbivores. First, several defences induced by plant–plant signals involve the recruitment of predators to the receiving plant. For example, a plant–plant signal increased nectar flow from extrafloral nectaries in lima bean, which commonly attract ants (Heil & Silva Bueno, 2007). In other cases plant–plant volatile signals increase VOC production in the receiving tissue, which can also attract predatory arthropods (Li, Holopainen, Kokko, Tervahauta, & Blande, 2012; Muroi et al., 2011) to neighbouring plants (Muroi et al., 2011). In other cases, the direct effect of a plant–plant signal on herbivores might cascade to population effects in other trophic levels. For example, plant–plant signals are expected to increase the spatial evenness of herbivore populations (Rubin, Ellner, Kessler, & Morrell, 2015), which could eventually alter how predators forage.

The substantial scientific effort into uncovering the physiological mechanisms that underlie plant–plant communication will surely feedback into interesting avenues for future ecological research. For example, many of the key questions about the consequences of plant–plant communication could best be addressed using mutant plants that are unable to produce, perceive, or respond to plant–plant signals. Such an approach has been very successful in documenting the consequences of numerous plant defences (Kessler & Baldwin, 2002). Currently, the mechanisms of VOC biosynthesis and release are fairly well known (Widhalm et al., 2015). Accordingly, 'mute' plant mutants exist that do not express particular VOC metabolites that have been implicated in plant–plant communication (Schuman, Barthel, & Baldwin, 2012). However, because VOCs probably benefit plants by mediating a great variety of ecological interactions, these mutants may not be the most useful plant manipulations for exploring the role of plant–plant communication (Baldwin, Halitschke, Paschold, von Dahl, & Preston, 2006). Moving forward, it will be particularly important to describe the mechanism of VOC perception and priming and to knock out mutants deficient in the specific function of VOC transfer and perception, leading to priming and defence actions in plants. Such 'deaf' plants could be used to unambiguously demonstrate the importance of plant—plant communication as a means of defence.

REFERENCES

Ali, M., Sugimoto, K., Ramadan, A., & Arimura, G. (2013). Memory of plant communications for priming anti-herbivore responses. *Science Reports, 3*, 1872.

Arimura, G., Huber, D. P. W., & Bohlmann, J. (2004). Forest tent caterpillars (*Malacosoma disstria*) induce local and systemic diurnal emissions of terpenoid volatiles in hybrid poplar (*Populus trichocarpa* x deltoides): cDNA cloning, functional characterization, and patterns of gene expression of (-)-germacrene D synthase, PtdTPS1. *Plant Journal, 37*, 603—616.

Arimura, G., Matsui, K., Koeduka, T., & Holopainen, J. K. (2016). Biosynthesis and regulation of plant volatiles, and their functional roles in ecosystem interactions and global environmental changes. In G. Arimura, & M. E. Maffei (Eds.), *Plant specialized metabolism: Genomics, biochemistry, and biological functions*. Boca Raton, FL: CRC Press/Taylor & Francis. pp. 185—237.

Arimura, G., Matsui, K., & Takabayashi, J. (2009). Chemical and molecular ecology of herbivore-induced plant volatiles: proximate factors and their ultimate functions. *Plant & Cell Physiology, 50*, 911—923.

Baldwin, I. T., Halitschke, R., Paschold, A., von Dahl, C. C., & Preston, C. A. (2006). Volatile signaling in plant-plant interactions: "talking trees" in the genomics era. *Science, 311*, 812—815.

Baldwin, I. T., & Schultz, J. C. (1983). Rapid changes in tree leaf chemistry induced by damage:. evidence for communication between plants. *Science, 221*, 277—279.

Felton, G. W., & Tumlinson, J. H. (2008). Plant-insect dialogs: complex interactions at the plant-insect interface. *Current Opinion in Plant Biology, 11*, 457—463.

Frost, C. J., Appel, H. M., Carlson, J. E., De Moraes, C. M., Mescher, M. C., & Schultz, J. C. (2007). Within-plant signalling via volatiles overcomes vascular constraints on systemic signalling and primes responses against herbivores. *Ecology Letters, 10*, 490—498.

Guenther, A. B., Hewitt, C. N., Erickson, D., Fall, R., Geron, C., Graedel, T., ... Zimmerman, P. (1995). A global model of natural volatile organic compound emissions. *Journal of Geophysical Research, 100*, 8873—8892.

Heil, M., & Adame-Alvarez, R. M. (2010). Short signalling distances make plant communication a soliloquy. *Biology Letters, 6*, 843—845.

Heil, M., & Karban, R. (2010). Explaining evolution of plant communication by airborne signals. *Trends in Ecology and Evolution, 25*, 137—144.

Heil, M., & Silva Bueno, J. C. (2007). Within-plant signaling by volatiles leads to induction and priming of an indirect plant defense in nature. *Proceedings of the National Academy of Sciences of the United States of America, 104*, 5467—5472.

Hilker, M., Kobs, C., Varama, M., & Schrank, K. (2002). Insect egg deposition induces *Pinus sylvestris* to attract egg parasitoids. *The Journal of Experimental Biology, 205*, 455—461.

Hughes, K. M., Pearse, I. S., Grof-Tisza, P., & Karban, R. (2015). Individual-level differences in generalist caterpillar responses to a plant-plant cue. *Ecological Entomology, 40*, 612—619.

Jaskiewicz, M., Conrath, U., & Peterhänsel, C. (2011). Chromatin modification acts as a memory for systemic acquired resistance in the plant stress response. *EMBO Reports, 12*, 50—55.

Karban, R. (2015). *Plant sensing and communication*. Chicago, IL, USA: University of Chicago Press.

Karban, R., & Baldwin, I. T. (1997). *Induced responses to herbivory*. Chicago: The University of Chicago Press.
Karban, R., & Shiojiri, K. (2009). Self-recognition affects plant communication and defense. *Ecology Letters, 12*, 502–506.
Karban, R., Shiojiri, K., Huntzinger, M., & McCall, A. C. (2006). Damage-induced resistance in sagebrush: volatiles are key to intra- and interplant communication. *Ecology, 87*, 922–930.
Karban, R., Shiojiri, K., & Ishizaki, S. (2010). An air transfer experiment confirms the role of volatile cues in communication between plants. *The American Naturalist, 176*, 381–384.
Karban, R., Shiojiri, K., Ishizaki, S., Wetzel, W. C., & Evans, R. Y. (2013). Kin recognition affects plant communication and defence. *Proceedings. Biological Sciences, 280*, 20123062.
Karban, R., Wetzel, W. C., Shiojiri, K., Ishizaki, S., Ramirez, S. R., & Blande, J. D. (2014). Deciphering the language of plant communication: volatile chemotypes of sagebrush. *New Phytology, 204*, 380–385.
Karban, R., Yang, L. H., & Edwards, K. F. (2014). Volatile communication between plants that affects herbivory: a meta-analysis. *Ecol Letters, 17*, 44–52.
Kessler, A., & Baldwin, I. T. (2002). Plant responses to insect herbivory: the emerging molecular analysis. *Annual Review of Plant Biology, 53*, 299–328.
Li, T., Holopainen, J. K., Kokko, H., Tervahauta, A. I., & Blande, J. D. (2012). Herbivore-induced aspen volatiles temporally regulate two different indirect defences in neighbouring plants. *Functional Ecology, 26*, 1176–1185.
Maffei, M., Camusso, W., & Sacco, S. (2001). Effect of *Mentha* x *piperita* essential oil and monoterpenes on cucumber root membrane potential. *Phytochemistry, 58*, 703–707.
Muroi, A., Ramadan, A., Nishihara, M., Yamamoto, M., Ozawa, R., Takabayashi, J., & Arimura, G. (2011). The composite effect of transgenic plant volatiles for acquired immunity to herbivory caused by inter-plant communications. *PLoS One, 6*, e24594.
Okada, K., Abe, H., & Arimura, G. (2015). Jasmonates induce both defense responses and communication in monocotyledonous and dicotyledonous plants. *Plant & Cell Physiology, 56*, 16–27.
Orrock, J. L., Sih, A., Ferrari, M. C., Karban, R., Preisser, E. L., Sheriff, M. J., & Thaler, J. S. (2015). Error management in plant allocation to herbivore defense. *Trends in Ecology and Evolution, 30*, 441–445.
Ozawa, R., Arimura, G., Takabayashi, J., Shimoda, T., & Nishioka, T. (2000). Involvement of jasmonate- and salicylate-related signaling pathways for the production of specific herbivore-induced volatiles in plants. *Plant & Cell Physiology, 41*, 391–398.
Pearse, I. S., Hughes, K., Shiojiri, K., Ishizaki, S., & Karban, R. (2013). Interplant volatile signaling in willows: revisiting the original talking trees. *Oecologia, 172*, 869–875.
Pearse, I. S., & Mescher, M. C. (2016). Information as a unifying theme of plant-animal interactions. *Current Opinion in Plant Biology, 32*, 69–76.
Pearse, I. S., Porensky, L. M., Yang, L. H., Stanton, M. L., Karban, R., Bhattacharyya, L., … Tanner, K. (2012). Complex consequences of herbivory and interplant cues in three annual plants. *PLoS One, 7*, e38105.
Rubin, I. N., Ellner, S. P., Kessler, A., & Morrell, K. A. (2015). Informed herbivore movement and interplant communication determine the effects of induced resistance in an individual-based model. *Journal of Animal Ecology, 84*, 1273–1285.
Scheiner, S. M., & Gurevitch, J. (2001). *Design and analysis of ecological experiments*. Oxford, UK: Oxford University Press.
Schmelz, E. A., Engelberth, J., Alborn, H. T., Tumlinson, J. H., 3rd, & Teal, P. E. A. (2009). Phytohormone-based activity mapping of insect herbivore-produced elicitors. *Proceedings of the National Academy of Sciences of the United States of America, 106*, 653–657.
Schuman, M. C., Barthel, K., & Baldwin, I. T. (2012). Herbivory-induced volatiles function as defenses increasing fitness of the native plant *Nicotiana attenuata* in nature. *eLife, 1*.

Scott-Phillips, T. C. (2008). Defining biological communication. *Journal of Evolutionary Biology, 21*, 387–395.
Sipura, M. (1999). Tritrophic interactions: willows, herbivorous insects and insectivorous birds. *Oecologia, 121*, 537–545.
Sugimoto, K., Matsui, K., Iijima, Y., Akakabe, Y., Muramoto, S., Ozawa, R., ... Takabayashi, J. (2014). Intake and transformation to a glycoside of (*Z*)-3-hexenol from infested neighbors reveals a mode of plant odor reception and defense. *Proceedings of the National Academy of Sciences of the United States of America, 111*, 7144–7149.
Trapp, S. C., & Croteau, R. B. (2001). Genomic organization of plant terpene synthases and molecular evolutionary implications. *Genetics, 158*, 811–832.
Unsicker, S. B., Kunert, G., & Gershenzon, J. (2009). Protective perfumes: the role of vegetative volatiles in plant defense against herbivores. *Current Opinion in Plant Biology, 12*, 479–485.
Veteli, T. O., Kuokkanen, K., Julkunen-Tiitto, R., Roininen, H., & Tahvanainen, J. (2002). Effects of elevated CO_2 and temperature on plant growth and herbivore defensive chemistry. *Global Change Biology, 8*, 1240–1252.
Widhalm, J. R., Jaini, R., Morgan, J. A., & Dudareva, N. (2015). Rethinking how volatiles are released from plant cells. *Trends in Plant Science, 20*, 545–550.
Zalucki, M. P., Clarke, A. R., & Malcolm, S. B. (2002). Ecology and behavior of first instar larval Lepidoptera. *Annual Review of Entomology, 47*, 361–393.

CHAPTER TWO

Allelopathy and the Role of Allelochemicals in Plant Defence

S. Latif*, G. Chiapusio[§,¶], L.A. Weston*
*Charles Sturt University, Wagga Wagga, NSW, Australia
[§]Université de Bourgogne Franche-Comté, UMR CNRS 6249, Montbéliard, France
[¶]Savoie Mont Blanc University INRA CARRTEL, Chambéry, France

Contents

Advances in Botanical Research, Volume 82
ISSN 0065-2296
http://dx.doi.org/10.1016/bs.abr.2016.12.001

Abstract

Allelopathy is described as the interference to plant growth resulting from chemical interactions among plants and other organisms mediated through release of plant-produced bioactive secondary metabolites referred to as allelochemicals. A number of mechanisms have been studied for the release of allelochemicals from various plant tissues including volatilization or leaching from aerial parts, exudation from roots and decomposition of plant residues in soil. Despite differences in biological activity and mode of action, related compounds commonly share similar biosynthetic pathways while some classes of metabolites can be produced using diverse biosynthetic pathways. Recently considerable research has also been undertaken to critically understand the role of allelochemicals in plant succession and plant invasion in native and nonnative ecosystem. In addition, numerous studies have been performed on the selection and utilization of weed suppressive crops and their residues for weed management in sustainable agriculture systems. A better understanding of allelochemical production with respect to plant defence strategies, both physical and chemical, may also allow us to better protect and manage developing crops, limit the spread of invasive weeds, preserve native plant stands and create strategies for allelochemical development and application as novel pesticides. The use of sensitive analytical techniques associated with performance of metabolomics in concert with other *omics* technologies has led to new advances in the identification of unique allelochemicals, the biosynthetic pathways associated with their production, their complex role(s) in the soil rhizosphere and their production as impacted by a changing climate. Identification of novel plant metabolites, including allelochemicals, may result in a source of biologically based pesticides through the provision of complementary structures for future synthesis and as an aid in the development of new molecular target sites.

1. INTRODUCTION

Plants and associated microflora collectively produce a vast assortment of compounds of diverse chemical nature, the majority of which do not appear to have a role in primary metabolism necessary for growth and development of the plant. These low organic weight compounds are referred to as plant secondary products or secondary metabolites. Certain classes of secondary metabolites called allelochemicals have stimulatory and/or inhibitory effects on the growth, health and behaviour of neighbouring plants (Rice, 1974). This phenomenon is referred to as allelopathy, a term first applied by Molisch (1937), broadly defined by some researchers as direct or indirect biochemical interactions among plants, and potentially also microorganisms, mediated through release of allelochemicals by plants (Weston & Duke, 2003).

The International Society of Allelopathy (IAS) defined allelopathy as chemical interactions among plants and other organisms, excluding herbivores (Weir, Park, & Vivanco, 2004). Given that most researchers studying allelopathy have focused on interactions among plants mediated by allelochemicals, we have chosen to describe plant–plant interactions in this chapter. However, it is important to note that microbes can both degrade and also activate secondary metabolites, in some cases rendering them more biologically active as plant growth inhibitors (Cipollini, Rigsby, & Barto, 2012). Secondary products exhibiting allelopathic potential are referred to as 'allelochemicals' (Whittaker and Feeny, 1971), however, both terms are not interchangeable. Moreover, allelopathic interference due to the action of allelochemicals differs from interference due to competition in which plants compete with neighbouring plants for soil resources including, water, space, light, gases and macro- and micronutrients necessary for growth (Weston & Duke, 2003). However, both competition and allelopathy generally result in growth reduction or interference with plant growth over time.

Allelopathy is not a new phenomenon and has been described for over 2000 years in the literature; in ancient manuscripts, plants were frequently reported to 'sicken the soil' and produce toxins that adversely affected the growth of other plants (Weston & Duke, 2003). The Greek botanist Theophrastus (300 BC) first noted that chickpeas exhausted the soil and inhibited weeds. Later, the Roman scholar Pliny described the toxicity of the walnut tree to neighbouring plants (Weir et al., 2004). In 1832, De Candolle performed the first well-described experiment to study toxicity associated with root exudates of allelopathic species (Singh, Batish, & Kohli, 2001). In recent years, following on from these foundational studies, research into the mechanisms associated with allelopathic interactions has increased exponentially. The latest studies on allelopathy have been critical in advancing our understanding of plant ecosystems and their drivers and have focused on the following: the impact of allelochemicals on successful invasion by plant invaders in the nonnative range; the role of allelochemicals in plant succession in dynamic ecosystems; the selection and utilization of weed suppressive crops and their residues for enhanced weed management; and the role of root exudates in driving rhizosphere interactions with plants and/or microbial symbionts.

This chapter aims to present a general overview of allelopathic interactions among plants and plant defence mechanisms, the biochemical classification of allelochemicals, allelochemical mode(s) of action and key allelochemical interactions occurring in the rhizosphere or plant–soil interface.

2. PLANT DEFENCE AND THE ROLE OF ALLELOCHEMICALS

Plants are sessile and therefore cannot easily move away from potential threats or towards beneficial entities. During the course of evolution, plants have developed both physical and chemical mechanisms of defence from pests and pathogens (Bernards, 2010). Traditionally resource competition has been considered as the single most important factor that influences the patterning of plant communities (Niklas & Hammond, 2013). However, recent research has described allelopathy as an important aspect of plant defence that impacts plant community diversity (Fernandez et al., 2013). In this process, plants release a diverse repertoire of low molecular weight secondary metabolites that are considered to interact with the surrounding environment by inhibiting the germination or growth of neighbouring plants (Ben, Jordan, & Osbourn, 2006; Fernandez et al., 2016). The majority of allelochemicals in the plant kingdom are found in vascular plants, but our knowledge of secondary metabolites in ancient terrestrial nonvascular plants such as mosses or liverworts has also increased in recent years. Allelochemicals can, therefore, play an important role in plant succession through their release by pioneer plants (i.e., Bryophytes) which contribute substantially to the accumulation of aboveground biomass, particularly in cold temperate biomes including boreal forests and peatlands (Chiapusio, Jassey, Hussain, & Binet, 2013; Michel, Burritt, & Lee, 2011).

Approximately 100,000 secondary metabolites have been identified to date in plants (Afendi et al., 2012; Croteau, Kutchan, & Lewis, 2000). A smaller number of these are described as bioactive allelochemicals and are generally classified as members of specific chemical families that include phenolics, terpenoids, glycosteroids and alkaloids (Ahuja, Kissen, & Bones, 2012). Plants can allocate large investments in carbon resources to synthesize, regulate and store secondary metabolites including allelochemicals involved in plant defence. However, many plants have evolved cost-effective strategies for production and recycling of these bioactive metabolites; the conversion of one product to another through plant metabolism may result in a completely different biological function. In some cases, a single compound or related family of compounds can have multiple functions from an ecological context. For example, strigolactones are stimulants of germination of parasitic plants while other lactones are potent germination inhibitors. Despite differences in biological activity and mode of action, related compounds commonly share similar biosynthetic pathways while some classes of metabolites including phenolics can be produced using diverse biosynthetic pathways and precursors (Neilson, Goodger, Woodrow, & Møller, 2013).

Plants have further evolved protection from potential predators, including herbivores and pathogens, through specific physical and chemical defence mechanisms. Plants possess specialized morphological structures including trichomes, spines and hairs that provide generalized physical defence. Specialized defence mechanisms include the production and release of bioactive metabolites, some of which are constitutive and found in almost every class of plants while others are synthesized or activated in response to biotic and abiotic stressors (Bartwal, Mall, Lohani, Guru, & Arora, 2013). Precursors which are formed immediately after pathogen attack from preexisting constituents are referred as phytoanticipins or more commonly as phytoalexins (Pedras, Yaya, & Glawischnig, 2011; VanEtten, Mansfield, Bailey, & Farmer, 1994). These low molecular mass metabolites are produced in host plants under diverse stressors and exhibit broad-spectrum inhibitory activity and extensive chemical diversity.

The development of microbial symbiosis is also crucial for the survival and growth of vascular plants. For example, mycorrhizal fungi form symbiotic associations with 90% of vascular plants and are also ubiquitously distributed in a wide range of ecosystems (Finlay, 2008; Read, Leake, & Perez-Moreno, 2004). These organisms often play fundamental roles in belowground processes because they are crucial for the mobilization of nitrogen and phosphorus from the soil to their host plant. These organisms are also sometimes involved in mediation of allelopathic interference as plant-produced allelochemicals, herbicides and other metabolites present in the soil rhizosphere are released into the rhizosphere and later taken up and transported to other plants by way of the endophyte's network of underground hyphal networks, sometimes referred to as the belowground 'superhighway' facilitating interplant transport of allelochemicals (Delaux, Sejalon-Delmas, Becard, & Ane, 2013; Souto, Pellissier, & Chiapusio, 2000).

2.1 Allelochemical Localization and Release Into the Environment

Secondary metabolites including allelochemicals are ubiquitous in nature and can be released over time from all plant tissues including leaves, stem, roots, flowers, seed, rhizomes, pollen, bark and buds (Weston & Duke, 2003). A number of mechanisms including volatilization or leaching from aerial parts, exudation from roots and decomposition of plant residues in

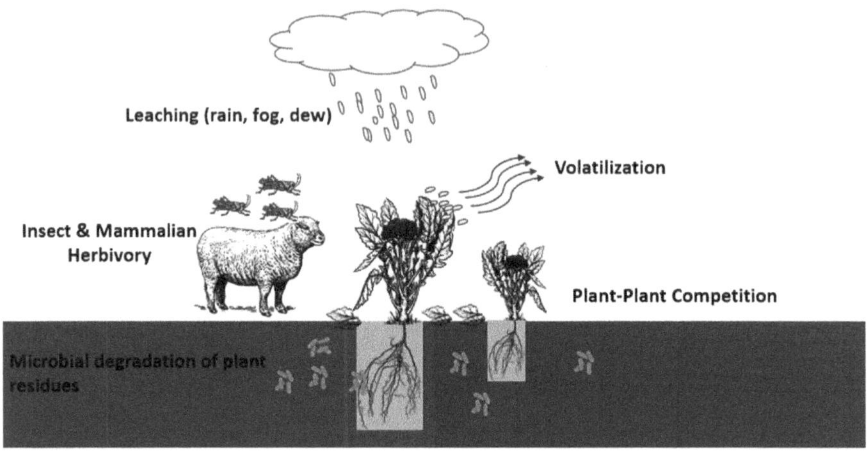

Figure 1 Chemical (leaching, volatilization) and biological (herbivory, microbial degradation, competition) mechanisms of release of allelochemicals from various plant tissues.

soil (Fig. 1) have been well characterized with respect to the release of allelochemicals over time in the environment (Cheng & Cheng, 2015; Putnam & Tang, 1986). Direct volatilization and precipitation, including rain, fog and dew, also play an important role in the solubilisation and release of allelochemicals from foliar parts of the plants, particularly those that possess glandular trichomes containing allelochemicals (Bais, Weir, Perry, Gilroy, & Vivanco, 2006; Inderjit & Duke, 2003). Trichomes are hair-like appendages on the surface of the plant leaf, stem or inflorescence. They provide a physical defence against mammalian herbivores upon ingestion, and potentially also chemical defences due to the unpleasant sensorial experience associated with the allelochemical release. In addition to a physical barrier, glandular trichomes have specialized cells which can synthesize allelochemicals as part of a constitutive chemical defence system (Champagne & Boutry, 2016).

3. CLASSIFICATION OF SECONDARY METABOLITES

Secondary metabolites including allelochemicals can logically be classified according to their carbon skeletal structure and type of functional groups; however, the most useful classification system is based on the biogenetic origin of metabolites (Fig. 2). As presented by Walton and Brown (1999) and Rice (1974) plant metabolites and allelochemicals can be further divided into three major groups: phenolics, terpenoids, and alkaloids.

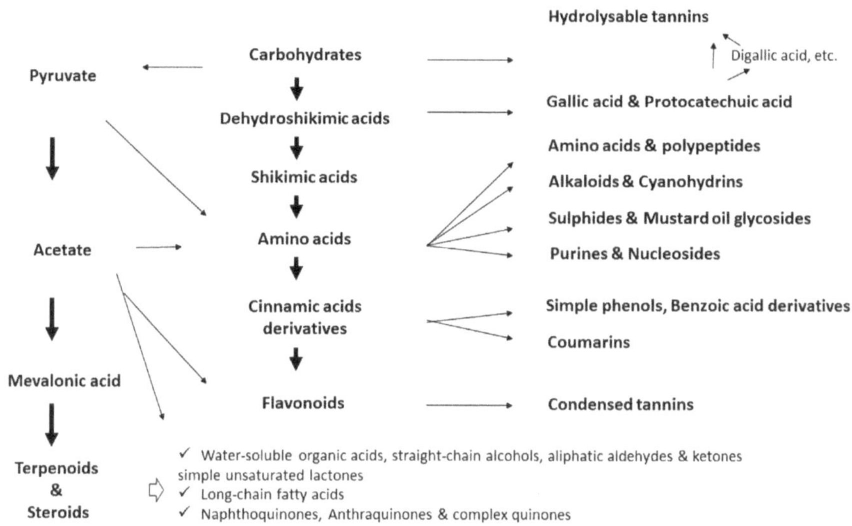

Figure 2 Major families of allelochemicals and their common biosynthetic pathways originating from numerous key precursors (Rice, 1974). Amino acids and polypeptides as well as purines and nucleosides are often categorized as primary metabolites due to their roles in primary biosynthetic pathways of proteins and DNA. However, some of these molecules are also bioactive in their own right and can serve as allelochemicals, thereby impacting higher plant growth.

3.1 Phenolic Compounds and Their Derivatives

Plant phenolics (Fig. 3) represent an extremely diverse group of organic compounds which share a common structure; they consist of an aromatic ring possessing at least one hydroxyl group and possibly other substituents (Stalikas, 2007). They are generally categorized as phenolic acids, flavonoids, stilbenes, coumarins, lignins and tannins (Cheynier, Comte, Davies, Lattanzio, & Martens, 2013; Dinelli et al., 2009). The term phenolic is descriptive as phenolics are secondary metabolites derived from the phenylpropanoid-acetate biosynthetic pathway (Croteau et al., 2000; Haig, 2008, pp. 63–104; Quideau, Deffieux, Douat-Casassus, & Pouysegu, 2011). Approximately 8000 naturally occurring secondary metabolites belonging to this group have been identified so far. They perform a broad array of structural and physiochemical roles *in planta* (Croteau et al., 2000). Phenolics are the most commonly reported metabolites known to play a role in defence mechanisms in higher plants; with variable toxicity, they target cellular functions at multiple sites (Haig, 2008, pp. 63–104). They also represent 40% of the structural matrix of plants in form of lignins

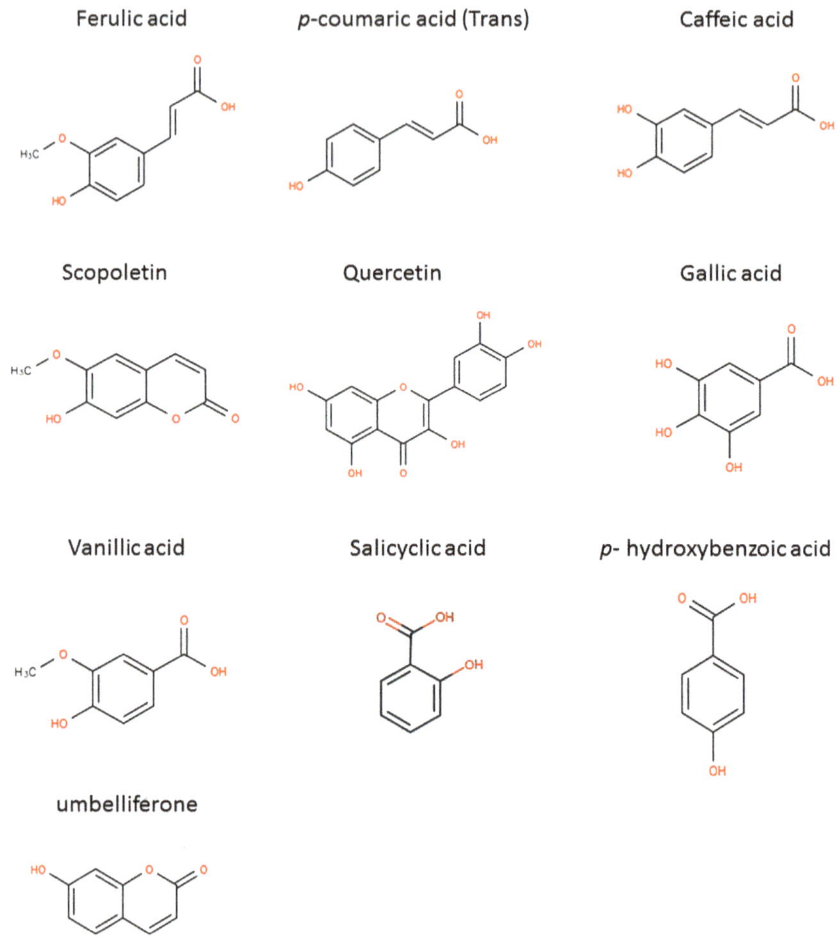

Figure 3 Some examples of diverse plant phenolics with known allelopathic properties.

and tannins, and through these molecules which degrade over time in the soil rhizosphere, maintain their abundance through the important ecological cycle of synthesis, transformation, release and decomposition (Dalton, 1999, pp. 57–74; Hattenschwiler & Vitousek, 2000).

The mechanism of allelopathy associated with phenolic compounds has been studied extensively and reports indicate that phenolics interfere with several key plant enzymes and physiological processes. For example, cinnamic and benzoic acids (Fig. 3) interfere with hormone activity, membrane permeability, photosynthesis, respiration and synthesis of organic compounds. However, there has been no evidence to date for phenolics impacting cell division or gene translation (Einhellig, 2004).

Perennial legumes including alfalfa (*Medicago sativa* L.) and clovers (*Trifolium* spp.) are known for their allelopathic properties following pasture establishment (Weston & Mathesius, 2013). Some species including alfalfa and clover can also prove autotoxic once they are established. These perennial plant stands tend to inhibit the growth of plants of the same species over time. Carlsen, Pedersen, Spliid, and Fomsgaard (2012) studied the phenomenon of phytotoxicity following clover establishment and reported that release of several secondary metabolites, including flavonoids and their glycosides, causes weed suppression and negative plant—plant interactions associated with replanting following legume establishment.

3.2 Terpenoids

Terpenoids are specific allelochemicals that are biosynthetically derived from the mevalonic acid and isopentenyl pyrophosphate pathways (Haig, 2008, pp. 63—104). The term 'terpene' or the preferred term 'terpenoid' comes from the German word 'terpentin' (terpentine) due to the fact that first reported terpenoid was isolated from turpentine (Herz, 1963). Naturally occurring, approximately 24,000 terpenoids are comprised of five carbon isoprene subunits (Fig. 4) linked together through the common head-to-head or less common head-to-tail linkages but some are also characterized by head-to-middle ring closure (Croteau et al., 2000). Based on the relative number of isoprene subunits possessed, they are classified into hemiterpenes (single five-carbon isoprene unit) monoterpenes (two isoprene units), sesquiterpenes (three isoprene units), diterpenes (four isoprene units), triterpenes (six isoprene units), tetraterpenes (eight isoprene units) and polyterpenes (with more than eight isoprene subunits). Terpenoids are typically volatile compounds having multiple biological activities in plants as signalling molecules, photoprotective agents, reproductive hormones but also allelochemicals (Croteau et al. 2000).

Monoterpenes, the main constituents of plant essential oils, are widely known for their strong inhibitory effects on plant growth and seedling germination (Haig, 2008, pp. 63—104). For example, the monoterpenes 1,4-cineole and 1,8-cineole (Fig. 5) are well studied as growth inhibitors

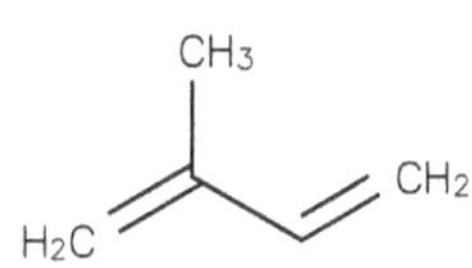

Figure 4 The structure of an isoprene unit, the backbone of terpenoids.

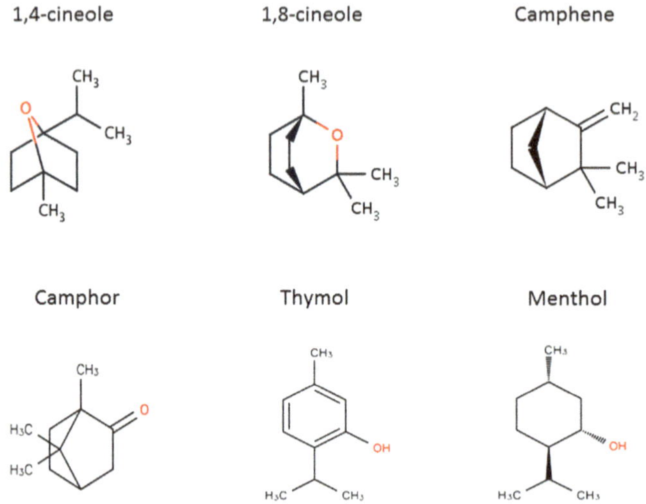

Figure 5 Some examples of plant monoterpenes with known allelopathic properties.

and potential candidates for herbicides (Duke et al., 2004). Despite structural similarities, these monoterpenes have very different modes of action where 1, 8-cineole affects all stages of mitosis and 1, 4-cineole causes growth abnormalities in shoots.

Sesquiterpenes and their associated metabolites along with the monoterpenes in essential oils have been well characterized with respect to their phytotoxic effects on plants. The sesquiterpene *β*-caryophyllene (Fig. 6) is

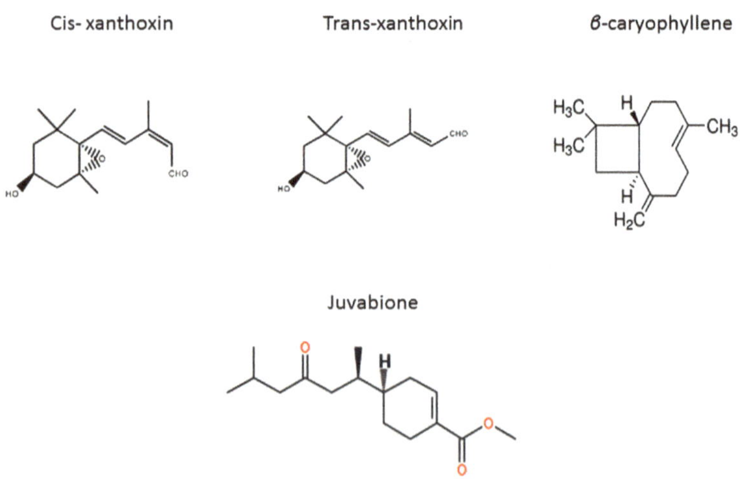

Figure 6 Some examples of plant sesquiterpenes with known allelopathic properties.

present in numerous plant volatiles and inhibits germination and seedling growth of *Brassica napus* L. and *Raphanus sativus* L. at very low concentrations (Wang, Peng, Zeng, Ding, & Xu, 2009). Geometric isomers of xanthoxin (Fig. 6) present in leaves of *Pueraria thunbergiana* L. (Sieb. and Zucc.) Benth., also known as the invasive kudzu vine, inhibited seedling root growth of numerous small-seeded annual broadleafs and grasses at very low concentrations (Kato-Noguchi, 2003; Rashid, Asaeda, & Uddin, 2010).

3.3 Alkaloids

Alkaloids (Fig. 7) are heterocyclic nitrogen-containing basic compounds of plant origin and are named accordingly, due to their alkaline chemical nature. Plant alkaloids predominate in four families of plants including the Asteraceae, Apocynaceae, Boraginaceae and Fabaceae (Haig, 2008, pp. 63–104).They have been important since antiquity due to their pharmacological properties and are among the largest group of secondary metabolites with approximately 20,000 compounds identified to date representing great structural biosynthetic diversity (Yang & Stöckigt, 2010). Based on their biosynthetic origin, alkaloids are classified into different classes; e.g., indole alkaloids are derived from tryptophan, pyrrolizidine alkaloids are derived from ornithine or arginine, and quinolizidine alkaloids are derived from lysine (Seigler, 1998, pp. 546–567).

Alkaloids are widely distributed across the plant kingdom and frequently reported to play key defensive roles in plant interactions against herbivores,

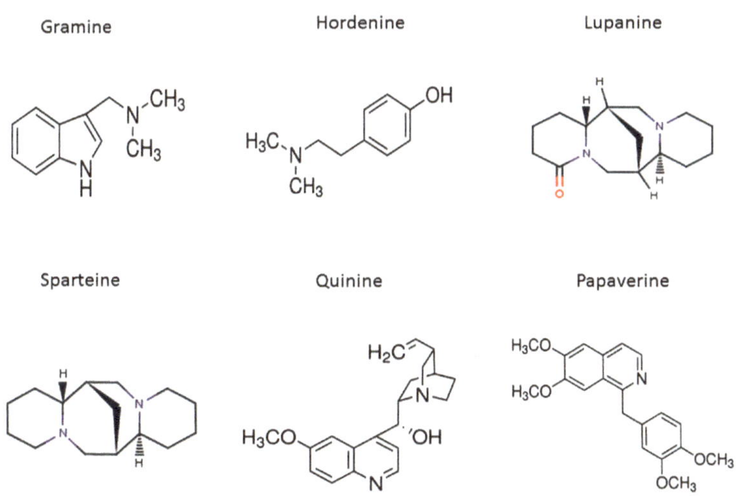

Figure 7 Some examples of plant alkaloids with known allelopathic properties.

microorganisms, fungi, and in some cases, neighbouring plants. Several naturally occurring pyrrolizidine alkaloids have been evaluated for hepatotoxicity to humans and grazing livestock (Quinn, Kessell, & Weston, 2014) and also deter herbivory as part of the plant's aboveground defence system. However, other alkaloids have in some cases shown activity as inhibitors of plant growth. Nicotine production in wild tobacco is upregulated after grazing or damage by herbivores; interestingly nicotine also possesses allelopathic activity against common annual weeds. Alkaloids are thought to inhibit plant growth by several mechanisms including interference with DNA, enzyme activity, protein biosynthesis and membrane integrity in developing plants (Wink, 2004). Lovett and Hoult (1995) studied defence mechanisms in barley (*Hordeum spp.)* through the release of the plant-produced alkaloids gramine and hordenine (Fig. 7) from living roots; these were shown to have allelopathic effects in seedling bioassays. The quinolizidine alkaloids studied by Wink (2004), including lupanine and sparteine, are produced by legumes and are thought to cause potent inhibition through interference with membrane permeability and protein synthesis.

3.4 Hydroxamic Acids of Benzoxazinoids: An Agronomic Case Study

Hydroxamic acids are phenol derivatives, known as cyclic 2-hydroxy-2H-1,4-benzoxazin-3-(4H)-ones, which are a basic skeleton of a broader family of bioactive plant secondary metabolites referred to as benzoxazinoids (Fig. 8). They have been well characterized by numerous scientists and have antimicrobial, antifungal, antifeedant and phytotoxic activities (Chiapusio et al., 2005, pp. 139–155).

Hydroxamic acids of benzoxazinoids present in cultivated cereals of the Graminaceae including maize, rye and wheat have been widely studied for their allelopathic potential and their abundance has been highly correlated with resistance to insects and other pathogens (Elek et al., 2013; Niemeyer, 2009). 2,4-Dihydroxy-7-methoxy-2H-1,4-benzoxazin-3-one (DIMBOA)

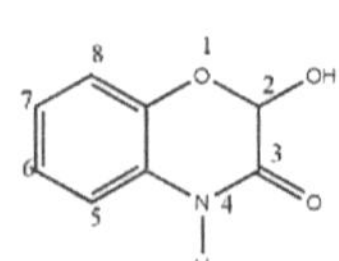

Figure 8 The structure of 2-hydroxy-2H-1, 4-benzoxazin-3-(4H)-ones, also known as hydroxamic acids of the benzoxazinoid family of allelochemicals.

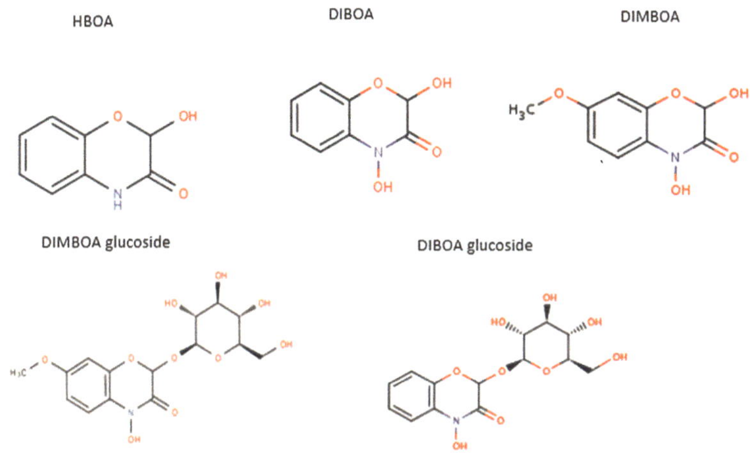

Figure 9 Common examples of benzoxazinoids with known phytotoxic properties.

and 2,4-dihydroxy-2H-1,4-benzoxazin-3-one (DIBOA) and their associated microbial conversion products (Fig. 9) are prominent hydroxamic acids with potent allelopathic effects on numerous broadleaf weeds and crops lettuce (*Lactuca sativa* L.) and wheat (*Triticum aestivum* L.) (Macías et al., 2003, pp. 239–252). They are typically released into soil from injured plant tissues or by degradation of crop residues after harvest. The glucosides of DIBOA and DIMBOA, together with their respective aglycones and degradation products released from rye (*Secale cereal* L.) root residues, are mainly responsible for allelopathic properties on developing broadleaf crops and weeds and several annual grasses (Schulz, Marocco, Tabaglio, Macias, & Molinillo, 2013).

4. ALLELOCHEMICAL MODE OF ACTION

One of the current challenges in allelopathy is to determine the specific mode(s) of action of allelochemicals in association with their diverse chemical nature and multiple target sites in higher plants. Various bioassays have been designed to carefully study the direct effects of allelochemicals on higher plant growth as well as evaluate their underlying mechanism of action. This can be a particularly difficult task, as bioassays for specific enzymes that may be potential target sites of allelochemicals are often challenging to devise and require prior knowledge of the structure of the catalytic or target site. In some cases, the actual target site of allelochemical action is not evident by visual observation of affected plant growth and/or morphology. In addition, the relevance of some in vitro bioassays with respect to the provision of conclusive evidence related to allelochemical mode of action has come under scrutiny, and the relationship of in vitro effects to those seen in a natural setting has also been questioned (Inderjit & Weston, 2000).

It has also been suggested that the relative concentration of allelochemicals typically released into the environment is often quite low and transient; therefore, the concentration of an allelochemical or complex mixtures of allelochemicals encountered in the soil or in the environment is often difficult to estimate or measure (Weir et al., 2004). Therefore, it is particularly important to use appropriate bioassays at relevant concentrations to assess the potential inhibitory effects of allelochemicals on higher plant growth. In many cases, it may be critical to use more than one bioassay to assess impacts on plant growth. For example, assays utilizing local soils, plant species of importance in natural systems and those providing uniform, repeatable results are critical for the development of useful data sets. Other investigators, including those assessing mode of action of synthetic and natural products such as allelochemicals in the agrochemical industry have developed robust models to assess plant growth using very sensitive assays. One can typically use several assay conditions to generate data including the use of an aqueous plant known as pondweed or duckweed (*Lemna spp.*) that is rapidly growing and uniformly reproduces additional plantlets over a one week period. The duckweed assay allows the testing of very low concentrations of allelochemicals as well as mixtures of metabolites, provided the compounds can be solubilized in an aqueous solution. In addition, various small pot assays using local soils and/or agar have been adopted to assess impacts both above- and belowground on shoots and roots, respectively (Blum, 2014, pp. 77–129). These small pot or container assays typically require a longer period to generate plant growth for assessment. Finally, a bioassay to assess direct impacts of decomposing plant residues has also been utilized to assess phytotoxicity associated with plant residues following a crop or a weed infestation using recovered plant tissue (Weston, Harmon, & Mueller, 1989). By use of appropriate assays for screening, a number of different mechanisms of action have been postulated and later confirmed using very specific assessment at the site of action; these include membrane permeability, water and nutrient uptake, respiration, photosynthesis, protein and nucleic acid synthesis, and growth regulation in susceptible plants (Einhellig, 1995).

4.1 Membrane Permeability

Many secondary metabolites including allelochemicals alter cell permeability and membrane function with exposure at adequate concentrations.

Exposure to such metabolites may result in leakage of cellular contents and consequently cell death through apoptosis and necrosis (Li, Wang, Ruan, Pan, & Jiang, 2010). This process eventually leads to tissue death and loss of specific function. For example, several phenolic compounds are reported to exhibit allelopathic effects by altering plasma membrane permeability. The compounds are easily able to cross cellular membranes either by diffusion or assisted transport mechanisms. Once membranes have experienced altered permeability, potassium channels are impacted and decreased permeability to chloride ions is often encountered. Yoon et al. (2000) and Singh, Batish, Kaur, Arora, and Kohli (2006) reported that the monoterpenoid, α-pinene, caused oxidative stress to the plasma membrane and disruption of membrane structural integrity which led to cell death. Chai, Ooh, Ooi, Chue, and Wong (2013) also found that concentrations of phenolic compounds greater than 1 mM and mimosine (a nonprotein amino acid) in *Leucaena leucocephala* (Lam.) leachates caused increased membrane permeability in leaf tissue of hyacinth (*Hyacinthus orientalis* L.). Membrane injury is typically assessed in vitro by measuring relative electrolyte leakage in treated plant culture bioassays in contrast to an untreated control.

4.2 Water and Nutrient Uptake

To further investigate the potential mode of action of allelochemicals released into the environment, it is important to focus upon the relative health and function of the plant's root system. In the living roots, allelochemicals can frequently affect the activity of Na^+/K^+ pumps involved in the absorption of ions across the plasma membrane. For example, Abenavoli, Lupini, Oliva, and Sorgonà (2010) found that nitrate uptake and membrane H^+-ATPase activity in *Zea mays* L. (corn) roots were inhibited by cinnamic acid, ferulic acid and *p*-coumaric acid. Franche, Lindström, and Elmerich (2009) also reported that the phenolic compounds cinnamic and ferulic acid reduced the nutrient uptake of P and Fe and thereby inhibited the growth of target plants. Phytotoxins produced by *Beta vulgaris* subspp. *vulgaris* (sugar beet) make zinc more available for zinc-sensitive crops such as beans (*Phaseolis vulgaris* L.) and corn when planted following sugar beet (Boawn, 1965).

4.3 Respiration

Allelochemicals released from donor plants can also seriously affect the growth of plants by impacting the process of respiration including electron transfer in the mitochondria, oxygen uptake, CO_2 generation and oxidative phosphorylation for ATP generation (Cheng & Cheng, 2015). For example, Abrahim, Takahashi, Kelmer-Bracht, and Ishii-Iwamoto (2003) studied the monoterpenes camphor, α-pinene and limonene and determined that each influences respiratory activity in hypocotyl mitochondria of various donor plants through different mechanisms i.e., mitochondrial uncoupling, electron flow in the cytochrome pathways and oxygen uptake, respectively. Czarnota, Paul, Dayan, Chandrashekhar, and Weston (2001) found that the inhibitor sorgoleone, exuded by living sorghum roots in copious quantities, is both an inhibitor of electron transport in photosystem II at the D1 binding protein and also inhibits mitochondrial electron transport by uncoupling electron flow in all plant species evaluated, including weeds and crops. Allelochemicals typically exhibit a stronger effect on seedling growth and germination in contrast to the growth of mature plants, and it has been suggested that one of the underlying mechanisms of germination and seedling growth inhibition is through frequent disruption of mitochondrial ability to carry out respiration (Weir et al., 2004).

4.4 Photosynthesis

Allelochemicals can significantly influence the process of photosynthesis in several ways, including disruption of electron flow in PS I and II, impacting the synthesis of photosynthetic pigments or stimulation of the decomposition of photosynthetic pigments. Consequently, decreased concentrations of photosynthetic pigments in plant shoots can lead to reduced synthesis of ATP, primarily in photosystem II (Cheng & Cheng, 2015; Fengzhi, Kai, Fengming, & Xuedong, 2004; Weir et al., 2004). For example, sorgoleone reacts at the D1 binding protein site in photosystem II and has shown strong competitive interference with electron transport (Einhellig, 1995; Gonzalez, Kazimir, Nimbal, Weston, & Cheniae, 1997). Weeds (*Amaranthus retroflexus* L.) that were resistant to PS II herbicide inhibitors by mutational changes at the D1 binding site were also resistant to the impact of sorgoleone at this site, providing further evidence for activity at the D1 binding protein in PS II (Nimbal, Yerkes, Weston, & Weller, 1996). Poonpaiboonpipat et al. (2013) observed that essential oil from *Cymbopogon citratus* (DC.) Stapf. (lemongrass) significantly decreased the concentration of chlorophyll a and b in the seedling barnyard grass, thereby reducing photosynthesis and subsequent plant growth. These findings suggest that essential oils have a direct impact on chlorophyll biosynthesis. Studies performed with the allelochemical citral, a volatile monoterpene found in many plants, has shown that citral is effective not only on inhibition of seedling metabolism through respiration and reduced photosynthesis but also on adult plants by inhibiting growth and altering the plant oxidative status (Graña, Sotelo, Díaz-Tielas, Reigosa, & Sánchez-Moreiras, 2013). This suggests that mode of action of certain allelochemicals is age and tissue dependent and also dependent on the site of uptake in the plant.

4.5 Protein and Nucleic Acid Synthesis and Growth Regulation

Allelochemicals can impact the growth of neighbouring plants by targeting key regulatory mechanisms affecting plant growth. These mechanisms include intercalation of DNA, inhibition of DNA polymerase I and inhibition of protein biosynthesis and have been associated with reduction in plant growth due to allelochemical exposure (Wink & Latz-Bruning, 1995). Phenolic allelochemicals can also interfere with the formation of nucleic acids, critical in cellular metabolism and gene expression. Li et al. (2010) determined that the phenolic compounds ferulic and cinnamic acid adversely affected total protein synthesis. Allelochemicals from donor plants have also been reported to regulate the expression of other defence genes in the target plant. For example, synthesis of enzymes which are involved in the biosynthesis of phenolic compounds in rice are upregulated in the presence of barnyard grass (*Echinochloa crus-galli* L.), a common weed in cultivated rice (He et al., 2012). In the case of seedling rice, the growth of rice with the weed competitor barnyardgrass results in an increase in production of the allelochemical momilolactone in rice (Kato-Noguchi, 2011). This suggests that stress caused by plant competition or interference due to allelopathy alters a target plant's ability to produce defence compounds and response may occur in a matter of hours to days as measured by enhanced allelochemical production. Similar phenolics have also been shown to oxidize indole-3-acetic acid, a hormone required for cell elongation in higher plants (Yang, Ye, Liao, & Yin, 2005).

5. LOCALIZATION AND TRANSPORT OF ALLELOCHEMICALS IN DONOR PLANTS

Plants have specialized tissues for synthesis and release of secondary metabolites in the environment including stomata and glandular trichomes in plant shoots, and root hairs, border cells, epidermis and periderm in roots. The root hair is a single cell extension of the root epidermis and plays an important role in complex interactions at root—soil interface (Weston, Ryan, & Watt, 2012). Root hairs have the ability to exude both low and high molecular weight organic molecules including ions, amino acids, growth regulators and allelochemicals (Table 1). Some compounds are synthesized within the root hair and directly exuded into the environment, such as sorgoleone (Weston et al., 2012) and shikonins which are directly exuded by *Echium plantagineum* L. seedlings (Zhu et al., 2016). Recently, the localization of red-coloured bioactive naphthoquinones (also known as shikonins) from the outer periderm of living roots of the highly invasive

Table 1 Diverse organic metabolites known to be released in root exudates of living plants (Bertin, Yang, & Weston, 2003)

Class of Compounds	Metabolites	Functions
Carbohydrates	Arabinose, glucose, raffinose, rhamnose, ribose, fructose, galactose, maltose, sucrose and xylose	Provide favourable environment for the growth of microorganisms
Amino acids and amides	All 20 proteinogenic amino acids, aminobutyric acid, homoserine, cystathionine, mugineic acid	Inhibit nematodes and root growth of different plant species
Aliphatic acids	Formic, acetic, butyric, isocitric, oxalic, fumaric, propionic, maleic, citric, tartaric, oxaloacetic, pyruvic, oxaloglutaric, glycolic, shikimic, acetonic, valeric, gluconic	Plant growth regulation and inhibition
Aromatic acids	*p*-hydroxybenzoic, caffeic, *p*-coumaric, ferulic, gallic, gentisic, protocatechuic, salicylic, sinapic, syringic	Stimulation depending on concentration
Miscellaneous phenolics	Flavanol, flavones, flavanones, anthocyanins, isoflavonoids	Plant growth inhibition or stimulation depending upon concentration
Fatty acids	Linoleic, linolenic, oleic, palmitic, stearic	Plant growth regulation
Sterols	Campestrol, cholesterol, sitosterol, stigmasterol	Plant growth regulation
Enzymes and miscellaneous		Unknown

weed *E. plantagineum* L. has also been studied; this plant accumulates these antimicrobial and phytotoxic naphthoquinones in the periderm of the primary taproot over time and also releases these molecules from seedling root hairs as red-coloured droplets. These localization studies were facilitated through the use of confocal microspectrofluorometry (Zhu et al., 2016).

In the shoot, allelochemicals can be directly exuded by glandular trichomes on the leaf surface, while in other cases they are translocated to the trichomes from other parts of the plant (Eom, Senesac, Tsontakis-Bradley, & Weston, 2005). In the case of catmint, *Nepeta x faassenii*, 21 volatile constituents were noted in the glandular exudates of the plant and proved to potently inhibit seed germination and seedling plant growth, when assays were conducted with foliage suspended in the proximity of germinating seedlings but not in contact with seedlings. This suggested that volatilization is important for dispersal and inhibition of plant growth processes. Three components, 2-(2-ethoxyethoxy)-ethanol, alloaromadendrene and chi-cadinene, were not only detected in both the volatile mixture and the methanolic extract, but also in an aqueous foliar extract that exhibited potential allelopathic activity on curly cress (*Lepidium sativum* L.) growth. Eom, Yang, and Weston (2006) studied the allelopathic potential of perennial groundcovers and found that foliar volatile compounds from *N. x faassenii* (catmint) significantly decreased the growth of *L. sativum* (curly cress) and also common roadside weeds under field conditions. Lipophilic or volatile compounds such as those produced by catmint can directly diffuse through the lipid bilayer while more polar compounds require specialized membrane-bound transport proteins to assist the transport process (Weston et al., 2012).

Once synthesized and accumulated in plant tissues, most allelochemicals are excreted from the plant to prevent autotoxicity or stored in vacuoles and small membranous vesicles or storage organs, such as the periderm, waxes and glands on the leaf surface or in extracellular spaces or compartments in plant tissues. They can also be transported to other plant tissues using diffusion or more likely specific transport proteins which are embedded in the plasma membrane and actively assist in the transport of a particular class of compounds, including allelochemicals (Fig. 10) (Weston et al., 2012).

5.1 Root Exudation of Allelochemicals

Root exudation is a common mechanism of release of allelochemicals from living plant roots, specifically through root hairs, which are individual cells

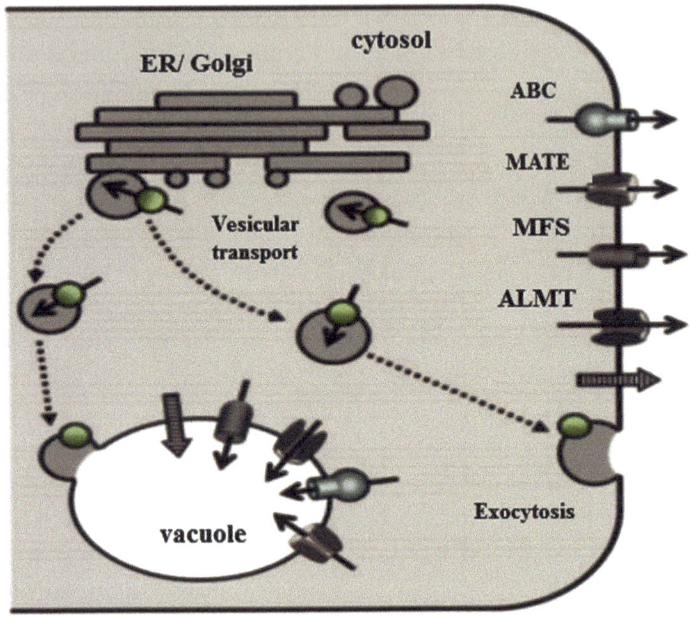

Figure 10 Transport of organic compounds in root cells through membrane-bound proteins including ATP binding cassette family (ABC), multidrug and toxic compound extrusion family (MATE), the major facilitator superfamily (MFS) and the aluminium-activated malate transport family (ALMT) (Weston et al., 2012).

associated with synthesis and excretion of bioactive metabolites, or by the actively growing tips of primary and secondary roots (Table 1) (Weston, Weston, & Hildebrand, 2013; Zhu et al., 2016). The soil-root interface or rhizosphere is considered as the site of greatest activity in the soil matrix as a myriad of organic compounds in varying concentrations are released from living roots, decomposing plant material and by associated microorganisms. The rhizosphere is an extremely complex environment to chemically survey due to actively changing the concentration of secondary metabolites and relative difficulty in extraction from soil matrix. Most recently, the development of silicone microprobe tubing and solid phase root zone extraction techniques have provided an opportunity to precisely profile nonpolar to moderately polar allelochemicals released from living plant roots in the soil (Zhu et al., 2016). In this case, a silicon tube is placed in the soil around the root surface and less polar molecules adhere to the silicone exterior of the tube and can be selectively removed by later solvent extraction (Weidenhamer, Boes, & Wilcox, 2009).

The composition of root exudates is complex and generally consists of carbon-containing compounds including polysaccharides (for example, arabinose, glucose, fructose, maltose), amino acids (for example, arginine asparagine, cysteine, glutamine), organic acids (for example, acetic, ascorbic, benzoic, ferulic acids), phenolic compounds, alkaloids, tannins, terpenoids, flavonoids, growth regulators, vitamins and nutrients and to a lesser extent, noncarbon compounds including ions, water and electrolytes (Table 1) (Baetz & Martinoia, 2014; Bertin et al., 2003). Once released, allelochemicals undergo physical, chemical and biological changes in the soil. Therefore, the biological activity of an allelochemical might be altered before reaching the potential target due to microbial degradation, oxidation or immobilization by irreversible binding to soil particles (Cheng, 1995).

Neumann and Römheld (2007) and Weston et al. (2013) described various mechanisms of root exudation including diffusion, vesicle transport and ion channels. A brief description of these mechanisms is given below.

5.2 Diffusion

Low molecular weight organic compounds such as sugars, amino acids, carboxylic acids and phenolics are released passively through a gradient of concentration between the cytoplasm of living root cells and the surrounding soil. This particular type of excretion depends on the physiological state of the root cells and the polarity of the organic compounds. The hydrophobic nature of plasma bilayer generally facilitates the transport of lipophilic compounds by diffusion.

5.3 Vesicle Transport

Transport of high molecular weight organic compounds (e.g., mucilage polysaccharides) is mediated through the formation of Golgi vesicles from secretory cells of the root cap. These cells are subsequently degenerated or sloughed off from the surface of the roots (Bertin et al., 2003). Enzymes such as acid phosphatase and peroxidase, synthesized at the level of membrane-bound polysomes, enter the lumen of endoplasmic reticulum using vectorial segregation (Neumann & Römheld, 2007). This exocytosis of vesicles depends on intracellular and extracellular calcium level.

5.4 Ion Channels

Exudation of some compounds (e.g., citrate, malate, oxalate) at higher concentrations, particularly under stressful conditions, cannot take place though simple diffusion in certain plant species (Neumann & Römheld, 2007). The use of patch clamp approaches and ion channel antagonist studies indicated the presence of ion channels on the plasma membrane of epidermal root cells involved in the release of malate and citrate in the *Triticum spp* (wheat) and corn (Weston et al., 2013).

6. FACTORS INFLUENCING THE RELEASE OF ALLELOCHEMICALS FROM THE PLANT

The nature and concentration of allelochemicals released by the plant into the environment, including the rhizosphere, is dependent on the plant itself and several other biotic and abiotic factors as shown in Fig. 11.

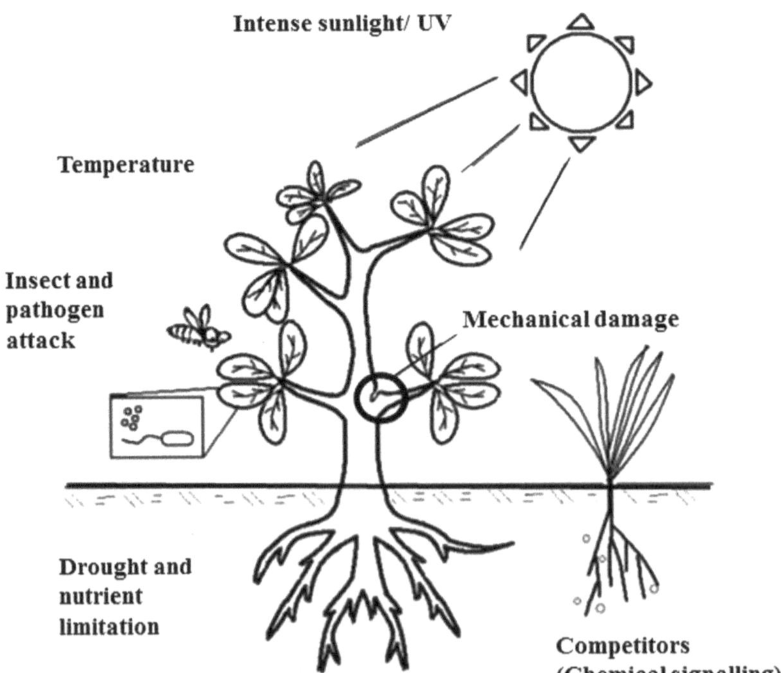

Figure 11 The release of allelochemicals in the environment under the influence of various factors (Albuquerque et al., 2011).

Plant factors include plant species and cultivar, plant age and the type of tissue under consideration for allelochemical production. Environmental factors regulating allelochemical production and release include pathogen infestation or physical injury or abiotic factors including drought, temperature, soil characteristics, rainfall, nutrient deficiency, irradiation, competitors and exposure to ultraviolet radiation (Brimecombe, Leij, & Lynch, 2001; Cseke & Kaufman, 1999; Einhellig, 1996; Iannucci, Fragasso, Platani, & Papa, 2013; Mahmood et al., 2013). A changing climate may also influence allelochemical production and this has been an area of particular interest in recent years (Jassey et al., 2013; Putten et al., 2013; Weston et al., 2013).

Taxonomically related species do not always release similar allelochemicals or concentrations of metabolites into the environment (Imatomi, Novaes, & Gualtieri, 2013). The release and dispersal of allelochemicals into the external environment are influenced by the chemical nature of allelochemicals including their respective molecular weight, polarity and concentration in the plant. Allelochemicals are most often released in mixtures in conjunction with other closely related metabolites. It is important to note that these metabolites can serve more than one biological role in the plant, and one compound can also defend its host from several unrelated competitors or predators (Macías, Molinillo, Varela, & Galindo, 2007). For example, phenolic compounds are known to stimulate and/or inhibit fungi in association with higher plants; certain fungi utilize phenyloxidase activities to metabolize and thus mitigate the toxicity of phenolic compounds produced by higher plants (Jassey, Chiapusio, Gilbert, Buttler, Toussaint, & Binet, 2011; Sinsabaugh, 2010).

The selection and use of bioassays to investigate the allelopathic potential of related metabolites must be carefully planned in any experimentation performed. Allelopathic activity may be seriously impacted by the mixture of metabolites present in complex mixtures, and therefore activity might be associated with complex molecular interactions including synergy, antagonism and enhanced effects in the presence of other metabolites (Albuquerque et al., 2011; Einhellig, 1996; Neilson et al., 2013). To estimate the bioactivity of allelochemicals, a dose/response study is essential to establish their potential effects on the environment. When studies utilize a dose significantly higher than the concentration(s) naturally present in the soil, the results are generally difficult to interpret from an allelopathic perspective. Moreover, the mode of action of allelochemicals can differ when applied at a dose well beyond that encountered in nature (Fujii & Hiradate, 2007).

7. ROLE(S) OF ALLELOCHEMICALS IN THE RHIZOSPHERE, IN NEIGHBOURING PLANTS AND OTHER ORGANISMS

The rhizosphere is a narrow zone of soil around living plant roots that is also inhabited by diverse groups of microorganisms; here plant roots compete with root systems of neighbouring plants and with other soilborne organisms, including bacteria and fungi (Weston et al., 2012). The rhizosphere itself is not easily defined by size but rather by its physical and biochemical properties, with a decreasing chemical gradient away from living plant roots. Newman, Fitter, Atkinson, Read, and Usher (1985) estimated that the roots of a typical higher plant can release 10–40% of photosynthetically fixed carbon in the form of organic and inorganic compounds into the rhizosphere. These root-produced products are collectively called rhizodeposits and include exudate, mucilage, sloughed off border cells and root cap cells (McNear Jr, 2013). As presented in Table 1, the majority of the root exudate consists of diverse low molecular weight organic compounds (Baetz & Martinoia, 2014).

Allelopathic interference in the rhizosphere is well documented under controlled conditions. However, allelopathy remains difficult to separate from competition among plants of different species for growth limiting nutrient resources under field conditions (Weston & Duke, 2003). However, recent advances have increased our knowledge about physiochemical properties of roots. Belowground interactions between roots and microbes are continuously occurring in the rhizosphere and are often difficult to study due to their presence in a complex soil matrix (Cipollini et al., 2012; Inderjit, Weston & Duke, 2005; McCully, 1999). There has been increasing evidence which suggests that root exudates provide a system of simultaneous communication with neighbouring plants, as well as with symbiotic and pathogenic organisms through biological and physical interactions (Fig. 11) (Bais, Park, Weir, Callaway, & Vivanco, 2004). Root exudates have been shown to regulate symbiotic relationships with microorganisms in the rhizosphere, protect against herbivory, alter the chemical and physiochemical properties of the soil and affect the growth of other plants in the rhizosphere (Mathesius & Watt, 2010; Nardi et al., 2000). Recently, secondary metabolites produced by plants have been shown to inhibit microbially mediated denitrification in the rhizosphere of invasive plants, thereby impacting growth and available N for the invasive community (Bardon et al., 2014). Effective symbioses between plants and microorganisms are generally facilitated by electrostatic and chemotactic responses created by the root and its exudates (Van West et al., 2002; Zheng & Sinclair, 1996).

Root exudates can also be associated with stimulation of plant growth, through the attraction of useful microbial communities to the plant rhizosphere having subsequent positive effects on plant growth or by direct effects of the plant-produced metabolites (Bais et al., 2006). The impact of allelochemicals in root exudates on neighbouring plant growth is dependent on the concentration of allelochemicals in the exudate, their relative rate of release, the vegetative stage of the target plant, as well as on the biotic and abiotic environmental conditions (Gniazdowska & Bogatek, 2005; Hill, Ngouajio, & Nair, 2006; Putnam, 1988). It is important to consider that allelochemicals can only mediate negative interference if they are released in sufficient quantities and persist long enough in the environment to remain active. Weidenhamer (1996) has demonstrated in controlled conditions that growth stage, biomass and density of target plants are important factors that alter the availability of allelochemicals released into the rhizosphere and should be considered when performing studies with allelochemicals. A greater density of receiver plants will thus require higher concentrations of allelochemicals provided by the donor to generate similar results in contrast to a reduced density. Therefore, interactions that are documented in one location may not be easily replicated in another natural setting due to varying conditions, including plant densities, encountered.

After release from plant tissue, other processes including leaching, oxidation, biodegradation and uptake by neighbouring plants can influence interference of allelochemicals upon target plants (Inderjit, 2001). Root exudates produced under the influence of chemical and biological elicitors may also act as antimicrobial agents against pathogenic bacteria in the rhizosphere (Zhu et al., 2016). Plant roots typically defend themselves against bacteria and fungi through the action of detachable border cells; these cells located at the tips of actively growing roots produce a vast assortment of bioactive secondary metabolites including allelochemicals. As an example, a newly germinating plant uses products generated by border cells to aid in plant defence, particularly during this sensitive period before the root system is well established in a field setting. Root border cells also produce extracellular DNA and associated proteins which can act as a trap for soilborne pathogens which are later negatively impacted by associated allelochemical production (Hawes et al., 2016). Oats produce avenicin, an important allelochemical and metabolite with potent activity against plant pathogenic organisms. This interaction has been well characterized by the Osbourn laboratory which has shown that triterpenes produced by oats play critical roles in defence against soil microbes and plant pathogens as well as abiotic stressors (Owatworakit et al., 2013). Interestingly, oats have diverged chemically from other cereal grasses which do not produce this family of compounds.

7.1 Tolerance to Allelochemicals

Some plant species are intrinsically insensitive to specific phytotoxins or have evolved mechanisms of defence from allelochemicals produced by the plant itself or from other species (Shitan, 2016). Plants producing allelochemicals as well as those absorbing them can resist toxicity by actively sequestering toxins in membrane-bound structures, vacuoles or vesicles, secreting the compounds immediately after absorption, or altering their chemical structure by primary and secondary metabolism (Bais et al., 2006). For example, *Polygonella myriophylla* (Small) Horton releases hydroquinone and benzoquinone allelochemicals in the form of arbutin, a glycoside of hydroquinone, which prevents autotoxicity (Weidenhamer & Romeo, 2004). Likewise, corn (*Z. mays* L.) produces inactive *N*−glycosylated forms of DIMBOA, DIBOA and BOA (Fig. 9), characteristic benzoxazolinones produced by several *Poaceae* spp (Schulz et al., 2016). Over millions of years, evolution has endowed selected plant species with the capacity to tolerate or metabolize commonly encountered phytotoxins. However, those species which do not encounter phytotoxins frequently are predicted to be less likely to develop resistance (Bais et al., 2006). A recent hypothesis suggests that greater success of invasive weeds is associated with their allelopathic potential to negatively affect native species by the production of unique allelochemicals or 'novel weapons' which can successfully interfere with the growth of native plant communities that are not well adapted to the presence of these bioactive metabolites (Callaway & Aschehoug, 2000).

8. METABOLIC PROFILING OF ALLELOCHEMICALS IN COMPLEX PLANT OR SOIL EXTRACTS OR MIXTURES

The biochemical interactions occurring in the rhizosphere are the least well characterized in all of the biotic zones studied in terrestrial ecology. Despite these challenges, new technological advancements in metabolite detection and identification are proving useful when studying complex systems and interactions, and solving ecological questions about regulation of allelochemicals and their roles in ecosystem function. Metabolic profiling is proving to be an important tool when studying these complex plant interactions and is utilized to study the plant metabolome (the total collection of primary and secondary products produced by a plant). By performing an extraction of plant tissue and evaluating the metabolome, one can assess the functional state of a biological system at a particular point in time (Roessner & Bacic, 2009).

Weston, Skoneczny, Weston, and Weidenhamer (2015) utilized a type of metabolomics referred to as metabolic profiling that is a set of analytical procedures designed to study targeted compounds of interest, including specific allelochemicals, in a biological system in response to a particular treatment. Metabolic profiling can provide a strong insight on the biochemical status of an organism so one might further understand complex interactions following the release of allelochemicals in the rhizosphere, or in some cases their intermediate metabolites (Putnam, 1988; Rice, 1974). It can also be used in conjunction with other *omics* technologies (proteomics and transcriptomics) to fully construct the biosynthetic pathways of targeted compounds of interest. Metabolic profiling of specific allelochemicals may also be employed to study possible mechanisms associated with invasion of weeds. Skoneczny et al. (2015) described significant upregulation of pyrrolizidine alkaloids and their N-oxides in the highly invasive species, *E. plantagineum* L., in contrast to the less successful invader, *Echium vulgare* L., in both field and laboratory settings, through the use of liquid chromatography coupled to mass spectrometry (LC-MS) for evaluation of a series of related pyrrolizidine alkaloids, including the toxins echiumine, leptanthine and echimidine.

Metabolomics is a high throughput approach which produces results in a relatively short period of time when analyzing large sample sets (Scognamiglio, D'Abrosca, Esposito, & Fiorentino, 2015). Progress in this field has been associated with the development of a diverse range of analytical platforms including gas and liquid column chromatography coupled with high-resolution mass spectrometry for targeted and nontargeted metabolic profiling of secondary metabolites and allelochemicals (Kim, Choi, & Verpoorte, 2010; Weston et al., 2015). More recently, the development of triple quadrupole MS or sensitive ion trap MS has facilitated the precise and accurate profiling for putative annotation of thousands of bioactive compounds present in very low quantity in complex matrices (Weston et al., 2015). A thorough description of how allelochemicals may be profiled by using metabolomics approaches is discussed in the review prepared by Weston et al., (2015). Most recently, metabolic profiling has also been applied to studies of the plant rhizosphere. One study was performed to evaluate the release of diverse flavonoids from clover (*Trifolium spp.*) in the soil rhizosphere when clovers have been utilized as cover crops, pointing to the role of the allelochemical kaempferol which persisted in the soil rhizosphere and was likely to be associated with consequent allelopathy in perennial stands of white clover (Weston & Mathesius, 2013). Additional studies to evaluate the role of a complex group of over 15 bioactive naphthoquinones, the shikonins, in plant defence and allelopathy in the rhizosphere, have also been facilitated by the use of sensitive metabolic profiling performed by UPLC QToF-MS (Zhu et al., 2016). Recovery of some of the more persistent allelochemicals including acetyl shikonin and shikonin in ppm levels in infested soils was also successful using LC/MS technologies. The increasing use of metabolomics to study metabolite dynamics and biochemical pathways in plants will no doubt result in an improved understanding of the presence and the role of allelochemicals in the soil environment.

After analysis by metabolomics, processing of substantial amounts of mass spectral data and subsequent characterization of metabolites are attempted (Smith et al., 2005). Although it is never possible to identify all plant metabolites in a complex mixture, and many remain currently unidentified or are undescribed. Such challenges are dealt with using efficient throughput computational and analysis tools which often involve data conversion, features detection, normalization and recently developed quality control of data processing (Sugimoto, Kawakami, Robert, Soga, & Tomita, 2012). The molecular features of the compounds present are normalized for baseline variation and aligned with available structural libraries. The METLIN library is one of the largest databases used for identification of natural products (Weston et al., 2015). Greater flexibility across its catalogues and multipurpose search options make it very convenient and widely used database (Smith et al., 2005).

9. CONCLUSIONS

Allelopathy, the study of complex plant–plant and plant–microbial interactions, is a very active field of research in the plant sciences. Recently, the number of researchers and research papers in the field has increased exponentially. Enhanced collaboration among soil scientists, chemists, ecologists, geneticists and molecular biologists will undoubtedly lead to new insights as to the factors regulating the production of allelochemicals in higher plants and how these metabolites impact invasion ecology, plant competition, plant interference, vegetation dynamics and crop production. The use of new analytical techniques associated with metabolomics in concert with other *omics* technologies has led to new advances in the identification of unique allelochemicals, the biosynthetic pathways associated with their production, their complex role(s) in the soil rhizosphere and their production as impacted by a changing climate. A better understanding of allelochemical production with respect to plant defence strategy may also allow us to better protect and manage developing crops, limit the spread of invasive weeds, preserve native plant stands, and create strategies for allelochemical development and application as novel pesticides.

Currently, increasing public awareness regarding safety and environmental issues associated with the use of herbicides, as well as profitable markets for organic commodities has resulted in greater emphasis on the development of natural product-based pesticide discovery programs (Duke et al., 2002; Weston & Duke, 2003). Identification of novel plant metabolites, including allelochemicals, may result in a source for future development of biologically based pesticides (Chiapusio et al., 2005, pp. 139–155), through the provision of complementary structures for synthetic compounds (Ghisalberti, 2007) and as an aid in the development of new molecular target sites (Duke et al., 2002).

REFERENCES

Abenavoli, M., Lupini, A., Oliva, S., & Sorgonà, A. (2010). Allelochemical effects on net nitrate uptake and plasma membrane H+-ATPase activity in maize seedlings. *Biologia Plantarum, 54*(1), 149–153.

Abrahim, D., Takahashi, L., Kelmer-Bracht, A., & Ishii-Iwamoto, E. (2003). Effects of phenolic acids and monoterpenes on the mitochondrial respiration of soybean hypocotyl axes. *Allelopathy Journal, 11*(1), 21–30.

Afendi, F. M., Okada, T., Yamazaki, M., Hirai-Morita, A., Nakamura, Y., Nakamura, K., ... Darusman, L. K. (2012). KNApSAcK family databases: integrated metabolite–plant species databases for multifaceted plant research. *Plant & Cell Physiology, 53*(2), 1–11.

Ahuja, I., Kissen, R., & Bones, A. M. (2012). Phytoalexins in defense against pathogens. *Trends in Plant Science, 17*(2), 73–90. http://dx.doi.org/10.1016/j.tplants.2011.11.002.

Albuquerque, M. B., Cavalcanti dos Santos, R., Lima, L. M., de Albuquerque Melo Filho, P., Custódio Nogueira, R. J. M., Gomes da Câmara, C. A., & de Rezende Ramos, A. (2011). Allelopathy, an alternative tool to improve cropping systems. A review. *Agronomy of Sustainable Development, 31*, 379–395.

Baetz, U., & Martinoia, E. (2014). Root exudates: the hidden part of plant defense. *Trends in Plant Science, 19*(2), 90–98. http://dx.doi.org/10.1016/j.tplants.2013.11.006.

Bais, H. P., Park, S.-W., Weir, T. L., Callaway, R. M., & Vivanco, J. M. (2004). How plants communicate using the underground information superhighway. *Trends in Plant Science, 9*(1), 26–32. http://dx.doi.org/10.1016/j.tplants.2003.11.008.

Bais, H. P., Weir, T. L., Perry, L. G., Gilroy, S., & Vivanco, J. M. (2006). The role of root exudates in rhizosphere interactions with plants and other organisms. *Annual Review of Plant Biology, 57*(1), 233–266. http://dx.doi.org/10.1146/annurev.arplant.57.032905.105159.

Bardon, C., Piola, F., Bellvert, F., Haichar, F. Z., Comte, G., Meiffren, G., … Poly, F. (2014). Evidence for biological denitrification inhibition (BDI) by plant secondary metabolites. *New Phytologist, 204*(3), 620–630. http://dx.doi.org/10.1111/nph.12944.

Bartwal, A., Mall, R., Lohani, P., Guru, S., & Arora, S. (2013). Role of secondary metabolites and brassinosteroids in plant defense against environmental stresses. *Journal of Plant Growth Regulation, 32*(1), 216–232.

Ben, F., Jordan, F., & Osbourn, A. (2006). First encounters: deployment of defence-related natural products by plants. *New Phytologist, 172*(2), 193–207.

Bernards, M. A. (2010). Plant natural products: a primer. *Canadian Journal of Zoology, 88*(7), 601–614.

Bertin, C., Yang, X., & Weston, L. A. (2003). The role of root exudates and allelochemicals in the rhizosphere. *Plant and Soil, 256*(1), 67–83. http://dx.doi.org/10.1023/a:1026290508166.

Blum, U. (2014). *Some issues and challenges when designing laboratory bioassays plant-plant allelopathic interactions II.* North Carolina, USA: Springer.

Boawn, L. C. (1965). Sugar beet induced zinc deficiency. *Agronomy Journal, 57*(5), 509.

Brimecombe, M. J., Leij, F. A., & Lynch, J. M. (2001). Nematode community structure as a sensitive indicator of microbial perturbations induced by a genetically modified *Pseudomonas fluorescens* strain. *Biology and Fertility of Soils, 34*(4), 270–275.

Callaway, R. M., & Aschehoug, E. T. (2000). Invasive plants versus their new and old neighbors: a mechanism for exotic invasion. *Science, 290*(5491), 521–523.

Carlsen, S. C., Pedersen, H. A., Spliid, N. H., & Fomsgaard, I. S. (2012). Fate in soil of flavonoids released from white clover (*Trifolium repens* L.). *Applied and Environmental Soil Science, 2012*, 1–10.

Chai, T.-T., Ooh, K.-F., Ooi, P.-W., Chue, P.-S., & Wong, F.-C. (2013). *Leucaena leucocephala* leachate compromised membrane integrity, respiration and antioxidative defence of water hyacinth leaf tissues. *Botanical Studies, 54*(1), 1–7. http://dx.doi.org/10.1186/1999-3110-54-8.

Champagne, A., & Boutry, M. (2016). Proteomics of terpenoid biosynthesis and secretion in trichomes of higher plant species. *Biochimica et Biophysica Acta (BBA) – Proteins and Proteomics, 1864*(8), 1039–1049. http://dx.doi.org/10.1016/j.bbapap.2016.02.010.

Cheng, H. H. (1995). Characterization of the mechanisms of allelopathy: modeling and experimental approaches. In Inderjit, K. M. M. Dakshini, & F. A. Einhellig (Eds.), *Allelopathy* (pp. 132–141). Washington, DC: American Chemical Society.

Cheng, & Cheng, Z. H. (2015). Research progress on the use of plant allelopathy in agriculture and the physiological and ecological mechanisms of allelopathy. *Frontiers in Plant Science, 6*, 1020.

Cheynier, V., Comte, G., Davies, K. M., Lattanzio, V., & Martens, S. (2013). Plant phenolics: recent advances on their biosynthesis, genetics, and ecophysiology. *Plant Physiology and Biochemistry, 72*, 1–20. http://dx.doi.org/10.1016/j.plaphy.2013.05.009.

Chiapusio, G., Gallet, C., Dobremez, J., Pellissier, F., Regnault-Roger, C., Philogène, B., & Vincent, C. (2005). *Allelochemicals: Tomorrow's herbicides?* Hampshire, UK: Intercept Ltd – Lavoisier Publ. Inc.

Chiapusio, G., Jassey, V. E. J., Hussain, M. I., & Binet, P. (2013). Evidences of bryophyte allelochemical interactions: the case of *Sphagnum*. In Z. A. Cheema, M. Farooq, & A. Wahid (Eds.), *Allelopathy: Current trends and future applications* (pp. 39–54). Springer Science & Business Media.

Cipollini, D., Rigsby, C. M., & Barto, E. K. (2012). Microbes as targets and mediators of allelopathy in plants. *Journal of Chemical Ecology, 38*(6), 714–727. http://dx.doi.org/10.1007/s10886-012-0133-7.

Croteau, R., Kutchan, T. M., & Lewis, N. G. (2000). Natural products (secondary metabolites). *Biochemistry and Molecular Biology of Plants, 24*, 1250–1319.

Cseke, L. J., & Kaufman, P. B. (1999). Regulation of metabolite synthesis in plants. In P. B. Kaufman, L. J. Cseke, S. Warber, J. A. Duke, & H. L. Brielmann (Eds.), *Natural products from plants* (pp. 91–122). Boca Raton, FL: CRC Press.

Czarnota, M. A., Paul, R. N., Dayan, F. E., Chandrashekhar, I. N., & Weston, L. A. (2001). Mode of action, localization of production, chemical nature, and activity of sorgoleone: a potent PSII inhibitor in Sorghum spp. Root exudates. *Weed Technology, 15*(4), 813–825.

Dalton, B. (1999). *The occurrence and behavior of plant phenolic acids in soil environments and their potential involvement in allelochemical interference interactions: Methodological limitations in establishing conclusive proof of allelopathy. Principles and practices in plant ecology: Allelochemical interactions*. Boca Raton, Florida: CRC Press.

Delaux, P. M., Sejalon-Delmas, N., Becard, G., & Ane, J. M. (2013). Evolution of the plant-microbe symbiotic 'toolkit'. *Trends in Plant Science, 18*(6), 298–304. http://dx.doi.org/10.1016/j.tplants.2013.01.008.

Dinelli, G., Carretero, A. S., Di Silvestro, R., Marotti, I., Fu, S., Benedettelli, S., … Gutiérrez, A. F. (2009). Determination of phenolic compounds in modern and old varieties of durum wheat using liquid chromatography coupled with time-of-flight mass spectrometry. *Journal of Chromatography A, 1216*(43), 7229–7240.

Duke, S. O., Dayan, F. E., Agnes, M. R., Schrader, K. K., Giovanni, A., Anna, O., & Romagni, J. G. (2002). Chemicals from nature for weed management. *Weed Science, 50*(2), 138–151.

Duke, S. O., Oliva, A., Macias, F. A., Galindo, J. C. G., Molinillo, J. M. G., & Cutler, H. G. (2004). *Mode of action of phytotoxic terpenoids*. Boca Raton, FL: CRC Press.

Einhellig, F. A. (1995). Mechanism of action of allelochemicals in allelopathy. In Inderjit, K. M. M. Dakshini, & F. A. Einhellig (Eds.), *Allelopathy* (pp. 96–116). Washington, DC: American Chemical Society.

Einhellig, F. A. (1996). Interactions involving allelopathy in cropping systems. *Agronomy Journal, 88*(6), 886–893.

Einhellig, F. A. (2004). Mode of allelochemical action of phenolic compounds. In F. A. Macías, J. C. G. Galindo, J. M. G. Molinillo, & H. G. Cutler (Eds.), *Allelopathy: Chemistry and mode of action of allelochemicals*. Boca Raton, Florida: CRC Press.

Elek, H., Smart, L., Martin, J., Ahmad, S., Gordon-Weeks, R., Welham, S., … Werner, C. (2013). The potential of hydroxamic acids in tetraploid and hexaploid wheat varieties as resistance factors against the bird-cherry oat aphid, *Rhopalosiphum padi*. *Annals of Applied Biology, 162*(1), 100–109.

Eom, S. H., Senesac, A. F., Tsontakis-Bradley, I., & Weston, L. A. (2005). Evaluation of herbaceous perennials as weed suppressive groundcovers for use along roadsides or in landscapes. *Journal of Environmental Horticulture, 23*(4), 198.

Eom, S. H., Yang, H. S., & Weston, L. A. (2006). An evaluation of the allelopathic potential of selected perennial groundcovers: foliar volatiles of catmint (*Nepeta × faassenii*) inhibit seedling growth. *Journal of Chemical Ecology, 32*(8), 1835–1848.

Fengzhi, W., Kai, P., Fengming, M., & Xuedong, W. (2004). Effects of cinnamic acid on photosynthesis and cell ultrastructure of cucumber seedlings. *Acta Horticulturae Sinica, 31*(2), 183–188.

Fernandez, C., Monnier, Y., Santonja, M., Gallet, C., Weston, L. A., Prévosto, B., … Bousquet-Mélou, A. (2016). The impact of competition and allelopathy on the trade-off between plant defense and growth in two contrasting tree species. *Frontiers in Plant Science, 7*, 594. http://dx.doi.org/10.3389/fpls.2016.00594.

Fernandez, C., Santonja, M., Gros, R., Monnier, Y., Chomel, M., Baldy, V., & Bousquet-Mélou, A. (2013). Allelochemicals of *Pinus halepensis* as drivers of biodiversity in mediterranean open mosaic habitats during the colonization stage of secondary succession. *Journal of Chemical Ecology, 39*(2), 298–311.

Finlay, R. D. (2008). Ecological aspects of mycorrhizal symbiosis: with special emphasis on the functional diversity of interactions involving the extraradical mycelium. *Journal of Experimental Botany, 59*(5), 1115–1126. http://dx.doi.org/10.1093/jxb/ern059.

Franche, C., Lindström, K., & Elmerich, C. (2009). Nitrogen-fixing bacteria associated with leguminous and non-leguminous plants. *Plant and Soil, 321*(1–2), 35–59.

Fujii, Y., & Hiradate, S. (2007). *Allelopathy: New concepts & methodology*. Hew Hamshire: USA Science Publishers.

Ghisalberti, E. L. (2007). Detection and isolation of bioactive natural products. In S. M. Colegate, & R. J. Molyneux (Eds.), *Bioactive natural products: Detection, isolation, and structural determination* (2nd ed., pp. 11–65). Boca Raton, FL: CRC Press.

Gniazdowska, A., & Bogatek, R. (2005). Allelopathic interactions between plants. Multi site action of allelochemicals. *Acta Physiologiae Plantarum, 27*(3B), 395–407.

Gonzalez, V. M., Kazimir, J., Nimbal, C., Weston, L. A., & Cheniae, G. M. (1997). Inhibition of a photosystem II electron transfer reaction by the natural product sorgoleone. *Journal of Agricultural and Food Chemistry, 45*(4), 1415–1421. http://dx.doi.org/10.1021/Jf960733w.

Graña, E., Sotelo, T., Díaz-Tielas, C., Reigosa, M. J., & Sánchez-Moreiras, A. M. (2013). The phytotoxic potential of the terpenoid citral on seedlings and adult plants. *Weed Science, 61*(3), 469–481. http://dx.doi.org/10.1614/WS-D-12-00159.1.

Haig, T. (2008). *Allelochemicals in plants. Allelopathy in sustainable agriculture and forestry*. Springer.

Hattenschwiler, S., & Vitousek, P. M. (2000). The role of polyphenols in terrestrial ecosystem nutrient cycling. *Trends in Ecology & Evolution, 15*(6), 238–243. http://dx.doi.org/10.1016/S0169-5347(00)01861-9.

Hawes, M., Allen, C., Turgeon, B. G., Curlango-Rivera, G., Minh, T. T., Huskey, D. A., & Xiong, Z. (2016). Root border cells and their role in plant defense. *Annual Review of Phytopathology, 355*(1), 1–16. http://dx.doi.org/10.1007/s11104-012-1218-3.

Herz, W. (1963). The organic constituents of higher plants, their chemistry and interrelationships. *Journal of the American Chemical Society, 85*(18), 2876. http://dx.doi.org/10.1021/ja00901a064.

He, H., Wang, H., Fang, C., Wu, H., Guo, X., Liu, C., ... Lin, W. (2012). Barnyard grass stress up regulates the biosynthesis of phenolic compounds in allelopathic rice. *Journal of Plant Physiology, 169*(17), 1747–1753.

Hill, E. C., Ngouajio, M., & Nair, M. G. (2006). Differential response of weeds and vegetable crops to aqueous extracts of hairy vetch and cowpea. *Horticulture Science, 41*(3), 695–700.

Iannucci, A., Fragasso, M., Platani, C., & Papa, R. (2013). Plant growth and phenolic compounds in the rhizosphere soil of wild oat (*Avena fatua* L.). *Frontiers in Plant Science, 4*(509), 1–7. http://dx.doi.org/10.3389/fpls.2013.00509.

Imatomi, M., Novaes, P., & Gualtieri, S. C. J. (2013). Interspecific variation in the allelopathic potential of the family Myrtaceae. *Acta Botanica Brasilica, 27*(1), 54–61.

Inderjit. (2001). Soil: environmental effects on allelochemical activity. *Agronomy Journal, 93*(1), 79–84.

Inderjit, & Duke, S. O. (2003). Ecophysiological aspects of allelopathy. *Planta, 217*(4), 529–539. http://dx.doi.org/10.1007/s00425-003-1054-z.

Inderjit, & Weston, L. (2000). Are laboratory bioassays for allelopathy suitable for prediction of field responses? *Journal of Chemical Ecology, 26*(9), 2111–2118. http://dx.doi.org/10.1023/A:1005516431969.

Inderjit, Weston, L. A., & Duke, S. O. (2005). Challenges, achievements and opportunities in allelopathy research. *Journal of Plant Interactions, 1*(2), 69–81. http://dx.doi.org/10.1080/17429140600622535.

Jassey, Chiapusio, G., Binet, P., Buttler, A., Laggoun-Defarge, F., Delarue, F., … Gilbert, D. (2013). Above- and belowground linkages in *Sphagnum peatland*: climate warming affects plant-microbial interactions. *Global Change Biology, 19*(3), 811–823. http://dx.doi.org/10.1111/gcb.12075.

Jassey, Chiapusio, G., Gilbert, D., Buttler, A., Toussaint, M. L., & Binet, P. (2011). Experimental climate effect on seasonal variability of polyphenol/phenoloxidase interplay along a narrow fen-bog ecological gradient in *Sphagnum fallax*. *Global Change Biology, 17*(9), 2945–2957. http://dx.doi.org/10.1111/j.1365-2486.2011.02437.x.

Kato-Noguchi, H. (2003). Allelopathic substances in *Pueraria thunbergiana*. *Phytochemistry, 63*(5), 577–580.

Kato-Noguchi, H. (2011). Barnyard grass-induced rice allelopathy and momilactone B. *Journal of Plant Physiology, 168*(10), 1016–1020.

Kim, H. K., Choi, Y. H., & Verpoorte, R. (2010). NMR-based metabolomic analysis of plants. *Nature Protocols, 5*(3), 536–549. http://dx.doi.org/10.1038/nprot.2009.237.

Li, Z.-H., Wang, Q., Ruan, X., Pan, C.-D., & Jiang, D.-A. (2010). Phenolics and plant allelopathy. *Molecules: a Journal of Synthetic Chemistry and Natural Product Chemistry, 15*(12), 8933–8952.

Lovett, J., & Hoult, A. (1995). Allelopathy and self-defense in barley. In Inderjit, K. M. M. Dakshini, & F. A. Einhellig (Eds.), *Allelopathy* (Vol. 582, pp. 170–183). Washington, DC: American Chemical Society.

Macías, F. A., Galindo, J. C., Molinillo, J. M., Cutler, H. G., Martínez, A., Reigosa, M., … González, L. (2003). *Mode of action of the hydroxamic acid BOA and other related compounds. Allelopathy: Chemistry and mode of action of allelochemicals*. CRC Press.

Macías, F. A., Molinillo, J. M. G., Varela, R. M., & Galindo, J. C. G. (2007). Allelopathy—a natural alternative for weed control. *Pest Management Science, 63*(4), 327–348. http://dx.doi.org/10.1002/ps.1342.

Mahmood, K., Khan, M. B., Song, Y. Y., Ijaz, M., Luo, S. M., & Zeng, R. S. (2013). UV-irradiation enhances rice allelopathic potential in rhizosphere soil. *Plant Growth Regulation, 71*(1), 21–29. http://dx.doi.org/10.1007/s10725-013-9804-9.

Mathesius, U., & Watt, M. (2010). Rhizosphere signals for plant–microbe interactions: implications for field-grown plants. *Progress in Botany, 72*, 125–161 (Springer, Berlin Heidelberg).

McCully, M. E. (1999). Roots in soil: unearthing the complexities of roots and their rhizospheres. *Annual Review of Plant Biology, 50*(1), 695–718.

McNear, D. H., Jr. (2013). The rhizosphere-roots, soil and everything in between. *Nature Education Knowledge, 4*(3), 1.

Michel, P., Burritt, D. J., & Lee, W. G. (2011). Bryophytes display allelopathic interactions with tree species in native forest ecosystems. *Oikos, 120*(8), 1272–1280. http://dx.doi.org/10.1111/j.1600-0706.2010.19148.x.

Molisch, H. (1937). *Der Einfluss einer Pflanze aufdie andere-Allelopathie* (pp. 158–160). Jena: Fischer.

Nardi, S., Concheri, G., Pizzeghello, D., Sturaro, A., Rella, R., & Parvoli, G. (2000). Soil organic matter mobilization by root exudates. *Chemosphere, 41*(5), 653–658.

Neilson, E. H., Goodger, J. Q. D., Woodrow, I. E., & Møller, B. L. (2013). Plant chemical defense: at what cost? *Trends in Plant Science, 18*(5), 250–258. http://dx.doi.org/10.1016/j.tplants.2013.01.001.

Neumann, G., & Römheld, V. (2007). The release of root exudates as affected by the plant physiological status. In P. Robert, V. Zeno, & N. Paolo (Eds.), *The Rhizosphere: Biochemistry and organic substances at the soil-plant interface* (pp. 23–72). NY: CRC.

Newman, E., Fitter, A., Atkinson, D., Read, D. J., & Usher, M. (1985). The rhizosphere: carbon sources and microbial populations. *Ecological Interactions in Soil: Plants, Microbes and Animals*, 107–121.

Niemeyer, H. M. (2009). Hydroxamic acids derived from 2-hydroxy-2H-1,4-benzoxazin-3(4H)-one: key defense chemicals of cereals. *Journal of Agricultural and Food Chemistry, 57*(5), 1677–1696. http://dx.doi.org/10.1021/jf8034034.

Niklas, K. J., & Hammond, S. T. (2013). Biophysical effects on plant competition and coexistence. *Functional Ecology, 27*(4), 854–864.

Nimbal, C. I., Yerkes, C. N., Weston, L. A., & Weller, S. C. (1996). Herbicidal activity and site of action of the natural product sorgoleone. *Pesticide Biochemistry and Physiology, 54*(1), 73–83. http://dx.doi.org/10.1006/pest.1996.0011.

Owatworakit, A., Townsend, B., Louveau, T., Jenner, H., Rejzek, M., Hughes, R. K., ... Osbourn, A. (2013). Glycosyltransferases from oat (*Avena*) implicated in the acylation of avenacins. *The Journal of Biological Chemistry, 288*(6), 3696–3704. http://dx.doi.org/10.1074/jbc.M112.426155.

Pedras, M. S. C., Yaya, E. E., & Glawischnig, E. (2011). The phytoalexins from cultivated and wild crucifers: chemistry and biology. *Natural Product Reports, 28*(8), 1381–1405.

Poonpaiboonpipat, T., Pangnakorn, U., Suvunnamek, U., Teerarak, M., Charoenying, P., & Laosinwattana, C. (2013). Phytotoxic effects of essential oil from *Cymbopogon citratus* and its physiological mechanisms on barnyardgrass (*Echinochloa crus-galli*). *Industrial Crops and Products, 41*, 403–407.

Putnam, A. R. (1988). Allelochemicals from plants as herbicides. *Weed Technology, 2*(4), 510–518.

Putnam, A. R., & Tang, C.-S. (1986). *The science of allelopathy*. Ithaca, NY, USA: John Wiley & Sons Inc.

Putten, W. H., Bardgett, R. D., Bever, J. D., Bezemer, T. M., Casper, B. B., Fukami, T., ... Schweitzer, J. A. (2013). Plant–soil feedbacks: the past, the present and future challenges. *Journal of Ecology, 101*(2), 265–276.

Quideau, S., Deffieux, D., Douat-Casassus, C., & Pouysegu, L. (2011). Plant polyphenols: chemical properties, biological activities, and synthesis. *Angewandte Chemie International Edition, 50*(3), 586–621.

Quinn, J. C., Kessell, A., & Weston, L. A. (2014). Secondary plant products causing photosensitization in grazing herbivores: their structure, activity and regulation. *International Journal of Molecular Sciences, 15*(1), 1441–1465. http://dx.doi.org/10.3390/ijms15011441.

Rashid, M. H., Asaeda, T., & Uddin, M. N. (2010). The allelopathic potential of kudzu (*Pueraria montana*). *Weed Science, 58*(1), 47–55.

Read, D. J., Leake, J. R., & Perez-Moreno, J. (2004). Mycorrhizal fungi as drivers of ecosystem processes in heathland and boreal forest biomes. *Canadian Journal of Botany-Revue Canadienne De Botanique, 82*(8), 1243–1263. http://dx.doi.org/10.1139/B04-123.

Rice, E. L. (1974). *Allelopathy*. Norman, Oklahoma: Acadamy Press.

Roessner, U., & Bacic, A. (2009). Metabolomics in plant research. *Australian Biochemistry, 40*, 9–20.

Schulz, M., Filary, B., Kühn, S., Colby, T., Harzen, A., Schmidt, J., ... Anders, N. (2016). Benzoxazolinone detoxification by N-Glucosylation: the multi-compartment-network of *Zea mays* L. *Plant Signaling & Behavior, 11*(1), e1119962. http://dx.doi.org/10.1080/15592324.2015.1119962.

Schulz, M., Marocco, A., Tabaglio, V., Macias, F. A., & Molinillo, J. M. (2013). Benzoxazinoids in rye allelopathy-from discovery to application in sustainable weed control and organic farming. *Journal of Chemical Ecology, 39*(2), 154–174.

Scognamiglio, M., D'Abrosca, B., Esposito, A., & Fiorentino, A. (2015). Metabolomics: an unexplored tool for allelopathy studies. *Journal of Allelochemical Interactions, 1*, 9–23.

Seigler, D. S. (1998). *Pyrrolizidine, quinolizidine, and indolizidine alkaloids. Plant secondary metabolism*. Springer.

Shitan, N. (2016). Secondary metabolites in plants: transport and self-tolerance mechanisms. *Bioscience, Biotechnology, and Biochemistry, 80*(7), 1283–1293. http://dx.doi.org/10.1080/09168451.2016.1151344.

Singh, H. P., Batish, D. R., Kaur, S., Arora, K., & Kohli, R. K. (2006). Alpha-Pinene inhibits growth and induces oxidative stress in roots. *Annals of Botany, 98*(6), 1261–1269. http://dx.doi.org/10.1093/aob/mcl213.

Singh, H. P., Batish, D. R., & Kohli, R. (2001). Allelopathy in agroecosystems: an overview. *Journal of Crop Production, 4*(2), 1–41.

Sinsabaugh, R. L. (2010). Phenol oxidase, peroxidase and organic matter dynamics of soil. *Soil Biology & Biochemistry, 42*(3), 391–404. http://dx.doi.org/10.1016/j.soilbio.2009.10.014.

Skoneczny, D., Weston, P. A., Zhu, X., Gurr, G. M., Callaway, R. M., & Weston, L. A. (2015). Metabolic profiling of pyrrolizidine alkaloids in foliage of two *Echium* spp. Invaders in Australia—a case of novel weapons? *International Journal of Molecular Sciences, 16*(11), 26721–26737.

Smith, C. A., Maille, G. O., Want, E. J., Qin, C., Trauger, S. A., Brandon, T. R., ... Siuzdak, G. (2005). METLIN: a metabolite mass spectral database. *Therapeutic Drug Monitoring, 27*(6), 747–751.

Souto, C., Pellissier, F., & Chiapusio, G. (2000). Allelopathic effects of humus phenolics on growth and respiration of mycorrhizal fungi. *Journal of Chemical Ecology, 26*(9), 2015–2023. http://dx.doi.org/10.1023/A:1005551912405.

Stalikas, C. D. (2007). Extraction, separation, and detection methods for phenolic acids and flavonoids. *Journal of Separation Science, 30*(18), 3268–3295.

Sugimoto, M., Kawakami, M., Robert, M., Soga, T., & Tomita, M. (2012). Bioinformatics tools for mass spectroscopy-based metabolomic data processing and analysis. *Current Bioinformatics, 7*(1), 96–108.

Van West, P.v., Morris, B., Reid, B., Appiah, A., Osborne, M., Campbell, T., Shepherd, S., & Gow, N. (2002). Oomycete plant pathogens use electric fields to target roots. *Molecular Plant-Microbe Interactions, 15*(8), 790–798.

VanEtten, H. D., Mansfield, J. W., Bailey, J. A., & Farmer, E. E. (1994). Two classes of plant antibiotics: phytoalexins versus "phytoanticipins". *The Plant Cell, 6*(9), 1191–1192. http://dx.doi.org/10.1105/tpc.6.9.1191.

Walton, N. J., & Brown, D. E. (1999). *Chemicals from plants: Perspectives on plant secondary products*. London: UK World Scientific.

Wang, R., Peng, S., Zeng, R., Ding, L. W., & Xu, Z. (2009). Cloning, expression and wounding induction of β-caryophyllene synthase gene from *Mikania micrantha* HBK and allelopathic potential of β-caryophyllene. *Allelopathy Journal, 24*, 35–44.

Weidenhamer, J. D. (1996). Distinguishing resource competition and chemical interference: overcoming the methodological impasse. *Agronomy Journal, 88*(6), 866–875.

Weidenhamer, J. D., Boes, P. D., & Wilcox, D. S. (2009). Solid-phase root zone extraction (SPRE): a new methodology for measurement of allelochemical dynamics in soil. *Plant and Soil, 322*(1), 177–186. http://dx.doi.org/10.1007/s11104-009-9905-4.

Weidenhamer, J. D., & Romeo, J. T. (2004). Allelochemicals of *Polygonella myriophylla*: chemistry and soil degradation. *Journal of Chemical Ecology, 30*(5), 1067–1082.

Weir, T. L., Park, S.-W., & Vivanco, J. M. (2004). Biochemical and physiological mechanisms mediated by allelochemicals. *Current Opinion in Plant Biology, 7*(4), 472–479. http://dx.doi.org/10.1016/j.pbi.2004.05.007.

Weston, L. A., & Duke, S. O. (2003). Weed and crop allelopathy. *Critical Reviews in Plant Sciences, 22*(3–4), 367–389. http://dx.doi.org/10.1080/713610861.

Weston, L. A., Harmon, R., & Mueller, S. (1989). Allelopathic potential of sorghum-sudangrass hybrid (sudex). *Journal of Chemical Ecology, 15*(6), 1855–1865.
Weston, L. A., & Mathesius, U. (2013). Flavonoids: their structure, biosynthesis and role in the rhizosphere, including allelopathy. *Journal of Chemical Ecology, 39*(2), 283–297. http://dx.doi.org/10.1007/s10886-013-0248-5.
Weston, L. A., Ryan, P. R., & Watt, M. (2012). Mechanisms for cellular transport and release of allelochemicals from plant roots into the rhizosphere. *Journal of Experimental Botany, 63*(9), 3445–3454. http://dx.doi.org/10.1093/jxb/ers054.
Weston, L. A., Skoneczny, D., Weston, P. A., & Weidenhamer, J. D. (2015). Metabolic profiling: an overview—new approaches for the detection and functional analysis of biologically active secondary plant products. *Journal of Allelochemical Interactions, 1*, 15–27.
Weston, P. A., Weston, L. A., & Hildebrand, S. (2013). Metabolic profiling in *Echium plantagineum* L: presence of bioactive pyrrolizidine alkaloids and napthoquinones from accessions across southeastern Australia. *Phytochemistry Reviews, 12*(4), 831–837. http://dx.doi.org/10.1007/s11101-013-9306-4.
Whittaker, R. H., & Feeny, P. P. (1971). Allelochemics: chemical interactions between species. *Science, 171*, 757–770.
Wink, M. (2004). Allelochemical properties of quinolizidine alkaloids. In F. A. Macias, J. C. G. Galindo, J. M. G. Molinillo, & H. Cutler (Eds.), *Allelopathy*. CRC Press.
Wink, M., & Latz-Bruning, B. (1995). Allelopathic properties of alkaloids and other natural products. In Inderjit, K. M. M. Dakshini, & F. A. Einhellig (Eds.), *Allelopathy: Organisms, processes and applications* (Vol. 582, pp. 117–126). Washington, DC: American Chemical Society.
Yang, L., & Stöckigt, J. (2010). Trends for diverse production strategies of plant medicinal alkaloids. *Natural Product Reports, 27*(10), 1469–1479.
Yang, Q., Ye, W., Liao, F., & Yin, X. (2005). Effects of allelochemicals on seed germination. *Chinese Journal of Ecology, 24*(12), 1459–1465.
Yoon, H. S., Moon, S. C., Kim, N. D., Park, B. S., Jeong, M. H., & Yoo, Y. H. (2000). Genistein induces apoptosis of RPE-J cells by opening mitochondrial PTP. *Biochemical and Biophysical Research Communications, 276*(1), 151–156. http://dx.doi.org/10.1006/bbrc.2000.3445.
Zheng, X., & Sinclair, J. (1996). Chemotactic response of *Bacillus megateriumstrain* B153-2-2 to soybean root and seed exudates. *Physiological and Molecular Plant Pathology, 48*(1), 21–35.
Zhu, X., Skoneczny, D., Weidenhamer, J. D., Mwendwa, J. M., Weston, P. A., Gurr, G. M., … Weston, L. A. (2016). Identification and localization of bioactive naphthoquinones in the roots and rhizosphere of Paterson's curse (*Echium plantagineum* L.), a noxious invader. *Journal of Experimental Botany, 67*(12), 3777–3788. http://dx.doi.org/10.1093/jxb/erw182.

CHAPTER THREE

Communication Between Host Plants and Parasitic Plants

P. Delavault[1], G. Montiel, G. Brun, J.-B. Pouvreau, S. Thoiron, P. Simier
Université de Nantes, Nantes, France
[1]Corresponding author: E-mail: philippe.delavault@univ-nantes.fr

Contents

Abstract

Parasitic plants are usually considered as botanical curiosities because of their shapes and colours. However, in some cases, they are proving to be terrible bioagressors in man-made ecosystems. Parasitic plants have indeed the capacity to connect intimately with other plants to exploit their resources (water, nutrients, growth regulators...) for their development. Thus, these plants are the result of an evolutionary transition from autotrophism to heterotrophism. The underlying process of this trophic exploitation, governed by a fine-tuned molecular dialogue between both partners, is an extraordinary example of adaptive plant biology operated by these parasitic organisms in the course of evolution. This transition is associated with remarkable morphological and physiological adaptations, and these adaptive strategies have led some parasitic plants to become deadly pests for crops against which no efficient control methods

Advances in Botanical Research, Volume 82
ISSN 0065-2296
http://dx.doi.org/10.1016/bs.abr.2016.10.006

are yet available, especially in terms of sustainability. This chapter highlights the actual knowledge regarding the molecular aspects of these intimate and intriguing communications between host plants and parasitic plants, with a special focus on root parasitic plants.

1. INTRODUCTION

Among the many parasitic plants, four biological types can be distinguished depending on their point of attachment to the host and on their capacity to carry out photosynthesis. Most of the parasitic plants invade the host root system (*Striga, Orobanche, Phelipanche, Tryphysaria*…), others, the epiphytic parasites, are attached to the aerial parts of their hosts, essentially stems but also leaves in some rare cases (*Cuscuta, Viscum*…). The latter, having no physical link with the soil, are completely dependent on their host for water and mineral supplies. Hemiparasitic plants (*Viscum, Striga, Tryphysaria*) are chlorophyllous and photosynthetically active and remain partially autotrophic for carbon. By contrast, the holoparasitic plants (*Orobanche, Phelipanche,* and *Cuscuta*) are achlorophyllous and then totally rely on their host plants to develop. Finally, it is also possible to differentiate facultative parasites (*Rhinanthus, Melampyrum, Euphrasia, Odontites, Tryphysaria*…) from obligate parasites (*Striga, Orobanche, Phelipanche, Cuscuta, Viscum*…). Facultative parasites, relatively frequent in natural ecosystems (meadows, moors…) are opportunist organisms. They can indeed grow and reproduce in the absence of a host plant. Nevertheless, in the vicinity of host plants, they adopt a parasitic lifestyle that provides them net benefits in terms of biomass production and reproductive success. In contrast, obligate parasitic plants, whether hemiparasites or holoparasites, must attach to a host after seed germination to complete their development cycle. For these species, the compulsory nature of parasitism is primarily explained by the absence of functional roots but also by the absence of chlorophyll in holoparasites.

Due to their impact on economically important crops, the biology of root parasitic weeds has been extensively studied. The hemiparasitic species of the *Striga* genus (witchweeds) impact cereal crop and are serious threats for food security in Africa (Parker, 2009). Members of the genera *Orobanche* and *Phelipanche* (broomrapes) also impact several important crops (legumes, sunflower, tomato, tobacco, oilseed rape…) in fields of Mediterranean and warm temperate areas of Europe, North Africa and Middle East (Parker, 2009). Interestingly, the establishment of the parasitic relationship between these parasites and their hosts relies on a fine-tuned exchange of signals. The first step of the relevant molecular dialogue between both partners takes place when the seeds of the parasite germinate in response to specific molecules exuded in the rhizosphere by host roots. These chemicals cues are often referred as germination stimulants (GS) most of which belonging to the strigolactones (SLs) (Butler, 1994; Yoneyama, Awad, Xie, Yoneyama, & Takeuchi, 2010). After germination, a second step of main importance takes place upon stimulation by other host-derived compounds leading to the development of a novel and specific organ, the haustorium, which attaches to and invades host tissues to establish a physical and physiological continuum with the host for translocation of resources (Joel & Losner-Goshen, 1994; Kuijt, 1969). Once this bridge is established and functional, the parasite develops outside the infected roots, hence causing major damages that severely and irrevocably reduce crop yields up to a total loss under high infestation levels. Although the molecular events governing these key developmental stages are considered as points of vulnerability and targets for new control methods, they still remain extremely elusive. This chapter reviews the molecular mechanisms governing the communication between plants and root parasitic weeds.

2. THE ORIGINAL LIFE CYCLE OF OBLIGATE ROOT PARASITIC PLANTS

Compared to autotrophic angiosperms, the life cycle of obligate root parasitic plants such as witchweeds and broomrapes is unique. This is due to its high degree of trophic specialization and its synchronization with that of the host (Westwood, 2013). This illustrates the need for a close dialogue between both partners (Fig. 1). This life cycle is divided into two distinct phases: the first starts with seed imbibition, followed by the germination and the development of the radicle, operated, thanks to the mobilization of seed reserves. Nevertheless, this energy independence hides a strong host dependency since a chemical signal produced by the roots of the host and exuded into the rhizosphere is essential for seed germination of these parasitic plants (Yoder, 2001). Indeed, the germination is a two-step process corresponding first to a conditioning period, thought to be required for the acquisition of the sensitivity to the GS, followed by the chemical stimulation of the germination itself that ends with the radicle protrusion (Lechat et al., 2012). During the conditioning period, seeds are exposed to a moist environment and suitable temperatures depending on the species

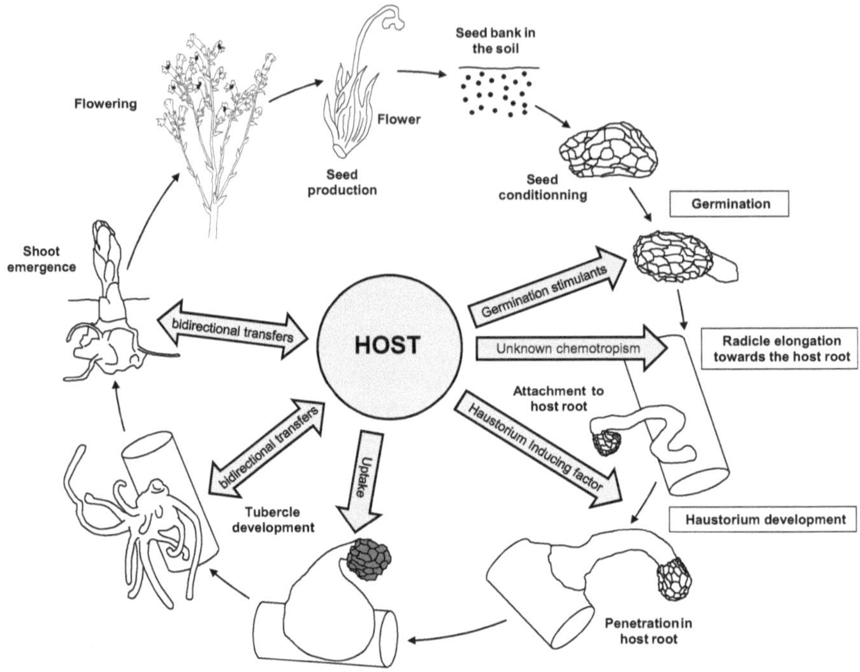

Figure 1 Life cycle of a parasitic plant and chemical communication between the parasite and its host.

(Linke, Sauerborn, & Saxena, 1989, p. 42). Once conditioned, seeds are still unable to germinate without the stimulation by GS. This mechanism implies that the parasite seeds are at the vicinity of the host roots (3–5 mm) on which they will eventually attach. During germination, the seed produces a radicle also called 'germ tube' of few millimetres long that elongates towards the host root, probably thanks to a positive chemotropism generated by the host roots, whose chemical nature has not yet been elucidated (Bouwmeester, Matusova, Zhongkui, & Beale, 2003). Seeds of *Phelipanche ramosa* and *Orobanche cumana* exposed to root exudates of tobacco and sunflower, respectively, release indole-3-acetic acid even before development of the germ tube, suggesting that auxin may be involved in the germination process (Slavov, van Onckelen, Batchvarova, Atanassov, & Prinsen, 2004). When the radicle reaches the host root, its apical cells differentiate into papillae (broomrapes) or root hair (witchweeds) structures producing yet unidentified carbohydrates that allow parasite adhesion to the host root (Joel & Losner-Goshen, 1994). This step of the parasite life cycle constitutes the transition to the parasitic phase.

The second phase begins with the development of the haustorium and its connection to the vascular system of the host. The mature haustorium, also called tubercle, acts as a foreign supernumerary and dominant sink organ for the host plant, then expressing genes notably involved in host-derived sucrose transport and utilization (Draie et al., 2011; Péron et al., 2012), and strongly accumulating auxin, potassium, amino-acids, mannitol, hexoses and starch (Abbes, Kharrat, Delavault, Chaïbi, & Simier, 2009). Hence, the tubercle serves as a transient storage organ for the subsequent growth of a subterranean shoot and a flowering spike after emergence. The survival of the parasite after germination is therefore highly dependent on this attachment step which thus represents the most critical step in the development cycle. The low success of attachment to the host roots is however offset by the very high rate of reproduction among these parasitic species, since only one parasitic plant can produce several tens of thousands of seeds (Joel et al., 2007). This reproductive strategy guarantees the sustainability of these species. Indeed, once the haustorium is fully differentiated, the parasitic plant diverts water, nutrients and hormones necessary for its development. The *Striga* differentiates a thin whitish stem that develops leaves with chlorophyll after emergence from the soil. Then the degree of heterotrophy for carbon decreases progressively in the parasitic plant due to the development of a photosynthetically active leafy shoot (Pageau, Simier, Robins, Naulet, & Fer, 1998). By contrast, the *Orobanche* first develops a tubercle, then a scaly underground stem and finally an achlorophyllous flowering spike after emergence from the soil.

Striga and *Orobanche* have extremely variable degrees of self-pollination according to the species, but they nevertheless have the common property of producing a multitude of tiny seeds (200—300 μm) that are easily disseminated by several means (Castejón-Munõz et al., 1991).

3. THE HOST-DEPENDENT SEED GERMINATION OF OBLIGATE ROOT PARASITIC PLANTS

Among the multiple physiological adaptations that allowed the transition from autotrophism to heterotrophism, the particular germination process of root parasitic plants is probably the most intriguing one. It is commonly known that seeds of autotrophic plants are able to germinate in appropriate conditions including temperature, light, moisture and oxygen (Finkelstein, Reeves, Ariizumi, & Steber, 2008). Nevertheless, seeds of obligate root parasites belonging to the Orobanchaceae family including broomrapes and witchweeds obligatory require host-derived signals, the GS, to germinate. Beyond their role as potent GS for parasitic plants, SLs have also been characterized as host recognition signals for arbuscular mycorrhizal fungi, by inducing oxidative metabolism, spore germination and hyphal branching required for fungi to invade host roots (Akiyama, Matsuzaki, & Hayashi, 2005; Besserer, Bécard, Jauneau, Roux, & Séjalon-Delmas, 2008). Despite their ecological interest as rhizospheric signalling molecules, information on their signalling pathways has for a long while been impaired due to a lack of genetic and genomic resources. It is only 2008 that SLs have been classified as new endogenous plant hormones involved in the inhibition of shoot branching (Gomez-Roldan et al., 2008; Umehara et al., 2008). This highlight was the beginning of rapid progress on deciphering strigolactone biosynthesis and signalling pathways in nonparasitic plants, hence providing many cues to analyze the strigolactone-dependent seed germination in parasitic plants.

3.1 The Germination Stimulants

Strigol was the first known active compound which stimulated germination of *Striga lutea* seeds, and it was purified 50 years ago from the root exudates of cotton, a nonhost plant (*Gossypium hirsutum*) (Cook, Whichard, Turner, Wall, & Egley, 1966) (Fig. 2). Many structural analogues of this compound have been discovered since then, and they are all part of the strigolactone family (Butler, 1994). SLs stimulate the germination of most parasitic plants

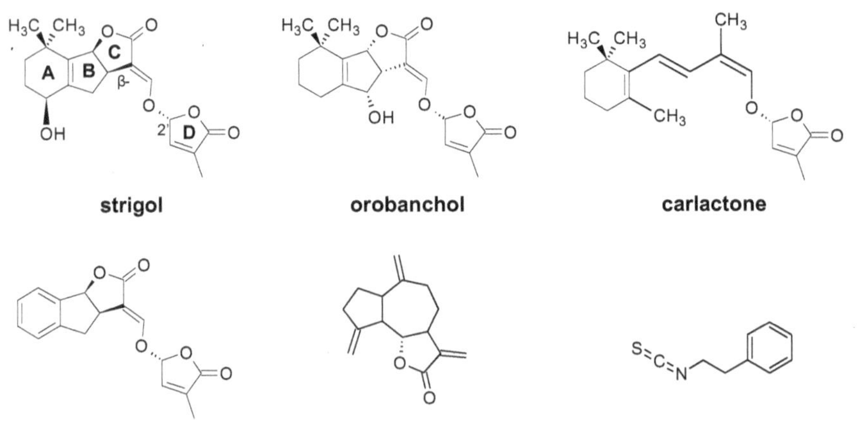

Figure 2 Chemical structure of some germination stimulants.

and were identified in root exudates of monocotyledons and dicotyledons plants (Ćavar, Zwanenburg, & Tarkowski, 2014; Yoneyama, Ruyter-Spira, & Bouwmeester, 2013). These compounds share a common structure, composed a tricyclic lactone (ABC part) linked to a butenolide group (the D-ring) by an enol ether bridge, which has been characterized in more than 15 SLs. Strigolactones were then divided in two classes according to the C-ring orientation: β-oriented for strigol-type and α—oriented for orobanchol-type. Furthermore, most of SLs present a C-2′ R stereo centre on the D-ring as its precursor carlactone (Seto et al., 2014). Substituent addition or modification on A or B rings are a source of diversity for SL structures. On the contrary C-D parts of SLs are mostly similar and carry GS activity (Ćavar et al., 2014). Interestingly, host plants have been shown to exude different SLs with low active concentrations on parasite seed germination (from 10^{-7} to 10^{-12} M), as exemplified by Fabaceae where up to four active SLs are detected (Auger et al., 2012; H.I. Kim, et al., 2014; Pouvreau et al., 2013; Xie et al., 2009; Yoneyama et al., 2008). GR24, a synthetic structural analogue of SLs, was designed and is now used as a standard for most researches on SL activities (Johnson, Rosebery, & Parker, 1976; Johnson et al., 1981) (Fig. 2).

SLs are biosynthesized by the carotenoid pathway (Matusova et al., 2005). From β-carotene, three enzymes sequentially synthesize carlactone (Alder et al., 2012; Seto et al., 2014): a 9-cis/all-*trans*-β-carotene isomerase (D27) and two carotenoid cleavage dioxygenase (CCD7 and CCD8) (Gomez-Roldan et al., 2008; Umehara et al., 2008). Carlactone only presents the A- and D-rings of the canonical SL and is converted in SLs by activities of cytochrome P450 homologues of MORE AXILLARY GROWTH 1 (MAX1) in rice (Zhang et al., 2014). In *Arabidopsis*, conversion of carlactone to SLs or carlactone derivative compounds without B- and C-rings (carlactonic acid and methyl carlactonoate), involves MAX1 and LATERAL BRANCHING OXIDOREDUCTASE (Abe et al., 2014; Brewer et al., 2016), but the active and mobile forms of SLs remain uncertain (Brewer et al., 2016). SLs are mainly synthesized in roots but also in shoots (Al-Babili & Bouwmeester, 2015; Ruyter-Spira, Al-Babili, van der Krol, & Bouwmeester, 2013), and their transport in the plant and their exudation from the roots to the rhizosphere involve the exporter PLEIOTROPIC DRUG RESISTANCE 1 (PDR1), first identified in *Petunia* (Kretzschmar et al., 2012; Sasse et al., 2015). Based on the fact that plant architecture is an important feature that controls plant survival and fitness but is also agronomically relevant for many crop species, many studies have been carried out to decipher what molecular cues underlie this trait. Genetic studies conducted since the 1990s led to the discovery of few genes involved in the repression of axillary shoot morphology, including DECREASED APICAL DOMINANCE (DAD) genes in *Petunia*, RAMOSUS (RMS) genes in pea (Napoli, Beveridge, & Snowden, 1999) and DWARF (D) genes in rice (Arite et al., 2007; Ishikawa et al., 2005). Further genetic evidences in *Arabidopsis* highlighted analogue implications of MAX genes, of which mutations induced similar repression of axillary growth as observed for mutations at RMS and DAD loci (Stirnberg, van De Sande, & Leyser, 2002). MAX1, MAX3 and MAX4 are required for the production of a carotenoid-derived signal that moves acropetally to repress shoot branching (Booker et al., 2004; Snowden et al., 2005; Sorefan et al., 2003), while MAX2 seems to mediate the response to this signal in the shoot. The *MAX2* gene encodes an F-box leucine-rich repeat protein (Stirnberg et al., 2002), that is part of a Skp1-Cullin1-F-box (SCF) complex to suppress shoot branching (Stirnberg, Furner, & Leyser, 2007). SCF complexes are specific ubiquitin E3 ligases that catalyze polyubiquitination of proteins targeted to proteasome 26S for degradation (Cardozo & Pagano, 2004). The involvement of members of this family, such as TRANSPORT INHIBITOR RESPONSE1 (TIR1), CORONATINE INSENSITIVE1 (COI1) and SLEEPY1 (SLY1), was already described in auxin, jasmonate and gibberellin signalling pathways, respectively, showing thus that the branching inhibitor signal shares a common mechanism with other major plant hormones (Chini et al., 2007; Dharmasiri, Dharmasiri, & Estelle, 2005; Dill, Thomas, Hu, Steber, & Sun, 2004; Stirnberg et al., 2007; Thines et al., 2007).

Apart from SLs, others classes of plant secondary metabolites have been also identified as GS of root parasitic plants. However, they are generally less active than SLs and to date they are shown to be involved in specific parasitic plant—host plant interactions. For example, dehydrocostus lactone (DCL) (Fig. 2), a guaianolide sesquiterpene lactone, is exuded by sunflower roots and specifically stimulates the seed germination of the sunflower broomrape, *O. cumana* (Joel, Chaudhuri, Plakhine, Ziadna, & Steffens, 2011). This compound has low or no activity on other parasitic plants. Similarly, *P. ramosa* germinates in response to 2-phenylethyl-isothiocyanate (2-PEITC) (Fig. 2) and others isothiocyanates (ITCs), which are glucosinolate-breakdown products present in the rhizosphere of its host, the oilseed rape (Auger et al., 2012). ITCs are produced through myrosinase activity in Brassicaceae, and they are released upon wounding (Ishida, Hara, Fukino, Kakizaki, & Morimitsu, 2014). Nevertheless, gluconisolates can also be exuded in the rhizosphere of Brassicaeae (Schreiner, Krumbein, Knorr, & Smetanska, 2011) and potentially interact with microorganisms (Bressan et al., 2009). The origin of the myrosinase activity required to produce ITCs needs therefore to be clarified (Gimsing, Sørensen, Tovgaard, Jørgensen, & Hansen, 2006). This raises thus the intriguing question of the significance of the rhizosphere microbiome in the establishment of the particular interaction between *P. ramosa* and oilseed rape (Auger et al., 2012).

In host rhizosphere, parasite seeds can thus germinate in response to complex mixtures of GS with, in some cases, a redundancy of signals with the different SLs and classes of GS. For example, in the rhizosphere of sunflower, *O. cumana* seeds perceive the presence of the host, thanks to heliolactone, a methyl carlactonoate-like compound (Ueno et al., 2014) and dehydrocostus lactone signals (Joel et al., 2011).

3.2 The Key Components of the Strigolactone Signalling Pathway

Most of the molecular actors shown so far to participate in the SL signalling pathways have been identified using SL- and karrikin-insensitive *Arabidopsis* mutants. Karrikins (KARs) are small molecules produced by burning vegetation which stimulate seed germination of autotrophic plant species following fire. This allows seeds that should have stayed dormant in the soil to rapidly germinate to recolonize landscapes (Nelson, Flematti, Ghisalberti, Dixon, & Smith, 2012). Evidences of a common signalling pathway between KARs and SLs came firstly with the central role of the F-box protein MAX2, which integrates both signals during *Arabidopsis* seed germination. Indeed, primary dormant wild-type seeds of *Arabidopsis* show enhanced germination upon GR24 or KAR treatments, whereas *max2* seeds exhibit high levels of dormancy and could not be rescued by neither SLs nor KARs (Nelson et al., 2011). MAX2 orthologs in rice and *Petunia* interact with the α/β hydrolase D14 (DWARF14) that is able to bind and hydrolyze SLs (Hamiaux et al., 2012; Zhou et al., 2013). Further investigations conducted on rice and *Arabidopsis* identified that MAX2 orthologs interact with D53 (DWARF53) and members of the SUPPRESSOR OF MAX2 1 LIKE (SMXL) family (consisting in eight members, SMAX1 and SMXL2–8), respectively. These D53/SMXL proteins seem to interact in different physiological processes with MAX2 upon GR24 or KAR treatment, highlighting a higher level of complexity considering SLs and KARs signalling (Jiang, Liu, et al., 2013; Stanga, Smith, Briggs, & Nelson, 2013; Soundappan et al., 2015; Wang et al., 2015; Zhou et al., 2013). Computational analyses identified that D53/SMXL proteins contain ETHYLENE-RESPONSIVE ELEMENT BINDING FACTOR-ASSOCIATED AMPHIPHILIC REPRESSION (EAR) motifs that are domains required for interaction with TOPLESS/TOPLESS RELATED (TPL/TPR) proteins (Causier, Ashworth, Guo, & Davies, 2012; Jiang, Liu, et al., 2013). Given the implication of TPL/TPR in other signalling pathways, this information strongly suggests a potential crosstalk between SLs, KARs and other hormones.

A last critical element of the discrimination of KARs and SLs signalling is based on the characterization of KARRIKIN INSENSITIVE2 (KAI2), an α/β hydrolase closely related to D14 (Nelson et al., 2011; Waters et al., 2012). This protein is required for responses to KARs during seed germination and seedling development (Scaffidi et al., 2014), and it shares the same catalytic triade Asp-Ser-His as its ortholog D14 (Kagiyama et al., 2013). Moreover KAI2 also functions in a MAX2/D53-dependent manner (Waters, Scaffidi, Sun, Flematti, & Smith, 2014), though its pathway is not recruited in vivo to canonical structures but to unnatural SL analogues (Scaffidi et al., 2014). Interestingly, studies on *Physcomitrella patens* (bryophytes) and algae showed that these early diverging plants contain canonical structures of SLs and are able to respond to application of exogenous SL analogues, though they do not have D14- or D14-LIKE proteins (Proust et al., 2011; Delaux et al., 2012). Both the similar structures and common signalling pathways between D14 and KAI2 receptors suggest that KAI2 could also respond to SLs under some circumstances.

3.3 Towards a Rhizosphere Signalling Paradigm?

By comparison with autotrophic plants, some points still remain to be elucidated in parasitic plants regarding (1) their ability or not to produce SLs, (2) the signalling pathway leading to their germination and (3) whether SLs could play other roles than germination in their development.

The answer to the first question came from the identification in *Striga hermonthica,* of an ortholog of MAX2 that is able to restore SL-dependent germination in *Arabidopsis max2* mutant (Liu et al., 2014). Moreover, this study also highlighted that extracts from *Striga* grown in vitro contained unidentified molecules able to stimulate the germination of *P. ramosa* (Liu et al., 2014). Genomic analyses in *Phelipanche aegyptiaca* (a very close species to *P. ramosa*) and *S. hermonthica* revealed that these parasitic plants contain a unique *D14* gene, while exhibiting multiple *KAI2* genes (Conn et al., 2015). Phylogenetic analyses group *KAI2* sequences in three different clades: a 'divergent' clade containing several members that are parasite-specific, included an 'intermediate' clade with a unique *S. hermonthica* sequence and a 'conserved' clade represented by a single copy of the canonical Angiosperm-type *KAI2* (Conn et al., 2015). *Arabidopsis kai2* mutants were complemented with *KAI2* genes from both parasitic plants, showing that two *KAI2* paralogs respond to natural SLs. A similar study exclusively conducted on *S. hermonthica* showed that six divergent *KAI2* genes encode proteins able to respond to natural or synthetic SLs, many of them allowing to 'sense' SLs far below micromolar concentrations (Toh et al., 2015; Tsuchiya et al., 2015). This means that the perception system of parasitic plants became more complex through receptor-encoding gene duplications, suggesting an adaptation of parasites to perceive a wide range of stimulants at low concentrations in the rhizosphere. As parasitic plants can also respond to ITC and DCL (Auger et al., 2012; Joel et al., 2011), it is questionable whether the perception system may have been diversified to allow a seed response to non-SL-type signals?

As a matter of fact, despite many studies having deciphered the SL-perception system during seed germination, little is still known about the key effectors involved in SL-dependent seed dormancy release. Few studies highlighted the role of SLs in the decrease of abscisic acid (ABA) biosynthesis and the increase of gibberellic acid (GA) biosynthesis during *Arabidopsis* germination (Nelson et al., 2009; Toh et al., 2012). These results are consistent with the role of ABA and GA in control of seed dormancy (Finkelstein et al., 2008). In the parasite *P. ramosa,* GR24 strongly triggers the upregulation of an ABA-hydroxylase encoding gene, *PrCYP707A1*, leading to an ABA catabolism, a *sine qua none* condition to induce germination (Fig. 3) (Lechat et al., 2012). This is consistent with the unequivocal involvement of CYP707 A enzymes to break seed dormancy in autotrophic plants (Nambara & Marion-Poll, 2005). This SL-dependent upregulation of *PrCYP707A1* is strictly controlled by the global DNA methylation status of parasitic seeds and is correlated to an extensive and site-specific DNA demethylation in the promoter of this gene during the conditioning period (Lechat et al., 2015). This result thus provides new insights regarding the link between epigenetic processes and SL signalling. Interestingly, TPL proteins can in some cases recruit histone deacetylases (HDA) that are major actors of epigenetic mechanisms. Besides, it has been shown that a brassinosteroid-mediated HDA activity can control *Arabidopsis* seed germination (Ryu, Cho, Bae, & Hwang, 2014). It is thus tempting to propose that parasitic seed germination may be controlled by epigenetic processes activated by host-derived compounds such as SLs and GS (Fig. 3).

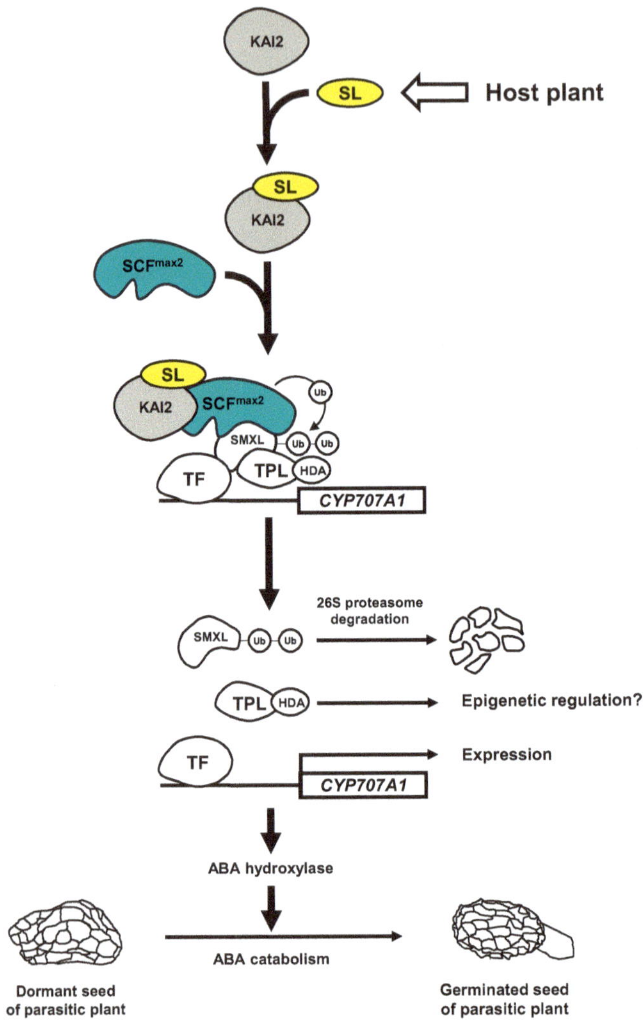

Figure 3 Putative strigolactone signalling pathway during seed germination of root parasitic plants.

4. THE HOST-DEPENDENT HAUSTORIUM DEVELOPMENT IN ROOT PARASITIC PLANTS

During the course of evolution, some angiosperms have acquired the ability to form a physical connection allowing them to 'feed on another plant species' and then became parasites of other higher plants. In some cases, this has led to the loss of autotrophy as a result of the loss of photosynthesis capacity, yet one of the main features of life in most plants. The occurrence of a specialized endophytic organ called haustorium (from the Latin word *haurire,* 'to draw') has been the major event for the evolutionary transition to a parasitic lifestyle. The haustorium is a common organ to all the parasitic plants and serves as a structural and physiological bridge allowing parasites to uptake water and nutrients from host plant vascular system. This specific multicellular organ allows the adhesion of the parasite to host roots, invades the host root cortex and then gives rise to searching invasive cells that elongate between host cells, thanks to the secretion of an arsenal of plant cell wall degrading enzymes, such as cellulases, polygalacturonases, pectine methyl esterases and peroxidases (Ben-Hod, Losner, Joel, & Mayer, 1993; Joel & Losner-Goshen, 1994; Losner-Goshen, Portnoy, Mayer, & Joel, 1998; Mitsumasu, Seto, & Yoshida, 2015; Véronési, Bonnin, Calvez, Thalouarn, & Simier, 2007). In compatible plant—parasite interactions, haustorial cells reach the host's vascular tissues and connect to them for water and nutrients uptake (Hibberd & Jeschke, 2001). Although haustorium formation is a common feature among parasitic plants, some phenotypic variations can be observed not only in the early steps of this process between hemi- and holoparasites but also between facultative and obligate root parasites.

4.1 Facultative Orobanchaceae

In the facultative hemiparasitic Orobanchaceae, *Agalinis purpurea* and *Tryphisaria versicolor,* haustoria develop laterally close to the root elongation zone. The early haustorial structure takes a hairy shape whose development does not durably alter the normal root meristem activity and thus the root growth (Albrecht, Yoder, & Phillips, 1999; Baird & Riopel, 1983; Riopel & Timko, 1995). Interestingly, a single *T. versicolor* root can form multiple lateral haustoria, and their development starts as early as 4 h after exposure to host roots or host root exudates, with the swelling of the root tip, a proliferation of epidermal hairs close to the tip and a proliferation of underlying cortical cells within 24 h (Bandaranayake et al., 2010; Bandaranayake & Yoder, 2013; Jamison & Yoder, 2001). Haustorial hairs are morphologically similar to root hairs and allow the adhesion to host roots (Baird & Riopel, 1985).

4.2 Obligate Orobanchaceae

Striga sp., which are obligate hemiparasitic plants, also develop haustorial hairs, but they are in this case located at the terminus of the root apex. In these species, haustorium ontology starts, as early as 6 h after exposure to host roots or host root—derived molecules, with the radial swelling of the meristem whose normal activity is thus altered. The distal cells of the meristem then enlarge to produce a large bulbous root tip from which the haustorial hairs will emerge (O'Malley & Lynn, 2000). The incorporation of H_3-thymidine into newly replicated DNA dramatically drops within an hour showing that the observed swelling is mainly the result of a redirection of cellular expansion rather than cell division (Keyes, O'Malley, Kim, & Lynn, 2000). Twenty-two hours after induction, the haustorial hairs have extended, and development of the early haustorium (attachment-competent organ) is complete (O'Malley & Lynn, 2000). In obligate holoparasitic plants such as members of the broomrape family (*Orobanche* and *Phelipanche* genera), haustorium ontology also starts at the apex of the root with the cessation of radicle elongation and the swelling of radicle tip, indicating thus a similar redirection of cellular extension from longitudinal to radial. Nevertheless, in these plants, epidermal cells then differentiate to form short bump-shaped extensions instead of haustorial hairs at the apex of root tip. These structure are described as papillae which serve as anchoring organs as they are covered with carbohydrate secretion that serve to tie the haustorium to the host root (Joel & Losner-Goshen, 1994).

4.3 Haustorium Inducing Factors or How the Host Triggers Haustorium Formation

Early haustorium formation is a crucial step in the parasitic mechanism, and it is triggered after chemical or mechanical stimulation. Although some parasites such as broomrapes were thought to form haustoria spontaneously or only from mechanical stimulation via contact with the host root, the ability to monitor haustorium formation in vitro allowed the identification of host-derived factors that display haustorium-inducing activity. The biochemical fractionation of host exudates allowed the identification of so-called haustorium-inducing factors (HIFs), which mainly belong to the phenolic acid family. The xenognosin A and B identified in extracts from *Astralagus* are flavonoids which trigger haustorium formation in *A. purpurea,* and the 2,6 dimethoxy-1,4-benzoquinone (DMBQ) isolated from *Sorghum* roots is effective for both *T. versicolor* and *S. hermonthica* haustorium formation (Chang & Lynn, 1986; Lynn et al., 1981) (Fig. 4). DMBQ is a common product of lignin biosynthesis and degradation released by host cells upon host peroxidase activity triggered by hydrogen peroxide production at the tip of *Striga* radicle (Keyes et al., 2000). This model thus explains how the parasite is able to extract HIFs and to develop a haustorium upon contact with the host root (Keyes et al., 2007) (Fig. 5). Interestingly exogenous

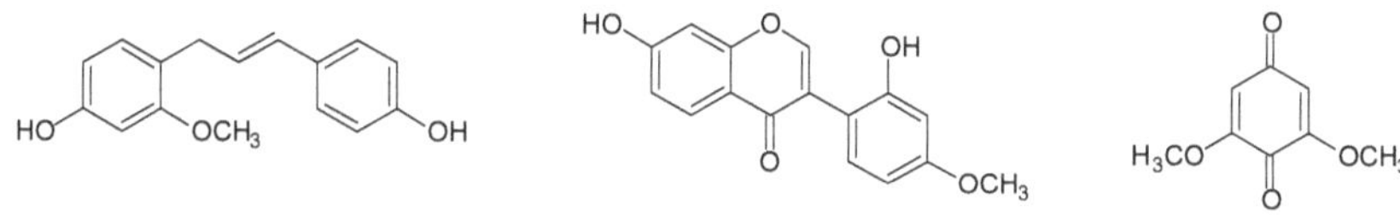

Figure 4 Chemical structure of haustorium inducing factors.

treatment with phytohormones such as indole-3-acetic acid (IAA) were also shown to trigger haustorial hair formation in *T. versicolor* but with differences in the degree and size of swelling and by the precise and synchronous location of the haustorial hairs with respect to the root tip (Tomilov, Tomilova, Abdallah, & Yoder, 2005). Nevertheless, an IAA treatment was shown to enhance haustorial hair formation in *T. versicolor* when low concentrations of DMBQ were applied, and a study by Ishida et al. (2016) showed that the auxin biosynthesis gene *YUCC3* encoding a Flavin monooxygenase was upregulated in cells of the haustorium apex and haustorial hairs of the facultative hemiparasitic plant *Phtheirospermum japonicum*. Moreover, overexpression of *YUCC3* in *P. japonicum* roots triggered haustorium formation and IAA levels were shown to be increased in cells of the root segment

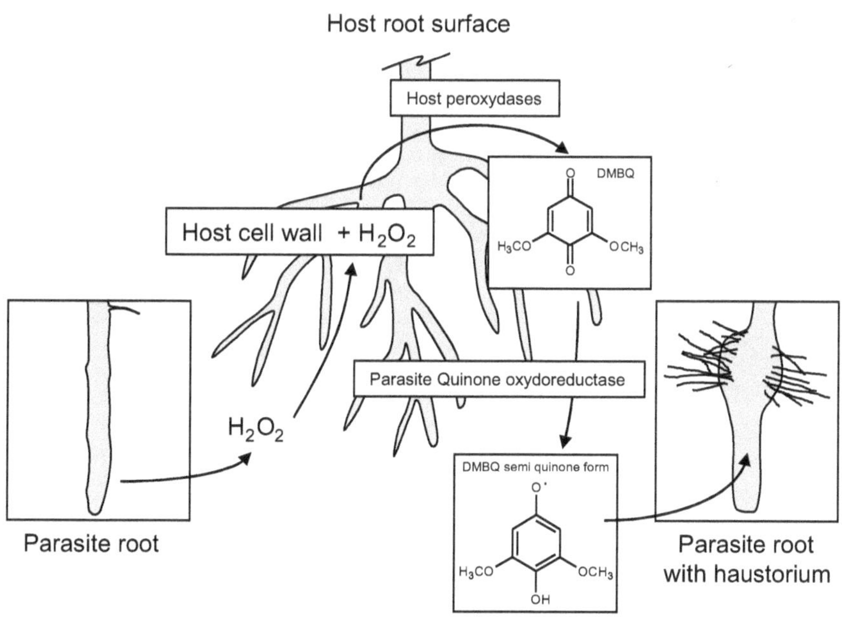

Figure 5 Haustorium formation in root hemiparasitic plants.

containing the haustorium formation site after DMBQ treatment (Ishida et al., 2016). Taken together these results thus suggest that the auxin biosynthetic and regulatory pathways may be recruited in the parasite during early haustorium formation. Interestingly, although none of the known host-derived HIFs have been shown to be effective in the holoparasitic plants *O. cumana*, *P. aegyptiaca* and *P. ramosa* (broomrapes), and no specific HIF has been identified yet, recent findings showed that exogenous treatments with the mycotoxins sphaeropsidones trigger early haustorium formation in *O. cumana* and *Orobanche crenata* (Fernández-Aparicio et al., 2016). These molecules are not related in structure with DMBQ which thus suggests that other receptors and signalling pathways may be required for early haustorium formation in broomrapes.

DMBQ recognition involves a modification of redox potential, in which semiquinone intermediates are formed as transitions between the quinone and phenolic states of DMBQ to trigger a redox-sensitive signal transduction pathway (Keyes et al., 2000) (Fig. 5). This redox cycling requires the activity of quinone oxydoreductases (EC 1.6.5) and two quinone oxydoreductase encoding genes (*TvQR1* and *TvQR2*) that were interestingly shown to be transcriptionally upregulated in *T. versicolor* roots in response to DMBQ (Bandaranayake et al., 2010). *TvQR1* is upregulated upon contact with the host root and when *Triphysaria* roots were transformed with inhibitory RNAi constructions targeting *TvQR1*, the number of haustoria was reduced on these roots, suggesting that TvQR1 is one of the actor of the DMBQ-dependent haustorium signalling pathway (Bandaranayake et al., 2010; Matvienko, Torres, & Yoder, 2001).

5. HOST-PARASITE COMMUNICATIONS DURING AND AFTER HAUSTORIUM FORMATION

In compatible plant–parasite interactions, haustorial cells reach the host's vascular tissues and associate with them for the recruitment of water, nutrients, growth regulators and macromolecules (proteins, nucleic acids...). Various feeding strategies can occur simultaneously in parasitic plants (Hibberd & Jeschke, 2001). The primary mechanism consists in a solute uptake from host apoplast to haustorial parenchymal cells and implies necessarily selective membrane carriers. Additionally, only after host contact, haustorial cortical cells differentiate into xylem cells opposite host xylem vessels, then forming the usual haustorial structure called the xylem bridge (Heide-Jorgensen & Kuijt, 1995). The implication of host signals in this event is assumed but still needs clarification (Tomilov et al., 2005). As such, the gene *PrSus1* is implied in xylem element maturation in the parasitic plant *P. ramosa* and the control of its expression by host-derived auxin has been demonstrated (Péron et al., 2012). Direct connections between host and parasite xylems cells through cross-species perforations (pits) or open conductive tubes (osculae) have been observed in host—hemiparasite interactions, then driving an intensive and nonselective spoliation of water and nutrients (minerals and organic compounds to a less extent) from host, thanks to the ability of hemiparasitic plants to maintain more negative water potential than hosts, which is generated by high and uncontrolled transpiration rates (Jiang, Jeschke, & Hartung, 2003). Similarly, symplastic connexions throughout interspecific plasmodesmata between haustorial parenchymal cells and host cells in *Cuscuta* or between host and parasite phloem elements in *Orobanche* and *Phelipanche*, correspond to a fundamental strategy of development in holoparasitic plants. Opening and functioning of these particular plasmodesmata have been evidenced, thanks to translocation experiments with labelled compounds, phloem-mobile dyes and virus (Birschwilks, Haupt, Hofius, & Neumann, 2006; Gaudin et al., 2014; Haupt, Oparka, Sauer, & Neumann, 2001). In spite of this symplastic phloem continuity resulting in nonselective nutrient transfers from host to parasite, simultaneous selective apoplastic transfers within the haustorium through specific membrane transporters cannot be ruled out. Indeed, genes specifically induced during haustorium formation in *Cuscuta* include genes related to cell wall modification, pathogen or stress response and nutrient transport (Ranjan et al., 2014). In addition, the strategy of phloem unloading in the parasitic plant has also to be considered as a regulatory step of host-derived nutrient transfers. Indeed, for example in the *Brassica napus*—*P. ramosa* interaction, the apoplastic pathway for phloem unloading dominates in the parasitic plant while host and parasite phloem elements are symplastically interconnected in the haustorium (Péron T., personal communication). Thus, downstream of the haustorium, specific membrane transporters, notably sucrose transporters (SUT) and SWEET (Chen, 2014) should be implied in phloem unloading of host-derived nutrients into sink organs of the parasitic plant.

Some macromolecules including proteins and nucleic acids are also translocated though the haustorium (Aly et al., 2011) and then correspond to potential regulators of host defence and immunity and parasite evolution. In *Orobanche*, xylem-mobile fluorescent-dye-labelled dextrans were used to demonstrate the bidirectional movement of macromolecules due to xylem continuity. Similarly, the possibility of a direct protein transfer throughout phloem plasmodesmata has been evidenced in *Orobanche* and *Cuscuta* by using (green fluorescent protein) GFP-producing host plants (Aly et al., 2011; Haupt et al., 2001; Smith, Mescher, & De Moraes, 2013). Studies demonstrating that transferred proteins function in the recipient host or parasite are rare. Nevertheless experiments in *Cuscuta* parasitizing transgenic soybean which produces a glufosinate-detoxifying enzyme demonstrate that this protein is transferred into the parasitic plant and then give herbicide tolerance to *Cuscuta* (Jiang, Qu, Li, & Doohan, 2013). Transcriptomic analyses in *Cuscuta* underline also bidirectional and selective transfers of several thousands of mRNAs from different host plants, whose quantity and identity vary according to the host plant (G. Kim et al., 2014; Kim & Westwood, 2015). Transfers of mRNAs should occur through plasmodesmata between host parenchyma and searching hyphae of *Cuscuta* (David-Schwartz, Runo, Townsley, Machuka, & Sinha, 2008; LeBlanc, Kim, Patel, Stromberg, & Westwood, 2013). Evidence that transferred mRNAs are translated to proteins or processed into silencing signals in the recipient plant is currently missing. Interestingly, hosts plants genetically transformed for producing silencing RNA molecules (siRNA) can affect targeted gene expression in parasites showing different feeding strategies, like *Tryphysaria* as xylem feeder (Bandaranayake et al., 2012; Bandaranayake & Yoder, 2013; Tomilov, Tomilova, Wroblewski, Michelmore, & Yoder, 2008) and *Cuscuta* or *Phelipanche* as phloem feeders (Alakonya et al., 2012; Aly et al., 2012, 2014). This raises another interesting question on the potential pathways of small RNA movements within the haustorium and on their significance in host–parasite communications. Practically, gene silencing RNA based strategies targeted against vital parasite genes are promising for engineering resistance in host plants against siRNA-feeding parasites.

Host–parasite communications include also direct DNA transfers from host to parasite, thanks to the haustorium which establishes direct connections. So, parasitic plants are excellent examples to study horizontal gene transfers (HGT) in plants (Davis & Xi, 2015; Yoshida, Maruyama, Nozaki, & Shirasu, 2010). Indeed HGT involving mainly mitochondrial DNA and nuclear genes to a lesser extent have been observed in 10 of the 11 parasitic lineages to date, revealing that such HGT correspond to frequent evolutionary events in parasitic plants. For example, HGT concerns up to 41% of mitochondrial genes in Rafflesiaceae (Xi et al., 2013), and some of the transgenes keep synteny with their hosts, suggesting that the native genes were displaced via homologous recombination. Concerning the mechanisms involved in HGT, vectors are under discussion (fungi, bacteria, virus, transposons). Direct uptake of DNA rather than mRNA is usually reported in parasitic plants, as supported by the presence of introns and the inclusion of cytosine to uracil RNA editing sites in mitochondrial transgenes. Another understanding question concerns the functionality of transgenes in parasitic plants. While most of the acquired mitochondrial DNAs are introns or pseudogenes in recipient parasitic plants, some nuclear transgenes representing a wide range of cellular functions are actively transcribed in their recipient parasites (Xi et al., 2012; Zhang et al., 2013). The suspected key evolutionary advantages of these functional transgenes in parasitic plants still need clarification.

6. HOST RESPONSE TO PARASITIC PLANT (BROOMRAPE) INFESTATION

In this type of interaction between two plants, one being a pathogen and the other being attacked, one deploys a strategy to invade, the other to defend itself. As observed in many pathogen/plant interactions, a complex molecular dialogue takes place between both protagonists. In the *Arabidopsis/P. ramosa* pathosystem, this dialogue starts as early as the first or second hours of contact between the germinated broomrape seeds and the host roots. Indeed, *Arabidopsis* was shown to respond to the parasite attack by eliciting defence mechanisms at an early stage of parasitism before *P. ramosa* has fully attached to the host roots, thus suggesting the involvement of yet unknown signalling molecules (Vieira Dos Santos, Letousey, Delavault, & Thalouarn, 2003). Throughout the interaction, in response to physical, mechanical, chemical and/or enzymatic aggressions performed by the parasite, the host defends itself by activating signalling pathways involving the phytohormones jasmonic acid (JA), salicylic acid (SA) and ethylene (ET). The nature and intensity of these defence mechanisms depend on the host species and the parasite species. For instance, *P. ramosa* induces both the JA and the ET signalling pathways in *Arabidopsis*, while SA-mediated response is not triggered (Vieira Dos Santos, Delavault, Letousey, & Thalouarn, 2003). In contrast, sunflower exhibits upregulation of SA responsive genes upon *O. cumana* infection (Letousey et al., 2007). Studies based on the observed degrees of resistance, together with histological observations and analyses of defence related gene, have demonstrated the existence of several resistance mechanisms whose nature is still largely unknown. Defensive responses such as physical or chemical barriers that prevent parasite intrusion and development can be activated at three different stages of the parasite–host interaction: preattachment of the radicle, prehaustorial stage or posthaustorial stage (Pérez-de-Luque et al., 2009). Indeed, early resistance mechanisms are part of the basal defence triggered during the preattachment phase and include low induction of seed germination by host roots (Labrousse, Arnaud, Serieys, Bervillé, & Thalouarn, 2001), but also exudation of germination inhibitors and/or phytoalexins (compounds of a phenolic nature) (Echevarría-Zomeño, Pérez-de-Luque, Jorrín, & Maldonado, 2006; Serghini, Pérez de Luque, Castejon-Munoz, GarciaTorres, & Jorrin, 2001). Later, prehaustorial resistance mechanisms take place from the first contact of the parasite to the host root until the establishment of vascular connections through haustorium development (Pérez-de-Luque et al., 2009). This includes formation of an encapsulation layer and cell wall reinforcement, through accumulation of lignin and suberin, protein cross-linking, or callose depositions (Echevarría-Zomeño et al. 2006; Letousey et al., 2007). Posthaustorial resistance is expressed once the parasite has developed vascular connections with the host. This mainly leads to parasite necrosis due to host vessel occlusion by mucilage, or host production of unidentified host-degraded products or production of toxic compounds, as observed in sunflower (Labrousse et al., 2001; Letousey et al., 2007; de Zelicourt et al., 2007). In the latter case, it was shown that the roots of a resistant sunflower genotype (LR1) exhibited a strong induction of the *HaDef1* gene encoding a small defence protein (Letousey et al., 2007). It has been suggested that the corresponding gene product, a defensin, may be the toxic compound leading to the parasite necrosis (de Zelicourt et al., 2007).

7. CONCLUSIONS

The world of parasitic plants is very diverse and represents a rich source of models that may be of interest for the understanding of how plants communicate with other plants and evolve together. Over the years, a great deal of study highlighted the chemical nature of certain molecules involved in the communication between the parasites and their hosts, especially in the case of the root parasitic plants. Whether it is for seed germination, haustorium development or host response, the diversity of the involved molecules should be emphasized. However, most of the experiments that conducted to the discovery of these signals have been carried out in artificial conditions in which the impact of the rhizosphere has unfortunately never been taken into account. Yet, it is undeniable that the microbial community of this part of the soil must be a major actor in the communication between plant partners. In the future, more emphasis needs to be placed on the role of the rhizosphere in the establishment of the plant—parasitic interactions, notably if the microbial communities are proved to be good targets for specific and efficient control methods.

REFERENCES

Abbes, Z., Kharrat, M., Delavault, P., Chaïbi, W., & Simier, P. (2009). Nitrogen and carbon relationships between the parasitic weed *Orobanche foetida* and susceptible and tolerant faba bean lines. *Plant Physiology and Biochemistry, 47*, 153—159.

Abe, S., Sado, A., Tanaka, K., Kisugi, T., Asami, K., Ota, S., ... Nomura, T. (2014). Carlactone is converted to carlactonoic acid by MAX1 in *Arabidopsis* and its methyl ester can directly interact with AtD14 in vitro. *Proceedings of the National Academy of Sciences of the United States of America, 111*, 18084—18089.

Akiyama, K., Matsuzaki, K., & Hayashi, H. (2005). Plant sesquiterpenes induce hyphal branching in arbuscular mycorrhizal fungi. *Nature, 435*, 824—827.

Alakonya, A., Kumar, R., Koenig, D., Kimura, S., Townsley, B., Runo, S., ... Sinha, N. (2012). Interspecific RNA interference of SHOOT MERISTEMLESS-like disrupts *Cuscuta pentagona* plant parasitism. *The Plant Cell, 24*(7), 3153—3166.

Al-Babili, S., & Bouwmeester, H. J. (2015). Strigolactones, a novel carotenoid-derived plant hormone. *Annual Review of Plant Biology, 66*, 161—186.

Albrecht, H., Yoder, J., & Phillips, D. (1999). Flavonoids promote haustoria formation in the root parasite *Triphysaria versicolor*. *Plant Physiology, 119*, 585—592.

Alder, A., Jamil, M., Marzorati, M., Bruno, M., Vermathen, M., Bigler, P., ... Al-Babili, S. (2012). The path from beta-carotene to carlactone, a strigolactone-like plant hormone. *Science, 335*, 1348—1351.

Aly, R., Cholakh, H., Joel, D. M., Leibman, D., Steinitz, B., Zelcer, A., ... Gal-On, A. (2012). Gene silencing of mannose 6-phosphate reductase in the parasitic weed *Orobanche aegyptiaca* through the production of homologous dsRNA sequences in the host plant. *Plant Biotechnology Journal*, 7(6), 487—498.

Aly, R., Dubey, N. K., Yahyaa, M., Abu-Nassar, J., & Ibdah, M. (2014). Gene silencing of CCD7 and CCD8 in *Phelipanche aegyptiaca* by tobacco rattle virus system retarded the parasite development on the host. *Plant Signaling & Behaviours, 9*(8), e29376.

Aly, R., Hamamouch, N., Abu-Nassar, J., Wolf, S., Joel, D. M., Eizenberg, H., ... Westwood, J. H. (2011). Movement of protein and macromolecules between host plants and the parasitic weed *Phelipanche aegyptiaca* Pers. *Plant Cell Report, 30*, 2233—2241.

Arite, T., Iwata, H., Ohshima, K., Maekawa, M., Nakajima, M., Kojima, M., ... Kyozuka, J. (2007). DWARF10, an RMS1/MAX4/DAD1 ortholog, controls lateral bud outgrowth in rice. *The Plant Journal, 51*, 1019—1029.

Auger, B., Pouvreau, J.-B., Pouponneau, K., Yoneyama, K., Montiel, G., Le Bizec, B., ... Simier, P. (2012). Germination stimulants of *Phelipanche ramosa* in the rhizosphere of *Brassica napus* are derived from the glucosinolate pathway. *Molecular Plant-Microbe Interactions, 25*, 993—1004.

Baird, W. V., & Riopel, J. L. (1983). Experimental studies of the attachment of the parasitic angiosperm *Agalinis purpurea* to a host. *Protoplasma, 118*, 206—218.

Baird, W. V., & Riopel, J. L. (1985). Surface characteristics of root and haustorial hairs of parasitic scrophulariaceae. *Botanical Gazette, 146*, 63—69.

Bandaranayake, P. C. G., Filappova, T., Tomilov, A., Tomilova, N. B., Jamison-McClung, D., Ngo, Q., … Yoder, J. I. (2010). A single-electron reducing quinone oxidoreductase is necessary to induce haustorium development in the root parasitic plant *Triphysaria*. *The Plant Cell, 22*, 1404–1419.

Bandaranayake, P. C. G., Tomilov, A., Tomilova, N. B., Ngo, Q. A., Wickett, N., dePamphilis, C. W., & Yoder, J. I. (2012). The *TvPirin* gene is necessary for haustorium development in the parasitic plant *Triphysaria versicolor*. *Plant Physiology, 158*(2), 1046–1053.

Bandaranayake, P. C. G., & Yoder, J. I. (2013). Trans-specific gene silencing of acetyl-CoA carboxylase in a root-parasitic plant. *Molecular Plant-Microbe Interactions, 26*, 575–584.

Ben-Hod, G., Losner, D., Joel, D. M., & Mayer, A. M. (1993). Pectin methyl-esterase in calli and germinating seeds of *Orobanche aegyptiaca*. *Phytochemistry, 32*, 1399–1402.

Besserer, A., Bécard, G., Jauneau, A., Roux, C., & Séjalon-Delmas, N. (2008). GR24, a synthetic analog of strigolactones, stimulates the mitosis and growth of the arbuscular mycorrhizal fungus *Gigaspora rosea* by boosting its energy metabolism. *Plant Physiology, 148*, 402–413.

Birschwilks, M., Haupt, S., Hofius, D., & Neumann, S. (2006). Transfer of phloem-mobile substances from the host plants to the holoparasite *Cuscuta* sp. *Journal of Experimental Botany, 57*, 911–921.

Booker, J., Auldridge, M., Wills, S., McCarty, D., Klee, H., & Leyser, O. (2004). MAX3/CCD7 is a carotenoid cleavage dioxygenase required for the synthesis of a novel plant signaling molecule. *Current Biology, 14*, 1232–1238.

Bouwmeester, H. J., Matusova, R., Zhongkui, S., & Beale, M. H. (2003). Secondary metabolite signalling in host–parasitic plant interactions. *Current Opinion in Plant Biology, 6*, 358–364.

Bressan, M., Roncato, M.-A., Bellvert, F., Comte, G., Haichar, F. Z., Achouak, W., & Berge, O. (2009). Exogenous glucosinolate produced by *Arabidopsis thaliana* has an impact on microbes in the rhizosphere and plant roots. *The ISME Journal, 3*, 1243–1257.

Brewer, P. B., Yoneyama, K., Filardo, F., Meyers, E., Scaffidi, A., Frickey, T., … Beveridge, C. A. (2016). LATERAL BRANCHING OXIDOREDUCTASE acts in the final stages of strigolactone biosynthesis in *Arabidopsis*. *Proceedings of the National Academy of Sciences of the United States of America, 113*, 6301–6306.

Butler, L. G. (1994). Chemical communication between the parasitic weed *Striga* and its crop host. *Allelopathy, 582*, 158–168.

Cardozo, T., & Pagano, M. (2004). The SCF ubiquitin ligase: insights into a molecular machine. *Nature Reviews. Molecular Cell Biology, 5*, 739–751.

Castejón-Munõz, M., Romero-Munoz, F., & Garciá-Torres, L. (1991). *Orobanche cernua* seed dispersal through sunflower achenes. *Helia, 14*, 51–54.

Causier, B., Ashworth, M., Guo, W., & Davies, B. (2012). The TOPLESS interactome: a framework for gene repression in *Arabidopsis*. *Plant Physiology, 158*, 423–438.

Ćavar, S., Zwanenburg, B., & Tarkowski, P. (2014). Strigolactones: occurrence, structure, and biological activity in the rhizosphere. *Phytochemistry Reviews, 14*, 691–711.

Chang, M., & Lynn, D. G. (1986). The haustorium and the chemistry of host recognition in parasitic angiosperms. *Journal of Chemical Ecology, 12*, 561–579.

Chen, L.-Q. (2014). SWEET sugar transporters for phloem transport and pathogen nutrition. *New Phytologist, 201*, 1150–1155.

Chini, A., Fonseca, S., Fernandez, G., Adie, B., Chico, J. M., Lorenzo, O., … Solano, R. (2007). The JAZ family of repressors is the missing link in jasmonate signalling. *Nature, 448*, 666–671.

Conn, C. E., Bythell-Douglas, R., Neumann, D., Yoshida, S., Whittington, B., Shirasu, K., … Nelson, D. C. (2015). Convergent evolution of strigolactone perception enabled host detection in parasitic plants. *Science, 349*, 540–543.

Cook, C. E., Whichard, L. P., Turner, B., Wall, M. E., & Egley, G. H. (1966). Germination of witchweed (*Striga lutea* lour.): isolation and properties of a potent stimulant. *Science, 154*, 1189–1190.

David-Schwartz, R., Runo, S., Townsley, B., Machuka, J., & Sinha, N. (2008). Long-distance transport of mRNA via parenchyma cells and phloem across the host–parasite junction in *Cuscuta*. *New Phytologist, 179*, 1133–1141.

Davis, C. C., & Xi, Z. (2015). Horizontal gene transfer in parasitic plants. *Current Opinion in Plant Biology, 26*, 14–19.

Delaux, P.-M., Xie, X., Yimme, R. E., Puech-Pagès, V., Dunand, C., Lecompte, E., … Séjalon-Delmas, N. (2012). Origin of strigolactones in the green lineage. *The New Phytologist, 195*, 857–871.

Dharmasiri, N., Dharmasiri, S., & Estelle, M. (2005). The F-box protein TIR1 is an auxin receptor. *Nature, 435*, 441–445.

Dill, A., Thomas, S. G., Hu, J., Steber, C. M., & Sun, T. (2004). The *Arabidopsis* F-box protein SLEEPY1 targets gibberellin signaling repressors for gibberellin-induced degradation. *The Plant Cell, 16*, 1392–1405.

Draie, R., Péron, T., Pouvreau, J. B., Véronési, C., Jégou, S., Delavault, P., … Simier, P. (2011). Invertases involved in the development of the parasitic plant *Phelipanche ramosa*: characterization of the dominant soluble acid isoform, PrSAI1. *Molecular Plant Pathology, 12*, 638–652.

Echevarría-Zomeño, S., Pérez-de-Luque, A., Jorrín, J., & Maldonado, A. M. (2006). Pre-haustorial resistance to broomrape (*Orobanche cumana*) in sunflower (*Helianthus annuus*): cytochemical studies. *Journal of Experimental Botany, 57*, 4189–4200.

Fernández-Aparicio, M., Masi, M., Maddau, L., Cimmino, A., Evidente, M., Rubiales, D., & Evidente, A. (2016). Induction of haustorium development by sphaeropsidones in radicles of the parasitic weeds *Striga* and *Orobanche*. A structure-activity relationship study. *Journal of Agricultural and Food Chemistry, 64*.

Finkelstein, R., Reeves, W., Ariizumi, T., & Steber, C. (2008). Molecular aspects of seed dormancy. *Annual Review of Plant Biology, 59*, 387–415.

Gaudin, Z., Cerveau, D., Marnet, N., Bouchereau, A., Delavault, P., Simier, P., & Poureau, J. B. (2014). Robust method for investigating nitrogen metabolism of 15N labeled amino acids using AccQ•Tag ultra performance liquid chromatography-photodiode array-electrospray ionization-mass spectrometry: application to a parasitic plant-plant interaction. *Analytical Chemistry, 86*(2), 1138–1145.

Gimsing, A. L., Sørensen, J. C., Tovgaard, L., Jørgensen, A. M. F., & Hansen, H. C. B. (2006). Degradation kinetics of glucosinolates in soil. *Environmental Toxicology and Chemistry, 25*, 2038–2044.

Gomez-Roldan, V., Fermas, S., Brewer, P. B., Puech-Pagès, V., Dun, E. A., Pillot, J.-P., … Rochange, F. (2008). Strigolactone inhibition of shoot branching. *Nature, 455*, 189–194.

Hamiaux, C., Drummond, R. S. M., Janssen, B. J., Ledger, S. E., Cooney, J. M., Newcomb, R. D., & Snowden, K. C. (2012). DAD2 is an α/β hydrolase likely to be involved in the perception of the plant branching hormone, strigolactone. *Current Biology, 22*, 2032–2036.

Haupt, S., Oparka, K. J., Sauer, N., & Neumann, S. (2001). Macromolecular trafficking between *Nicotiana tabacum* and the holoparasite *Cuscuta reflexa*. *Journal of Experimental Botany, 152*, 173–177.

H1eide-Jorgensen, H. S., & Kuijt, J. (1995). The haustorium of the root parasite *Triphysaria* (Scrophulariaceae), with special reference to xylem bridge ultrastructure. *American Journal of Botany, 82*, 782–797.

Hibberd, J. M., & Jeschke, W. D. (2001). Solute flux into parasitic plants. *Journal of Experimental Botany, 52*, 2043–2049.

Ishida, M., Hara, M., Fukino, N., Kakizaki, T., & Morimitsu, Y. (2014). Glucosinolate metabolism, functionality and breeding for the improvement of Brassicaceae vegetables. *Breeding Science, 64*, 48–59.

Ishida, J. K., Wakatake, T., Yoshida, S., Takebayashi, Y., Kasahara, H., Wafula, E., ... Shirasu, K. (2016). Local auxin biosynthesis mediated by a YUCCA Flavin monooxygenase regulates haustorium development in the parasitic plant *Phtheirospermum japonicum. The Plant Cell, 28.*

Ishikawa, S., Maekawa, M., Arite, T., Onishi, K., Takamure, I., & Kyozuka, J. (2005). Suppression of tiller bud activity in tillering dwarf mutants of rice. *Plant & Cell Physiology, 46*, 79–86.

Jamison, D. S., & Yoder, J. I. (2001). Heritable variation in quinone-induced haustorium development in the parasitic plant *Triphysaria. Plant Physiology, 125*, 1870–1879.

Jiang, F., Jeschke, W. D., & Hartung, W. (2003). Water flows in the parasitic association *Rhinanthus* minor-*Hordeum vulgare. Journal of Experimental Botany, 54*, 1985–1993.

Jiang, L., Liu, X., Xiong, G., Liu, H., Chen, F., Wang, L., ... Li, J. (2013). DWARF 53 acts as a repressor of strigolactone signalling in rice. *Nature, 504*, 401–405.

Jiang, L., Qu, F., Li, Z., & Doohan, D. (2013). Inter-species protein trafficking endows dodder (*Cuscuta pentagona*) with a host specific herbicide-tolerant trait. *New Phytologist, 198*, 1017–1022.

Joel, D. M., Chaudhuri, S. K., Plakhine, D., Ziadna, H., & Steffens, J. C. (2011). Dehydrocostus lactone is exuded from sunflower roots and stimulates germination of the root parasite *Orobanche cumana. Phytochemistry, 72*, 624–634.

Joel, D. M., Hershenhorn, J., Eizenberg, H., Aly, R., Ejeta, G., Rich, P. J., ... Rubiales, D. (2007). Biology and management of weedy root parasites. *Horticultural Reviews, 33*, 267–349.

Joel, D. M., & Losner-Goshen, D. (1994). The attachment organ of the parasitic angiosperms *Orobanche cumana* and *O. aegyptiaca* and its development. *Canadian Journal of Botany, 72*, 564–574.

Johnson, A. W., Gowada, G., Hassanali, A., Knox, J., Monaco, S., Razavi, Z., & Rosebery, G. (1981). The preparation of synthetic analogues of strigol. *Journal of the Chemical Society, Perkin Transactions, 1*, 1734–1743.

Johnson, A. W., Rosebery, G., & Parker, C. (1976). A novel approach to *Striga* and *Orobanche* control using synthetic germination stimulants. *Weed Research, 16*, 223–227.

Kagiyama, M., Hirano, Y., Mori, T., Kim, S.-Y., Kyozuka, J., Seto, Y., ... Hakoshima, T. (2013). Structures of D14 and D14L in the strigolactone and karrikin signaling pathways. *Genes to Cells, 18*, 147–160.

Keyes, W. J., O'Malley, R. C., Kim, D., & Lynn, D. G. (2000). Signaling organogenesis in parasitic angiosperms: xenognosin generation, perception, and response. *Journal of Plant Growth Regulation, 19*, 217–231.

Keyes, W. J., Palmer, A. G., Erbil, W. K., Taylor, J. V., Apkarian, R. P., Weeks, E. R., & Lynn, D. G. (2007). Semagenesis and the parasitic angiosperm *Striga asiatica. The Plant Journal, 51*, 707–716.

Kim, G., LeBlanc, M. L., Wafula, E. K., de Pamphilis, C. W., & Westwood, J. H. (2014). Genomic-scale exchange of mRNA between a parasitic plant and its hosts. *Science, 345*(6198), 808–811.

Kim, G., & Westwood, J. H. (2015). Macromolecule exchange in *Cuscuta*– host plant interactions. *Current Opinion in Plant Biology, 26*, 20–25.

Kim, H. I., Kisugi, T., Khetkam, P., Xie, X., Yoneyama, K., Uchida, K., ... Yoneyama, K. (2014). Avenaol, a germination stimulant for root parasitic plants from *Avena strigosa. Phytochemistry, 103*, 85–88.

Kretzschmar, T., Kohlen, W., Sasse, J., Borghi, L., Schlegel, M., Bachelier, J. B., ... Martinoia, E. (2012). A *Petunia* ABC protein controls strigolactone-dependent symbiotic signalling and branching. *Nature, 483*, 341–344.

Kuijt, J. (1969). *The biology of parasitic flowering plants*. Berkeley, CA: University of California Press.

Labrousse, P., Arnaud, M.-C., Serieys, H., Bervillé, A., & Thalouarn, P. (2001). Several mechanisms are involved in resistance of *Helianthus* to *Orobanche cumana* Wallr. *Annals of Botany, 88*, 859–868.

LeBlanc, M., Kim, G., Patel, B., Stromberg, V., & Westwood, J. H. (2013). Quantification of tomato and *Arabidopsis* mobile RNAs trafficking into the parasitic plant *Cuscuta pentagona*. *New Phytologist, 200*, 1225–1233.

Lechat, M.-M., Brun, G., Montiel, G., Véronési, C., Simier, P., Thoiron, S., ... Delavault, P. (2015). Seed response to strigolactone is controlled by abscisic acid-independent DNA methylation in the obligate root parasitic plant, *Phelipanche ramosa* L. Pomel. *Journal of Experimental Botany, 66*, 3129–3140.

Lechat, M.-M., Pouvreau, J.-B., Péron, T., Gauthier, M., Montiel, G., Véronési, C., ... Delavault, P. (2012). *PrCYP707A1*, an ABA catabolic gene, is a key component of *Phelipanche ramosa* seed germination in response to the strigolactone analogue GR24. *Journal of Experimental Botany, 63*, 5311–5322.

Letousey, P., De Zelicourt, A., Vieira Dos Santos, C., Thoiron, S., Monteau, F., Simier, P., ... Delavault, P. (2007). Molecular analysis of sunflower resistance mechanisms to *Orobanche cumana*. *Plant Pathology, 56*, 536–546.

Linke, K. H., Sauerborn, J., & Saxena, M. C. (1989). Orobanche *field guide*. Hohenheim, Germany: University of Hohenheim.

Liu, Q., Zhang, Y., Matusova, R., Charnikhowa, T., Amini, M., Jamil, M., ... Bouwmeester, H. J. (2014). *Striga hermonthica* MAX2 restores branching but not the very low fluence response in the *Arabidopsis thaliana* max2 mutant. *The New Phytologist, 202*, 531–541.

Losner-Goshen, D., Portnoy, V. H., Mayer, A. M., & Joel, D. M. (1998). Pectolytic activity by the haustorium of the parasitic plant *Orobanche* L. (Orobanchaceae) in host roots. *Annals of Botany, 81*, 319–326.

Lynn, D. G., Steffens, J. C., Kamut, V. S., Graden, D. W., Shabanowitz, J., & Riopel, J. L. (1981). Isolation and characterization of the first host recognition substance for parasitic angiosperms. *Journal of the American Chemical Society, 103*, 1868–1870.

Matusova, R., Rani, K., Verstppen, F. W. A., Franssen, M. C. R., Beale, M. H., & Bouwmeester, H. J. (2005). The strigolactone germination stimulants of the plant-parasitic *Striga* and *Orobanche* spp. are derived from the carotenoid pathway. *Plant Physiology, 139*, 920–934.

Matvienko, M., Torres, M. J., & Yoder, J. I. (2001). Transcriptional responses in the hemiparasitic plant *Triphysaria versicolor* to host plant signals. *Plant Physiology, 127*, 272–282.

Mitsumasu, K., Seto, Y., & Yoshida, S. (2015). Apoplastic interactions between plants and plant root intruders. *Frontiers in Plant Science, 6*, 617.

Nambara, E., & Marion-Poll, A. (2005). Abscisic acid biosynthesis and catabolism. *Annual Review of Plant Biology, 56*, 165–185.

Napoli, C. A., Beveridge, C. A., & Snowden, K. C. (1999). Reevaluating concepts of apical dominance and the control of axillary bud outgrowth. *Current Topics in Developmental Biology, 44*, 127–169.

Nelson, D. C., Flematti, G. R., Ghisalberti, E. L., Dixon, K. W., & Smith, S. M. (2012). Regulation of seed germination and seedling growth by chemical signals from burning vegetation. *Annual Review of Plant Biology, 63*, 107–130.

Nelson, D. C., Riseborough, J.-A., Flematti, G. R., Stevens, J., Ghisalberti, E. L., Dixon, K. W., & Smith, S. M. (2009). Karrikins discovered in smoke trigger *Arabidopsis* seed germination by a mechanism requiring gibberellic acid synthesis and light. *Plant Physiology, 149*, 863–873.

Nelson, D. C., Scaffidi, A., Dun, E. A., Waters, M. T., Flematti, G. R., Dixon, K. W., ... Smith, S. M. (2011). F-box protein MAX2 has dual roles in karrikin and strigolactone signaling in *Arabidopsis thaliana*. *Proceedings of the National Academy of Sciences of the United States of America, 108*, 8897–8902.

O'Malley, R. C., & Lynn, D. G. (2000). Expansin message regulation in parasitic angiosperms: marking time in development. *The Plant Cell, 12*, 1455–1465.

Pageau, K., Simier, P., Robins, R., Naulet, N., & Fer, A. (1998). Carbon dependency of the hemiparasite *Striga hermonthica* on *Sorghum bicolor* determined by carbon isotopic and gas exchange analyses. *Australian Journal of Plant Physiology, 25*, 695–700.

Parker, C. (2009). Observations on the current status of *Orobanche* and *Striga* problems worldwide. *Pest Management Science, 65*, 453–459.

Pérez-de-Luque, A., Fondevilla, S., Pérez-Vich, B., Aly, R., Thoiron, S., Simier, P., ... Delavault, P. (2009). Understanding *Orobanche* and *Phelipanche*–host plant interactions and developing resistance. *Weed Research, 49*, 8–22.

Péron, T., Véronési, C., Mortreau, E., Pouvreau, J. B., Thoiron, S., Leduc, N., ... Simier, P. (2012). Role of the sucrose synthase encoding *PrSus1* gene in the development of the parasitic plant *Phelipanche ramosa* L. (Pomel). *Molecular Plant-Microbe Interactions, 25*, 402–411.

Pouvreau, J.-B., Gaudin, Z., Auger, B., Lechat, M.-M., Gauthier, M., Delavault, P., & Simier, P. (2013). A high-throughput seed germination assay for root parasitic plants. *Plant Methods, 9*, 32.

Proust, H., Hoffmann, B., Xie, X., Yoneyama, K., Schaefer, D., Yoneyama, K., ... Rameau, C. (2011). Strigolactones regulate protonema branching and act as a quorum sensing-like signal in the moss *Physcomitrella patens*. *Development, 138*, 1531–1539.

Ranjan, A., Ichihashi, Y., Farhi, M., Zumstein, K., Townsley, B., David-Schwartz, R., & Sinha, N. R. (2014). De novo assembly and characterization of the transcriptome of the parasitic weed dodder identifies genes associated with plant parasitism. *Plant Physiology, 166*, 1186–1199.

Riopel, J. L., & Timko, M. P. (1995). Haustorial initiation and differentiation. In *Parasitic plants* (pp. 39–79). New York: Chapman & Hall: JD Graves.

Ruyter-Spira, C., Al-Babili, S., van der Krol, S., & Bouwmeester, H. (2013). The biology of strigolactones. *Trends in Plant Science, 18*, 72–83.

Ryu, H., Cho, H., Bae, W., & Hwang, I. (2014). Control of early seedling development by BES1/TPL/HDA19-mediated epigenetic regulation of *ABI3*. *Nature Communications, 5*, 4138.

Sasse, J., Simon, S., Gübeli, C., Liu, G.-W., Cheng, X., Friml, J., ... Borghi, L. (2015). Asymmetric localizations of the ABC transporter PaPDR1 trace paths of directional strigolactone transport. *Current Biology, 25*, 647–655.

Scaffidi, A., Waters, M. T., Sun, Y. K., Skelton, B. W., Dixon, K. W., Ghisalberti, E. L., ... Smith, S. M. (2014). Strigolactone hormones and their stereoisomers signal through two related receptor proteins to induce different physiological responses in *Arabidopsis*. *Plant Physiology, 165*, 1221–1232.

Schreiner, M., Krumbein, A., Knorr, D., & Smetanska, I. (2011). Enhanced glucosinolates in root exudates of *Brassica rapa* ssp. rapa mediated by salicylic acid and methyl jasmonate. *Journal of Agricultural and Food Chemistry, 59*, 1400–1405.

Serghini, K., Pérez de Luque, A., Castejon-Munoz, M., GarciaTorres, L., & Jorrin, J. (2001). Sunflower (*Helianthus annuus* L.) response to broomrape (*Orobanche cernua* Loefl.) parasitism: induced synthesis and excretion of 7-hydroxylated simple coumarins. *Journal of Experimental Botany, 52*, 2227–2234.

Seto, Y., Sado, A., Asami, K., Hanada, A., Umehara, M., Akiyama, K., & Yamaguchi, S. (2014). Carlactone is an endogenous biosynthetic precursor for strigolactones. *Proceedings of the National Academy of Sciences of the United States of America, 111*, 1640–1645.

Slavov, I., van Onckelen, H., Batchvarova, R., Atanassov, A., & Prinsen, E. (2004). IAA production during germination of *Orobanche* spp. seeds. *Journal of Plant Physiology, 161*, 847–853.

Smith, J. D., Mescher, M. C., & De Moraes, C. M. (2013). Implications of bioactive solute transfer from hosts to parasitic plants. *Current Opinion in Plant Biology, 16*, 464–472.

Snowden, K. C., Simkin, A. J., Janssen, B. J., Templeton, K. R., Loucas, H. M., Simons, J. L., ... Klee, H. J. (2005). The Decreased apical dominance1/*Petunia hybrida* CAROTENOID CLEAVAGE DIOXYGENASE8 gene affects branch production and plays a role in leaf senescence, root growth, and flower development. *The Plant Cell, 17*, 746–759.

Sorefan, K., Booker, J., Haurogné, K., Goussot, M., Bainbridge, K., Foo, E., ... Leyser, O. (2003). MAX4 and RMS1 are orthologous dioxygenase-like genes that regulate shoot branching in *Arabidopsis* and pea. *Genes & Development, 17*, 1469–1474.

Soundappan, I., Bennett, T., Morffy, N., Liang, Y., Stanga, J. P., Abbas, A., ... Nelson, D. C. (2015). SMAX1-LIKE/D53 family members enable distinct MAX2-dependent responses to strigolactones and karrikins in *Arabidopsis*. *The Plant Cell, 27*, 3143–3159.

Stanga, J. P., Smith, S. M., Briggs, W. R., & Nelson, D. C. (2013). SUPPRESSOR OF MORE AXILLARY GROWTH2 1 controls seed germination and seedling development in *Arabidopsis*. *Plant Physiology, 163*, 318–330.

Stirnberg, P., Furner, I. J., & Leyser, O. (2007). MAX2 participates in an SCF complex which acts locally at the node to suppress shoot branching. *The Plant Journal, 50*, 80–94.

Stirnberg, P., van De Sande, K., & Leyser, O. (2002). MAX1 and MAX2 control shoot lateral branching in *Arabidopsis*. *Development, 129*, 1131–1141.

Thines, B., Katsir, L., Melotto, M., Niu, Y., Mandaokar, A., Liu, G., ... Browse, J. (2007). JAZ repressor proteins are targets of the SCF(COI1) complex during jasmonate signalling. *Nature, 448*, 661–665.

Toh, S., Holbrook-Smith, D., Stogios, P. J., Onopriyenko, O., Lumba, S., Tsuchiya, Y., ... McCourt, P. (2015). Structure-function analysis identifies highly sensitive strigolactone receptors in *Striga*. *Science, 350*, 203–207.

Toh, S., Kamiya, Y., Kawakami, N., Nambara, E., McCourt, P., & Tsuchiya, Y. (2012). Thermoinhibition uncovers a role for strigolactones in *Arabidopsis* seed germination. *Plant & Cell Physiology, 53*, 107–117.

Tomilov, A. A., Tomilova, N. B., Abdallah, I., & Yoder, J. I. (2005). Localized hormone fluxes and early haustorium development in the hemiparasitic plant *Triphysaria versicolor*. *Plant Physiology, 138*, 1469–1480.

Tomilov, A. A., Tomilova, N. B., Wroblewski, T., Michelmore, R., & Yoder, J. I. (2008). Trans-specific gene silencing between host and parasitic plants. *The Plant Journal, 56*, 389–397.

Tsuchiya, Y., Yoshimura, M., Sato, Y., Kuwata, K., Toh, S., Holbrook-Smith, D., ... Hagihara, S. (2015). Probing strigolactone receptors in *Striga hermonthica* with fluorescence. *Science, 349*, 864–868.

Ueno, K., Furumoto, T., Umeda, S., Mizutani, M., Takikawa, H., Batchvarova, R., & Sugimoto, Y. (2014). Heliolactone, a non-sesquiterpene lactone germination stimulant for root parasitic weeds from sunflower. *Phytochemistry, 108*, 122–128.

Umehara, M., Hanada, A., Yoshida, S., Akiyama, K., Arite, T., Takeda-Kamiya, N., ... Yamaguchi, S. (2008). Inhibition of shoot branching by new terpenoid plant hormones. *Nature, 455*, 195–200.

Véronési, C., Bonnin, E., Calvez, S., Thalouarn, P., & Simier, P. (2007). Activity of secreted cell wall modifying enzymes and expression of peroxidase-encoding gene following germination of *Orobanche ramosa*. *Biologia Plantarum, 51*, 391–394.

Vieira Dos Santos, C., Delavault, P., Letousey, P., & Thalouarn, P. (2003). Identification by suppression subtractive hybridization and expression analysis of *Arabidopsis thaliana*

putative defence genes during *Orobanche ramosa* infection. *Physiological and Molecular Plant Pathology, 62*, 297–303.
Vieira Dos Santos, C., Letousey, P., Delavault, P., & Thalouarn, P. (2003). Defense gene expression analysis of *Arabidopsis thaliana* parasitized by *Orobanche ramosa*. *Phytopathology, 93*, 451–457.
Wang, L., Wang, B., Jiang, L., Liu, X., Li, X., Lu, Z., … Li, J. (2015). Strigolactone signaling in *Arabidopsis* regulates shoot development by targeting D53-like SMXL repressor proteins for ubiquitination and degradation. *The Plant Cell, 27*, 3128–3142.
Waters, M. T., Nelson, D. C., Scaffidi, A., Flematti, G. R., Sun, Y. K., Dixon, K. W., & Smith, S. M. (2012). Specialisation within the DWARF14 protein family confers distinct responses to karrikins and strigolactones in *Arabidopsis*. *Development, 7*, 1285–1295.
Waters, M. T., Scaffidi, A., Sun, Y. K., Flematti, G. R., & Smith, S. M. (2014). The karrikin response system of *Arabidopsis*. *The Plant Journal, 79*, 622–631.
Westwood, J. H. (2013). The physiology of the established parasite-host association. In D. M. Joel, J. Gressel, & L. J. Musselman (Eds.), *Parasitic Orobanchaceae* (pp. 87–114). Berlin: Springer.
Xi, Z., Bradley, R. K., Wurdack, K. J., Wong, K., Sugumaran, M., Bomblies, K., … Davis, C. C. (2012). Horizontal transfer of expressed genes in a parasitic flowering plant. *BMC Genomics, 13*, 227.
Xie, X., Yoneyama, K., Kurita, J., Harada, Y., Yamada, Y., Takeuchi, Y., & Yoneyama, K. (2009). 7-Oxoorobanchyl acetate and 7-Oxoorobanchol as germination stimulants for root parasitic plants from flax (*Linum usitatissimum*). *Bioscience, Biotechnology, and Biochemistry, 73*, 1367–1370.
Xi, Z., Wang, Y., Bradley, R. K., Sugumaran, M., Marx, C. J., Rest, J. S., & Davis, C. C. (2013). Massive mitochondrial gene transfer in a parasitic flowering plant clade. *PLoS Genetics, 9*, e1003265.
Yoder, J. I. (2001). Host-plant recognition by parasitic Scrophulariaceae. *Current Opinion in Plant Biology, 4*, 359–365.
Yoneyama, K., Awad, A. A., Xie, X., Yoneyama, K., & Takeuchi, Y. (2010). Strigolactones as germination stimulants for root parasitic plants. *Plant & Cell Physiology, 51*, 1095–1103.
Yoneyama, K., Ruyter-Spira, C., & Bouwmeester, H. (2013). Induction of germination. In M. D. Joel, J. Gressel, & J. L. Musselman (Eds.), *Parasitic Orobanchaceae: Parasitic mechanisms and control strategies* (pp. 167–194). Berlin, Heidelberg: Springer Berlin Heidelberg.
Yoneyama, K., Xie, X., Sekimoto, H., Takeuchi, Y., Ogasawara, S., Akiyama, K., … Yoneyama, K. (2008). Strigolactones, host recognition signals for root parasitic plants and arbuscular mycorrhizal fungi, from *Fabaceae* plants. *New Phytologist, 179*, 484–494.
Yoshida, S., Maruyama, S., Nozaki, H., & Shirasu, K. (2010). Horizontal gene transfer by the parasitic plant *Striga hermonthica*. *Science, 328*, 1128.
de Zelicourt, A., Letousey, P., Thoiron, S., Campion, C., Simoneau, P., Elmorjani, K., … Delavault, P. (2007). Ha-DEF1, a sunflower defensine, induces cell death in *Orobanche* parasitic plant. *Planta, 226*, 591–600.
Zhang, Y., Fernandez-Aparicio, M., Wafula, E. K., Das, M., Jiao, Y., Wickett, N. J., … Depamphilis, C. W. (2013). Evolution of a horizontally acquired legume gene, albumin 1, in the parasitic plant *Phelipanche aegyptiaca* and related species. *BMC Evolutionary Biology, 13*, 48.
Zhang, Y., van Dijk, A. D. J., Scaffidi, A., Flematti, G. R., Hofmann, M., Charnikhova, T., … Bouwmeester, H. J. (2014). Rice cytochrome P450 MAX1 homologs catalyze distinct steps in strigolactone biosynthesis. *Nature Chemical Biology, 10*, 1028–1033.
Zhou, F., Lin, Q., Zhu, L., Ren, Y., Zhou, K., Shabek, N., … Wan, J. (2013). D14-SCFD3-dependent degradation of D53 regulates strigolactone signalling. *Nature, 504*, 406–410.

CHAPTER FOUR

Plant–Plant Communication Through Common Mycorrhizal Networks

L. Gilbert*, D. Johnson§,1
*James Hutton Institute, Aberdeen, United Kingdom
§University of Aberdeen, Aberdeen, United Kingdom
[1]Corresponding author: E-mail: d.johnson@abdn.ac.uk

Contents

Abstract

Mycorrhizal fungi can interconnect individual plants to form common mycorrhizal networks (CMNs). Recent work has shown that these networks can transport signals produced by plants in response to herbivore and pathogen infestation to neighbouring plants before they are themselves attacked. The speed of transfer to uninfested plants is such that the mechanism is likely to have measurable benefits for plant protection. At present, the nature of the signals and the degree of plant and fungal control of their delivery are unknown. Interplant communication of signals via mycorrhizal fungi means we must rethink our understanding of multitrophic interactions in nature, and the findings have led to several hypotheses concerning the evolutionary outcomes of the process. We urgently need to better understand the mechanisms of signal transfer in natural plant communities, and attempt to harness the formation of CMNs as a tool for sustainable pest management in agriculture.

Advances in Botanical Research, Volume 82
ISSN 0065-2296
http://dx.doi.org/10.1016/bs.abr.2016.09.001

1. INTRODUCTION

Mycorrhizal fungi colonize the roots of terrestrial plants in almost all families (Smith & Read, 2008) and are ubiquitous within natural and many agricultural ecosystems (Gosling, Hodge, Goodlass, & Bending, 2006). This is a symbiosis whereby the fungus gains carbon in the form of sugars from the plant, transported below ground very quickly after CO_2 is converted to sugar during photosynthesis (Johnson, Leake, Ostle, Ineson, & Read, 2002); meanwhile, the fungus provides benefits for plants through increased acquisition of nutrients, especially nitrogen and phosphorus, and water (Smith & Read, 2008), as well as improved resilience against biotic (Van Wees, Van der Ent, & Pieterse, 2008) and abiotic (Maya & Matsubara, 2013) stresses. The development of extraradical mycelium means that mycorrhizal fungi can form hyphal connections between the root systems of two or more host plants (Selosse, Richard, He, & Simard, 2006), forming 'common mycorrhizal networks' (CMNs) below ground between multiple plants, popularly referred to as the 'Wood-Wide Web' (Helgason, Daniell, Husband, Fitter, & Young, 1998).

Although we have known for many years that mycorrhizal fungi can form CMNs in the laboratory (e.g., Whittingham & Read, 1982), there remain few studies of their distribution and extent in nature because of the difficulty in unequivocally identifying such connections. Heap and Newman (1980) provided unequivocal evidence in pasture systems using buried slides from which images of hyphal connections could be analyzed, while more recent work across greater areas necessitates indirect measures using population genetics (e.g., Beiler, Durall, Simard, Maxwell, & Kretzer, 2010). Despite the limited field evidence, CMNs are likely to be ubiquitous in nature because most types of mycorrhizal fungi produce abundant extraradical mycelium (Leake et al., 2004). There are three broad categories of mycorrhizal fungi [ectomycorrhizal (ECM) fungi, ericoid mycorrhizal (ERM) fungi and arbuscular mycorrhizal (AM) fungi] and these are likely to differ in the extent and function of their CMNs. Most ECM fungi (with the exception of smooth contact type) produce extraradical mycelium and some 'long-distance' growth forms produce rhizomorphs that can grow many metres through soil giving them potential to connect many roots (Taylor, 2006). ERM fungi tend to have hyphae that are concentrated close to their hosts' roots, so CMNs formed by ERM fungi are likely to operate only at a localized spatial scale (Grelet, Johnson, Vrålstad, Alexander, & Anderson, 2010). The length of extraradical mycelia of AM and ECM fungi vary considerably, with typical estimates ranging from 10 to 100 m hyphae g^{-1} soil, or even up to hundreds of metres of hyphae per metre of root length (Leake et al., 2004).

CMNs can link not only plants of the same species, but also connect different species and it is clear that CMNs are crucial components of ecosystems. For example, CMNs promote 'network-enhanced bioactive zones' (Barto et al., 2011) and have key roles in facilitating nutrient transport and redistribution between individual plants of the same and of different species (Selosse et al., 2006; Song, Simard, Carroll, Mohn, & Zeng, 2015). Carbon and mineral nutrients can move throughout CMNs, although our understanding of the factors controlling the speed, direction and magnitude of these movements is poor. In some cases, net transfer of carbon has been reported between green plants (Simard et al., 1997), with plants under shade apparently being greater sinks for carbon. For green plants, the ecological significance of net transfers of carbon has been questioned (Robinson & Fitter, 1999). However, for achlorophyllous mycoheterotrophic plants, net transfer of carbon from neighbouring green plants via CMNs is unequivocal (e.g., McKendrick, Leake, & Read, 2000), being essential for plant growth and fitness.

One of the most exciting aspects of CMN functioning that has come to light recently is their role in transporting warning signals between plants (Fig. 1). A series of experiments has now demonstrated that CMNs facilitate signals that are sent in response to pest or pathogen attack in one plant, that

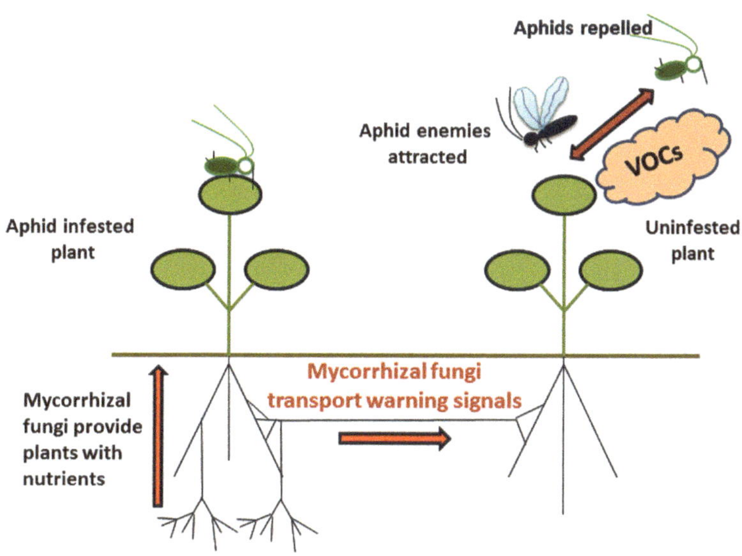

Figure 1 The interactions between plants, mycorrhizal fungi, aphids and parasitoid wasps in the context of plant to plant transfer of signals.

aid neighbouring plants that are connected via the CMN to the plant under attack to defend themselves against insect herbivores (Babikova et al., 2013; Babikova, Johnson, Bruce, Pickett, & Gilbert, 2013b; Song et al., 2014, 2015) and foliar necrotrophic fungi (Song et al., 2010). This new knowledge has opened the door for further research to understand the wider ecological roles of CMNs in nature.

This chapter details the discovery of these CMN-based warning signals and discusses the constraints and potential evolutionary and ecological consequences. It then suggests how CMNs could be manipulated to aid in crop pest control as part of the toolbox for promoting integrated pest management and sustainable agriculture.

2. EVIDENCE OF INTERPLANT SIGNALLING VIA COMMON MYCORRHIZAL NETWORKS

2.1 Background

The idea that CMNs could potentially send signals between plants has developed over the last few years due to various pieces of accruing evidence. For example, it is known that CMNs can transport mineral nutrients (Finlay & Read, 1986) and other resources such as carbon (Simard et al., 1997) between plants. Since fungal hyphae can transfer large molecules such as lipids, amino acids and sugars (e.g., Bago et al., 2002), it therefore seems reasonable to question whether they may also be able to transport chemicals that elicit plant defences against pests and pathogens. Further evidence had shown that semiochemicals can indeed move through liquids between hydroponically grown plants, thus stimulating receiving plants to elicit defences (Chamberlain et al., 2001). The mechanism for this was unknown and no CMNs were present; however, based on combining the above evidence of nutrient transfer via CMNs and semiochemical transfer through liquid, Dicke and Dijkman (2001) suggested that CMNs may be able to act as conduits for transfer of semiochemicals between connected plants.

2.2 Experimental Challenges

However, testing the hypothesis that CMNs can transport signals between plants is far from straightforward due to the need to experimentally control for the many potentially confounding factors. First, and perhaps most fundamental, is the simple problem of being sure that the plants are indeed connected by the same CMN. Analysis of genetic markers of fungi in roots may identify identical genotypes, from which we might assume the plants are connected by the same individual fungus; however, it is not possible to definitively determine if these might be disconnected genets (e.g., Beiler et al., 2010). It is not possible to merely compare plants that are grown with mycorrhizal fungi with those that are not mycorrhizal, because mycorrhizal colonization affects plant signalling pathways, causing a change in the quantity and type of volatile organic compounds (VOCs) emitted by the plant's leaves (Babikova, Gilbert, Bruce, et al., 2014; Babikova, Gilbert, Randall, et al., 2014; Fontana, Reichelt, Hempel, Gershenzon, & Unsicker, 2009; Pozo & Azcón-Aguilar, 2007) as well as dramatically affecting plant health, all of which change the plant's attractiveness to herbivores. It is often these VOCs and the attractiveness to herbivore pests that are analyzed to determine whether or not a plant has received the signal, so this consideration is fundamental. Similarly, comparisons of plants in the mycorrhizal condition versus plants that never form mycorrhizas, such as members of the *Brassicaceae*, are also inadequate because different plant species also produce different VOCs. Other confounding factors that must be controlled for include the release of exudates by both roots and mycorrhizal hyphae, and diffusion of the signal through the soil.

Despite these challenges, several experiments have now managed to convincingly demonstrate that CMNs can act as conduits for signals between interconnected plants and, by so doing, elicit defence responses in the receiving plants.

2.3 The Discovery of Common Mycorrhizal Networks–Based Signals

The first study (Song et al., 2010) to demonstrate CMN-based interplant signals used a biotrophic fungus (*Alternaria solani*) pathogenic on tomato plants that were grown with an AM fungus. To compare connected versus unconnected tomato plants, Song et al. (2010) (1) used waterproof membranes between mycorrhizal 'donor' plants (those that received the pathogenic fungus) and uninfested receivers that completely prevented diffusion or (2) made contrasts between donor and receiver plants that were both grown in the nonmycorrhizal condition. They therefore did not overcome all the experimental challenges listed above, such as teasing apart the effects exudation by mycorrhizal hyphae and diffusion of molecules through soil solution, but nevertheless showed striking results of disease resistance and activities of defensive enzymes in healthy neighbouring 'receiver' tomato plants when they were connected to an infested 'donor' tomato plant via a CMN.

The subsequent study to demonstrate CMN-based interplant signals used aphids infesting bean plants, all of which were grown with an AM fungus (Babikova et al., 2013). While this study could not completely determine that plants were connected to the same individual AM fungus (as explained above, this might be impossible), all bean plants were grown in the mycorrhizal state, and Babikova et al. (2013) were able to compare connected and nonconnected plants in two ways that controlled for root—root contact and soil diffusion. A central donor bean plant that received aphids was surrounded by four 'receiver' bean plants (that never received aphids), two of which were connected to the donor via a CMN, and two of which were in the 'unconnected' treatment group. Of particular novelty was the use of two 'unconnected' treatment groups: the first was to simply prevent contact using a 0.5 μm mesh through which fungal hyphae cannot grow, and a second 'unconnected' treatment grew the receiver plant inside a 40 μm mesh core that did allow hyphal contact and, when aphids were added to the donor plant, this core was rotated to snap the hyphal connections (Johnson, Leake, & Read, 2001). This neatly designed treatment group thus controlled for any effect of plants being connected to one another before the connection was broken. The above-ground parts of all plants were sealed in plastic bags when aphids were added to the donor, to prevent any aerial communication. Babikova et al. (2013) analyzed the VOCs and measured the attractiveness of the VOCs to aphids and parasitoid wasps (aphid enemies) of all plants. They found that connected receiver plants produced the same concoction of VOCs as the aphid-infested donors, even though they themselves were not infested, as well as being similarly repellent to aphids and attractive to parasitoid wasps. By contrast, unconnected beans maintained the same VOC composition as expected from uninfested plants, and were attractive to aphids and unattractive to parasitoids, which is the normal state of plants that have not induced their defences. This demonstrated clearly and unequivocally that CMNs can be conduits for aphid-warning signals between plants, and redefines our understanding of multitrophic interactions.

The third study to demonstrate CMN-based warning signals (Song et al., 2014) found that caterpillars on tomato plants induced defence genes in neighbouring tomato plants when they were connected by CMNs formed by an AM fungus. This experiment used tomato plant donors that had a jasmonate biosynthesis defective mutant that were not able to induce defence responses; accordingly there was no change in insect resistance in 'receiver' plants, suggesting that the jasmonate signalling pathway is required for CMN-based interplant communication.

The most recent study Song et al. (2015) to show CMN-based pest warning signals was particularly interesting because it, for the first time, showed that these signals can be mediated through ECM fungi rather than the previous experiments that used AM fungi, as well as the signals passing between different plant species, and it showed CMN-based signalling in response to mechanical (manual) damage as well as insect herbivore damage.

Song et al. (2015) used pairs of seedlings, a Douglas fir (*Pseudotsuga menziesii*) donor with just one ponderosa pine (*Pinus ponderosa*) receiver that was subject to one of three treatments: (1) grown together in bulk soil allowing both roots and hyphal contact, (2) receivers grown in 35 μm mesh allowing only hyphal contact and (3) receivers grown in 0.5 μm mesh allowing only diffusion through soil, i.e., no hyphal or root contact. Douglas fir donors were then either defoliated manually or suffered light damage by infestation by western spruce budworm (*Choristoneura occidentalis*). Both manual defoliation and insect damage of donors led to increased activity of defence enzymes (peroxidase, polyphenol oxidase and superoxide dismutase) in both the Douglas fir donors and the ponderosa pine receivers grown in the 35 μm mesh that allowed only hyphal contact. This suggests that ECM fungal CMNs can transfer warning signals between plant species, although it is so far not known why signalling did not appear to occur between plants that had both hyphal and root contact (receivers grown with no mesh barrier).

There is now unequivocal evidence that plants must induce signals that can be transported via CMNs between plants of the same and different species, through both AM and ECM fungi, and in response to aphids (suckers), caterpillars (chewers), manual/mechanical damage and fungal pathogens. However, we need a better understanding of the specificity of responses by both donor and receivers to different stimuli. Thus far, the key responses of receiver plants to a range of stimuli are production of methyl salicylate and genes related to jasmonate signalling pathways.

2.4 Speed of Signal Transfer

Now that we know that plants can receive warning signals via CMNs in response to pest and pathogen attack, the ecological and evolutionary implications are potentially large, and there may even be scope for us to utilize this signalling pathway to aid in crop pest control. However, the importance of the signals depends to a large extent on whether the signal, and the response to the signal by receiver plants, is rapid enough to change the dynamics between pest, host plant and fungus. Crucially, the signal (and the response to the signal) must be rapid enough to successfully deter herbivores before they can travel between the infested donor and the uninfested receiver plants. Johnson and Gilbert (2015, 2017) combined the evidence from the studies to date of the speed of the signal and response, i.e., the time from pest or pathogen addition in the donor to the response detected in the receiver plants, and concluded that the signal was detected, in terms of measurable response by receiver plants, within 24 h of the donor receiving the pest. Peak response in receivers occurred between 48 and 100 h after infestation of the donor (Fig. 2). These findings suggest that the CMN-based pest warning signals are indeed likely to be of importance to pest–plant–fungal dynamics, opening the field for further research into the wider ecological and evolutionary consequences of this mode of plant–plant communication. The length of time over which the signals are effective (i.e., how long the plants produce the defence response) is so far unknown, but this is also likely to affect strongly the wider ecosystem impact of CMN-based signalling.

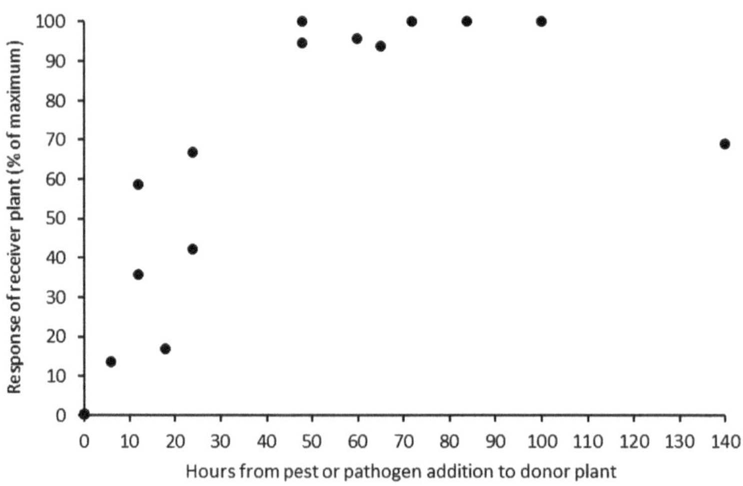

Figure 2 The speed of response of uninfested 'receiver' plants when connected via common mycorrhizal networks to neighbouring plants infested with different pests or damaged mechanically.

3. BIOPHYSICAL CONSTRAINTS OF SIGNAL TRANSFERS IN COMMON MYCORRHIZAL NETWORKS

While the impact of CMN-based warning signals will depend on their speed and the period of time over which their effects are manifested, at least as important is the physical distance over which they can travel and how many receiver plants they can affect. Experiments testing these factors are yet to be conducted, although Babikova et al. (2013) and Babikova, Johnson, Bruce, Pickett, and Gilbert (2013a) showed that signals can travel through CMNs of at least 20 cm in length. Depending on the nature of the signal, i.e., what substance is transported, there must be physical constraints to the length of hyphae through which a signal can travel. These constraints may be analogous to thermodynamic laws of liquids moving through tubes. The nature of the signal is unknown, but a chemical transport mechanism seems likely, given that sugars, lipids and amino acids can be exchanged between plants and mycorrhizal fungi (Smith & Read, 2008). However, identifying the molecular structure of signals will require extremely challenging controlled experiments. It is likely that a combination of advanced analytical chemistry and gene silencing approaches will be required, which would be facilitated through the use of model plant and fungi that have had their genomes sequenced. Alternatively, it is not impossible that action potentials, or electrical signals, are part of the mechanism, especially as it is known that action potentials are produced in plants in response to leaf mechanical damage (Mousavi, Chauvin, Pascaud, Kellenberger, & Farmer, 2013), chewing insects (Salvador-Recatala, Tjallingii, & Farmer, 2014) and in roots in response to AM fungi (Ayling, Smith, Smith, & Kolesik, 1997). Furthermore, action potentials are produced in fungi in response to detection of a nutrient resource (Brand & Gow, 2009).

Babikova et al. (2013b) hypothesized that signals may be able to be relayed down a chain of plants, as well as affecting receiver plants in direct contact with the original donor. If each receiver plant resends the signal onto the next plant, thus acting like a new donor, the signal will be able to travel much further than otherwise suggested from the physical constraints of narrow tube transport.

Another fundamental aspect of CMN-based signalling, which will affect how we think about the ecological and evolutionary implications, is which partner is in control of the signal? Because it is the donor plant that is attacked by the pest, and it is the receiver plant that receives and responds to the signal to protect itself from pest attack, it is easy to take a phytocentric point of view and think of the process in terms of a signal being sent to warn neighbours of pest attack, or that receiver plants may 'eaves-drop' and use the signal to protect themselves. However, it is also important to consider the mycocentric stand-point, as there is a possibility that the fungus controls the signal, since the signal travels through it. There are likely to be fitness consequences of these signals to both donor and receiver plants, as well as the fungi, so it could benefit either partner to try to control the signal for their own benefit.

4. EVOLUTIONARY CONSEQUENCES OF SIGNALLING VIA COMMON MYCORRHIZAL NETWORKS

It is easy to consider the fitness benefits of the CMN-based signal to the receiver plant because it can then respond by inducing defence chemicals, thus protecting itself against imminent pest attack. While eliciting such defence pathways can be costly to the plant (Suza et al., 2010), infestation by pests and disease is often extremely costly or even fatal (Babikova, Gilbert, Bruce, et al., 2014; Babikova, Gilbert, Randall, et al.,2014; Guerrieri & Digilio, 2008).

The benefit to the donor plant of sending the signal (if, indeed, it does actively 'send' the signal, from a phytocentric point of view) is more difficult to understand. However, if kin are in close proximity and likely to be connected to the same CMN, it might increase the donor's 'inclusive fitness' (that is, the fitness of its genepool including its family) to warn kin of imminent pest attack. Indeed, Karban, Shiojiri, Ishizaki, Wetzel, and Evans (2013) have shown that aerial signalling between plants via VOCs infers more benefit to close kin than to unrelated receiver plants.

Babikova et al. (2013b) took a mycocentric viewpoint and formulated hypotheses on the fitness consequences to the mycorrhizal fungi of sending the warning signals between plants. These hypotheses were based on the assumption that the key benefit to the fungus is to protect its source of carbon from its host plants. Babikova et al. (2013b) therefore hypothesized that (if there is a mechanism in place that allows fungi to allocate signal specifically) the fungus would preferentially allocate signals to those individual plants that are their most lucrative carbon source, or those plants that are best able to protect themselves from pest attack after receiving the signal (e.g., by producing the most effective VOCs), or to plants that they have exclusive rights to, in preference to plants that are also connected to competing mycorrhizal fungi. Johnson and Gilbert (2015) speculated further with this mycocentric viewpoint, and theorized that in stable ecosystems, such as natural grasslands or forests, where particular CMNs may persist long-term, there may be a higher cost to the fungus of losing an individual host plant to pests or disease and thus a greater benefit of transporting warning signals, than in unstable ecosystems with a rapid turnover of individual CMNs.

Insect herbivores are likely to suffer a cost of CMN-based signalling, because the response of plants is to produce defences against them, emitting VOCs that are repellent to the herbivores yet attractive to the herbivore's enemies (e.g., parasitoid wasps that lay eggs inside herbivores such as aphids). Gilbert and Johnson (2015) theorized on the effects of CMN-based plant–plant signalling on the evolutionary strategies of insect herbivores. Assuming that signals act against insect herbivores and thinking from an evolutionary stand-point, Gilbert and Johnson (2015) hypothesized that, if signals enable plants across a wide area to upregulate their defences (rather than just one receiver plant), there would be a selective advantage to insects of dispersing over wider areas to locate patches of plants that are not emitting defence chemicals. They similarly speculated that, if signals operate only between similar plant species, wide dispersal would be unnecessary if the herbivore is a generalist, so an alternative evolutionary strategy to dispersal could be for specialists to become generalists. By contrast, Gilbert and Johnson (2015) argued, if signals operate among multiple plant species, wide dispersal would be advantageous even for generalist insect herbivores.

Taking the insect–plant–fungi system as a whole, Johnson and Gilbert (2015) discussed how the specialization of this three-way interaction might affect the likelihood of CMN-based plant–plant signalling evolving at all. In particular they focussed on specialist versus generalist insect herbivores and the specificity of the signal (i.e., whether the signal operates within single-plant species only or over multiple plant species). They hypothesized that in an environment with high plant species diversity and where insect pests specialize on particular plant species, there should be selective pressure for species-specific signals to evolve. Conversely, they speculated that in single-plant-species environments or where insect herbivores are generalists (feeding on a wider range of plant species), generic signals would be beneficial to plants and fungi and so could evolve. Little is known about the specificity of the signalling pathways through CMNs. However, we do know that plant responses to herbivory tend not to be specific (Bidart-Bouzat & Kliebenstein, 2011), which suggests that there is no selective advantage of producing signals that are specific to particular herbivores.

5. POTENTIAL TO MANAGE COMMON MYCORRHIZAL NETWORKS TO AID CROP PROTECTION

Given that CMN-based plant–plant signals warn plants before the pest attack, there may be scope for developing these signals as part of the toolkit in the fight against agricultural pests and disease. While pesticides have helped hugely to increase crop yields throughout the 20th century (Pretty, 2008; Weisenburger, 1993), there is now an increasing need to find alternative sustainable ways to maintain and improve food security. Indeed, the EU has withdrawn 80% of assessed substances, with more currently being phased out (Karabelas, Plakas, Solomou, Drossou, & Sarigiannis, 2009), due to the growing recognition of the negative aspects to pesticide use. These include environmental pollution (Pretty, 2008), human health issues (Karabelas et al., 2009; Weisenburger, 1993), killing beneficial arthropods such as pollinators and natural enemies of pests (Campbell et al., 2000; Crane et al., 2006), and the development of acquired resistance to many pesticides by the pests (Bjornlund et al., 2000; Pimentel et al., 1991). Therefore, there is now a growing body of research into integrated pest management to encourage more sustainable crop management. Babikova et al. (2013, 2013b) first suggested that CMN-based plant–plant signals could be part of this toolkit. Clearly, for crops to self-protect in this way, CMNs need to connect the individual crop plants. Since deep ploughing and other forms of tilling destroy the fungal networks, this type of pest control could operate only in minimal till or no till systems, including perennials. Further research is needed on the extent to which soil management aids or inhibits CMN formation and signal transfer. It would also be essential to understand which crop species can send or receive signals, how far signals can travel and through how many individual crop plants. One idea proposed by Babikova et al. (2013) is that of 'sacrificial' donor plants in the crop that are subject to controlled pest infestation or mechanical injury, which then activates signals to protect the rest of the crop. A quantification of the cost of responding to such signals is needed, in case this unduly limits crop yield, and the timing would need to be optimized by first understanding the period over which receiver plants respond, which would need to correspond with periods of greatest pest risk.

6. CONCLUSIONS

Interplant signalling via CMNs is a recently discovered process that has wide implications for ecology, evolution and sustainable agriculture. We need to increase the types of experiments to consider more natural systems, including those with diverse assemblages of plants, fungi and insect herbivores. Moreover, we need to conduct detailed multidisciplinary mechanistic studies to identify the nature of the signals that are transported via CMNs. Finally, consideration of the ecological, evolutionary and agronomic effects of signalling via CMNs needs to be considered alongside the broader suite of ecological functions undertaken by mycorrhizal fungi to gain a more holistic understanding of fungal networks.

REFERENCES

Ayling, S. M., Smith, S. E., Smith, F. A., & Kolesik, P. (1997). Transport processes at the plant-fungus interface in mycorrhizal associations: physiological studies. *Plant and Soil, 196*, 305—310.

Babikova, Z., Gilbert, L., Bruce, T. J. A., Birkett, M., Caulfield, J. C., Woodcock, C., ... Johnson, D. (2013). Underground signals carried through common mycelial networks warn neighbouring plants of aphid attack. *Ecology Letters, 16*, 835—843. http://dx.doi.org/10.1111/ele.12115.

Babikova, Z., Gilbert, L., Bruce, T. J. A., Dewhirst, S., Pickett, J. A., & Johnson, D. (2014). Arbuscular mycorrhizal fungi and aphids interact by changing host plant quality and volatile emission. *Functional Ecology, 28*, 375—385. http://dx.doi.org/10.1111/1365-2435.12181.

Babikova, Z., Gilbert, L., Randall, K. C., Bruce, T. J. A., Pickett, J. A., & Johnson, D. (2014). Increasing phosphorus supply is not the mechanism by which arbuscular mycorrhiza increase attractiveness of bean (*Vicia faba*) to aphids. *Journal of Experimental Botany, 65*, 5231—5241. http://dx.doi.org/10.1093/jxb/eru283.

Babikova, Z., Johnson, D., Bruce, T. J. A., Pickett, J. A., & Gilbert, L. (2013). How rapid is aphid-induced signal transfer between plants via common mycelial networks? *Communicative & Integrative Biology, 6*, e25904.

Babikova, Z., Johnson, D., Bruce, T. J. A., Pickett, J. A., & Gilbert, L. (2013). Underground allies: how and why do mycelial networks help plants defend themselves? *BioEssays, 36*, 21—26. http://dx.doi.org/10.1002/bies.201300092.

Bago, B., Zipfel, W., Williams, R. M., Jun, J., Arreola, R., Lammers, P. J., ... Shachar-Hill, Y. (2002). Translocation and utilization of fungal storage lipid in the arbuscular mycorrhizal symbiosis. *Plant Physiology, 128*, 108—124.

Barto, K. E., Hilker, M., Müller, F., Mohney, F., Weidenhamer, J. D., & Rillig, M. C. (2011). The fungal fast lane: common mycorrhizal networks extend bioactive zones of allelochemicals in soils. *PLoS One, 6*, e27195.

Beiler, K. J., Durall, D. M., Simard, S. W., Maxwell, S. A., & Kretzer, A. M. (2010). Architecture of the wood-wide web: *Rhizopogon* spp. genets link multiple Douglas-fir cohorts. *New Phytologist, 185*, 543—553.

Bidart-Bouzat, M. G., & Kliebenstein, D. (2011). An ecological genomic approach challenging the paradigm of differential plant responses to specialist versus generalist insect herbivores. *Oecologia, 167*, 677—689.

Bjornlund, L., Ekelund, F., Christensen, S., Jacobsen, C. S., Krogh, P. H., & Johnsen, K. (2000). Interactions between saprotrophic fungi, bacteria and protozoa on decomposing wheat roots in soil influenced by the fungicide fenpropimorph (Corbel (R)): a field study. *Soil Biology & Biochemistry, 32*, 967—975.

Brand, A., & Gow, N. A. R. (2009). Mechanisms of hypha orientation of fungi. *Current Opinion in Microbiology, 12*, 350—357.

Campbell, P. J., Brown, K. C., Harrison, E. G., Bakker, F., Barrett, K. L., Candolfi, M. P., … Waltersdorfer, A. (2000). A hazard quotient approach for assessing the risk to non-target arthropods from plant protection products under 91/414/EEC: hazard quotient trigger value proposal and validation. *Journal of Pest Science, 73*, 117–124.
Chamberlain, K., Guerrieri, E., Pennacchio, F., Pettersson, J., Pickett, J. A., Poppy, G. M., … Woodcock, C. M. (2001). Can aphid-induced plant signals be transmitted aerially and through the rhizosphere? *Biochemical Systematics and Ecology, 29*, 1063–1074.
Crane, M., Norton, A., Leaman, J., Chalak, A., Bailey, A., Yoxon, M., … Fenlon, J. (2006). Acceptability of pesticide impacts on the environment: what do United Kingdom stakeholders and the public value? *Pest Management Science, 62*, 5–19.
Dicke, M., & Dijkman, H. (2001). Within-plant circulation of systemic elicitor of induced defence and release from roots of elicitor that affects neighbouring plants. *Biochemical Systematics and Ecology, 29*, 1075–1087.
Finlay, R. D., & Read, D. J. (1986). The structure and function of the vegetative mycelium of ectomycorrhizal plants .2. The uptake and distribution of phosphorus by mycelial strands interconnecting host plants. *New Phytologist, 103*, 157–165.
Fontana, A., Reichelt, M., Hempel, S., Gershenzon, J., & Unsicker, S. B. (2009). The effects of arbuscular mycorrhizal fungi on direct and indirect defense metabolites of *Plantago lanceolata* L. *Journal of Chemical Ecology, 35*, 833–843.
Gilbert, L., & Johnson, D. (2015). Plant mediated "apparent effects" between mycorrhiza and insect herbivores. *Current Opinion in Plant Biology, 26*, 100–105. http://dx.doi.org/10.1016/j.pbi.2015.06.008.
Gosling, P., Hodge, A., Goodlass, G., & Bending, G. D. (2006). Arbuscular mycorrhizal fungi and organic farming. *Agriculture, Ecosystems & Environment, 113*, 17–35.
Grelet, G.-A., Johnson, D., Vrålstad, T., Alexander, I. J., & Anderson, I. C. (2010). New insights into the mycorrhizal *Rhizoscyphus ericae* aggregate: spatial structure and co-colonization of ectomycorrhizal and ericoid roots. *New Phytologist, 188*, 210–222.
Guerrieri, E., & Digilio, M. C. (2008). Aphid–plant interactions: a review. *Journal of Plant Interactions, 3*, 223–232.
Heap, A. J., & Newman, E. I. (1980). Links between roots by hyphae of vesicular-arbuscular mycorrhizas. *New Phytologist, 85*, 169–171. http://dx.doi.org/10.1111/j.1469-8137.1980.tb04457.x.
Helgason, T., Daniell, T. J., Husband, R., Fitter, A. H., & Young, J. P. W. (1998). Ploughing up the wood-wide web? *Nature, 394*, 431.
Johnson, D., & Gilbert, L. (2015). Interplant signalling through hyphal networks. *New Phytologist, 205*, 1448–1453. http://dx.doi.org/10.1111/nph.13115.
Johnson, D., & Gilbert, L. (2017). Mycorrhizal fungal networks as plant communication systems. In J. Dighton, & J. F. White (Eds.), *The fungal community: Its organization and role in the ecosystem* (4th ed., pp. 539–548). CRC Taylor & Francis (in press).
Johnson, D., Leake, J. R., Ostle, N., Ineson, P., & Read, D. J. (2002). In situ $^{13}CO_2$ pulse-labelling of upland grassland demonstrates a rapid pathway of carbon flux from arbuscular mycorrhizal mycelium to the soil. *New Phytologist, 153*, 327–334.
Johnson, D., Leake, J. R., & Read, D. J. (2001). Novel in-growth core system enables functional studies of grassland mycorrhizal mycelial networks. *New Phytologist, 152*, 555–562.
Karabelas, A. J., Plakas, K. V., Solomou, E. S., Drossou, V., & Sarigiannis, D. A. (2009). Impact of European legislation on marketed pesticides — a view from the standpoint of health impact assessment studies. *Environment international, 35*, 1096–1107.
Karban, R., Shiojiri, K., Ishizaki, S., Wetzel, W. C., & Evans, R. Y. (2013). Kin recognition affects plant communication and defence. *Proceedings of the Royal Society B-Biological Sciences, 280*, 20123062.

Leake, J. R., Johnson, D., Donnelly, D. P., Muckle, G., Boddy, L., & Read, D. J. (2004). Networks of power and influence: the role of mycorrhizal mycelium in controlling plant communities and agroecosystem functioning. *Canadian Journal of Botany, 82*, 1016–1045.

Maya, M. A., & Matsubara, Y. (2013). Influence of arbuscular mycorrhiza on the growth and antioxidative activity in cyclamen under heat stress. *Mycorrhiza, 23*, 381–390.

McKendrick, S. L., Leake, J. R., & Read, D. J. (2000). Symbiotic germination and development of myco-heterotrophic plants in nature: transfer of carbon from ectomycorrhizal *Salix repens* and *Betula pendula* to the orchid *Corallorhiza trifida* through shared hyphal connections. *New Phytologist, 145*, 539–548.

Mousavi, S. A. R., Chauvin, A., Pascaud, F., Kellenberger, S., & Farmer, E. E. (2013). Glutamate receptor-like genes mediate leaf-to-leaf wound signalling. *Nature, 500*, 422–425.

Pimentel, D., Mclaughlin, L., Zepp, A., Lakitan, B., Kraus, T., Kleinman, P., ... Selig, G. (1991). Environmental and economic-effects of reducing pesticide use – a substantial reduction in pesticides might increase food costs only slightly. *Bioscience, 41*, 402–409.

Pozo, M. J., & Azcón-Aguilar, C. (2007). Unravelling mycorrhiza-induced resistance. *Current Opinion in Plant Biology, 10*, 393–398.

Pretty, J. (2008). Agricultural sustainability: concepts, principles and evidence. *Philosophical Transactions of the Royal Society B-Biological Sciences, 363*, 447–465.

Robinson, D., & Fitter, A. (1999). The magnitude and control of carbon transfer between plants linked by common mycorrhizal network. *Journal of Experimental Botany, 50*, 9–13.

Salvador-Recatala, V., Tjallingii, W. F., & Farmer, E. E. (2014). Real-time, in vivo intracellular recordings of caterpillar-induced depolarization waves in sieve elements using aphid electrodes. *New Phytologist, 203*, 674–684.

Selosse, M.-A., Richard, F., He, X. H., & Simard, S. W. (2006). Mycorrhizal networks: des liaisons dangereuses? *Trends in Ecology and Evolution, 21*, 621–628.

Simard, S. W., Perry, D. A., Jones, M. D., Myrold, D. D., Durall, D. M., & Molina, R. (1997). Net transfer of carbon between ectomycorrhizal tree species in the field. *Nature, 388*, 579–582.

Smith, S. E., & Read, D. J. (2008). *Mycorrhizal symbiosis* (3rd ed.). London: Academic Press.

Song, Y. Y., Simard, S. W., Carroll, A., Mohn, W. W., & Zeng, R. S. (2015). Defoliation of interior Douglas fir elicits carbon transfer and stress signalling to ponderosa pine neighbors through ectomycorrhizal networks. *Scientific Reports, 5*, 8495.

Song, Y. Y., Ye, M., Li, C. Y., He, X. H., Zhu-Salzman, K. Y., Wang, R. L., ... Zeng, R. S. (2014). Hijacking common mycorrhizal networks for herbivore-induced defence signal transfer between tomato plants. *Scientific Reports, 4*, 3915.

Song, Y. Y., Zeng, R. S., Xu, J. A. F., Li, J., Shen, X. A., & Yihdego, W. G. (2010). Interplant communication of tomato plants through underground common mycorrhizal networks. *PLoS One, 5*, e13324.

Suza, W. P., Avila, C. A., Carruthers, K., Kulkarni, S., Goggin, F. L., & Lorence, A. (2010). Exploring the impact of wounding and jasmonates on ascorbate metabolism. *Plant Physiology & Biochemistry, 48*, 337–350.

Taylor, A. F. S. (2006). Common mycelial networks: life-lines and radical addictions. *New Phytologist, 169*, 6–8.

Van Wees, S. C. M., Van der Ent, S., & Pieterse, C. M. J. (2008). Plant immune responses triggered by beneficial microbes. *Current Opinion in Plant Biology, 11*, 443–448.

Weisenburger, D. D. (1993). Human health-effects of agrichemical use. *Human Pathology, 24*, 571–576.

Whittingham, J., & Read, D. J. (1982). Vesicular-arbuscular mycorrhiza in natural vegetation systems .3. Nutrient transfer between plants with mycorrhizal interconnections. *New Phytologist, 90*, 277–284. http://dx.doi.org/10.1111/j.1469-8137.1982.tb03259.x.

PART TWO

Plant Communication With Microbes

CHAPTER FIVE

Plant Communication With Associated Microbiota in the Spermosphere, Rhizosphere and Phyllosphere

P. Lemanceau*,[1], M. Barret[§], S. Mazurier*, S. Mondy*, B. Pivato*, T. Fort[¶], C. Vacher[¶]

*Agroécologie, AgroSup Dijon, INRA, University of Bourgogne Franche-Comté, Dijon, France
[§]IRHS, INRA Site d'Angers Beaucouzé, Beaucouzé, France
[¶]BIOGECO, INRA, University of Bordeaux, Pessac, France
[1]Corresponding author: E-mail: philippe.lemanceau@inra.fr

Contents

Advances in Botanical Research, Volume 82
ISSN 0065-2296
http://dx.doi.org/10.1016/bs.abr.2016.10.007

Abstract

Plants are surrounded with microorganisms whose abundance is promoted by the release of plant organic compounds and by the presence of niches favourable to microbial development and activities. These microorganisms thrive in three main plant compartments, i.e., spermosphere, rhizosphere and phyllosphere, which are interconnected. They are recruited from the environment (soil, atmosphere) and from the mother plant via the seed. Plants indeed modulate the composition and activities of the hosted microbial populations through complex communication trackways relying on trophic interactions and/or molecular signalization. The tuning of these interactions by the plant favours beneficial microbial populations and activities while depressing deleterious ones, which have a major impact on plant growth and health. This review presents the current knowledge of the plant communication with associated microorganisms in the spermosphere, rhizosphere and phyllosphere and of plant and microbial traits involved. Possible prospects of application of this knowledge for monitoring plant–microbe interactions in agroecological systems with reduced chemical inputs are discussed.

1. INTRODUCTION

In contrast with animals which have developed the ability to escape biotic and abiotic stresses, plants cannot move and have therefore evolved strategies to cope with them. These strategies rely in large part on the plant-associated microbiota. Long standing coevolution process has indeed led to reciprocal beneficial interactions between plants and microorganisms. The first evidence for such association dates back to 400 million years ago with the discovery of fossils of plant–fungus associations, including arbuscular mycorrhiza ancestors (Redecker, Kodner, & Graham, 2000). These associated microorganisms are heterotrophs or even biotrophs and are supported by the primary production of host plant photosynthesis. In turn, they contribute to the nutrition and health of the host plant. The stability of the association depends on the benefits generated for each partner.

Plants are surrounded with microorganisms whose abundance is promoted by the release of plant organic compounds and the presence of niches favourable to microbial development and activities. These microorganisms thrive in three main plant compartments, i.e., spermosphere, rhizosphere and phyllosphere, which are interconnected. They are recruited from the environment (soil, atmosphere) and from the mother plant via the seed. Plants indeed modulate the composition and activities of the hosted microbial populations through complex communication trackways relying on trophic interactions and/or molecular signalization. The tuning of these interactions by the plant favours beneficial microbial populations and activities while depressing deleterious ones.

The close connection of microorganisms with plants, and more generally with eukaryotes, has recently led to the emergence of the holobiont concept, defined as the host plus all of its associated microbes (Theis et al., 2016; Vandenkoornhuyse, Quaiser, Duhamel, Le Van, & Dufresne, 2015). The study of the holobiont has been recently boosted by the development of molecular tools allowing us to characterize the taxonomic and functional diversity of microbial communities, but also their location and interaction with the host at the molecular and cellular levels.

Below we review current knowledge of the plant communication with associated microorganisms in the spermosphere, rhizosphere and phyllosphere and of plant and microbial traits involved. Possible prospects of application of this knowledge for monitoring plant–microbe interactions in agroecological systems are further discussed.

2. PLANT COMMUNICATION WITH ASSOCIATED MICROORGANISMS IN THE SPERMOSPHERE

2.1 Acquisition of the Seed Microbiota

Seeds, defined here as sexually derived structures of spermatophytes, are not only decisive for transmission of genetic material from one plant generation to another but are also strongly linked to the vertical transmission of multiple microorganisms ranging from plant-growth promoting agents to plant pathogens (Darrasse et al., 2010; Quesada-Moraga, Lopez-Diaz, & Landa, 2014; Truyens, Weyens, Cuypers, & Vangronsveld, 2015). These microorganisms are associated with different seed compartments, such as the embryo (e.g., the embryonic axis and the cotyledons), the storages tissues (e.g., endosperm and/or perisperm) and the seed coat, representing distinct microhabitats. The colonization of these microhabitats is strongly dependent of the seed transmission pathway employed by microorganisms. To date, three main seed transmission pathways have been documented: (1) the internal pathway, in which microorganisms colonize developing seeds via the xylem or nonvascular tissue of the mother plant; (2) the floral pathway that represents microbial colonization of developing seeds through the stigma and (3) the external pathway, which corresponds to colonization of mature seeds via contact of the seed with microorganisms located on fruits or threshing residues (Maude, 1996, Fig. 1). Microorganisms transmitted by the internal and floral pathways are usually found in all seed compartments, while those transmitted through the external pathway are almost exclusively associated with the seed coat (Singh & Mathur, 2004). As the external pathway is more permissive than the internal or floral pathway,

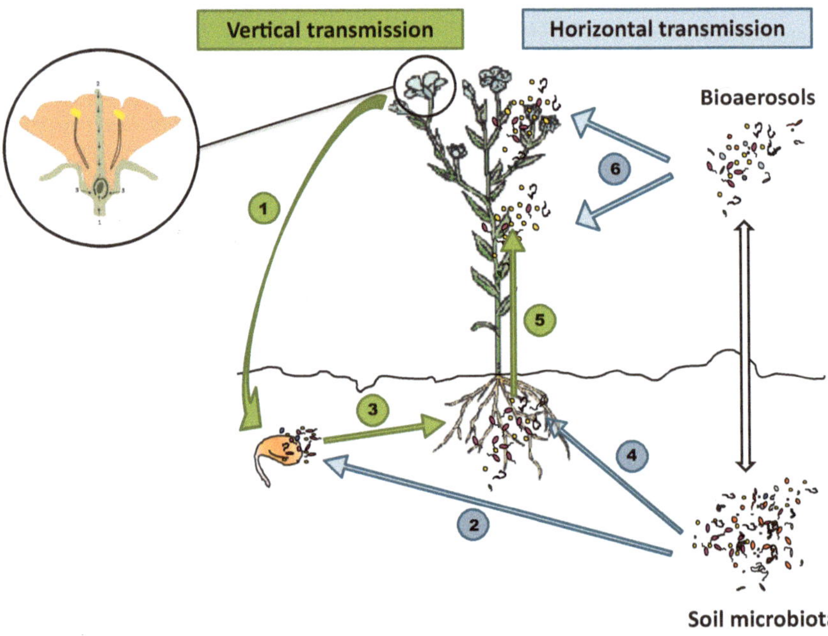

Figure 1 Schematic diagram representing the sources of microorganisms associated with the different plant compartments (spermosphere, rhizosphere and phyllosphere) and the continuum between them. Microorganisms in the spermosphere are recruited from flowers fruits and seeds (1) through internal, floral or external pathway and from soilborne communities (2) via trophic and signal communication; microorganisms in the rhizosphere are recruited from the spermosphere (3), and from soilborne communities via trophic and signal communication (4); microorganisms of the phyllosphere originate from the seed and rhizosphere (3, 5) but mostly from airborne communities upon their ability to adhesion to the plant surface and to resist to biotic and abiotic stresses (6). *Flower scheme adapted from Maude, R.B. (1996).* Seedborne diseases and their control: Principles and practice. *Wallinford, Oxon, UK: CAB International.*

microbial communities associated with the seed coat are usually more diverse than those associated with the endosperm or the embryo (Barret, Guimbaud, Darrasse, & Jacques, 2016).

2.2 Dynamics of the Seed Microbiota During Germination

Seed-associated microorganisms may either act as ephemeral colonizers of the seed habitat (seedborne microorganisms) or persist during the different plant developmental stages (seed-transmitted microorganisms). With the exception of microorganisms located within the embryo, the persistence of microbial plant interactions is strongly dependent of the outcome of intermicrobial competition occurring in the area of soil under the influence of germinating seeds (ranging from seed imbibition to radicle protrusion). This area of 'elevated microbial activity around a germinating seed' has been coined spermosphere (Verona, 1958) in analogy to the rhizosphere. During seed imbibition, multiple compounds such as carbohydrates, fatty acids, organic acids and amino acids are released to the surrounding soil (Nelson, 2004; Schiltz et al., 2015). The nature and quantity of these exudates are not only linked to the plant species but also to abiotic factors such as soil type and temperature (Nelson, 2004; Schiltz et al., 2015). The release of these organic compounds into the soil creates a zone densely populated with microorganisms that compete for resources and space. To protect itself against the myriad of microorganisms in its vicinity, seeds release antimicrobial compounds during the germination process such as flavonoids and peptides modulating microbial growth (Scarafoni et al., 2013; Terras et al., 1995).

Given the high competition for resources and space occurring within the spermosphere, only microorganisms with specific traits succeed in colonizing germinating seeds. The first set of traits important for successful spermosphere competence is related to trophic interactions. A prompt consumption of resources represents a major competitive advantage and explains why copiotrophic microorganisms are usually enriched during germination and seedling emergence (Barret et al., 2015). Chemotaxis-based motility towards seed exudates such as amino acids and organics acids has also been reported for multiple spermophere bacterial taxa including *Bacillus*, *Pseudomonas* and *Rhizobium* (Gamliel & Katan, 1992; Zheng & Sinclair, 1996). In addition to chemotactic ability, microbial populations with a versatile metabolic potential are favoured in the spermosphere. For instance, proteomics data suggest that transport and biosynthesis of 15 different amino acids are required for an optimal growth of *Salmonella enterica* on alfalfa seedling exudates (Kwan, Pisithkul, Amador-Noguez, & Barak, 2015). Moreover, carbohydrate catabolism is decisive for successful bacterial colonization as illustrated by the impaired proliferation of *Enterobacter cloacae* in the spermosphere of various plant species after mutation of the phosphofructokinase gene (*pfk*A) (Roberts, Dery, Yucel, & Buyer, 2000).

Attachment to seeds represents another crucial trait for spermosphere colonization by commensal (Deflaun, Marshall, Kulle, & Levy, 1994; Espinosa-Urgel, Salido, & Ramos, 2000) and pathogenic (Darsonval et al., 2009) bacteria. The attachment can be either mediated by flagellin (Deflaun et al., 1994) or by secreted bacterial adhesins (Darsonval et al., 2009; Espinosa-Urgel et al., 2000). Since these adhesins are involved in biofilm formation and are regulated by quorum-sensing (QS), transition from planktonic to sessile phenotype appears to be necessary for spermosphere colonization. This has been illustrated in several reports showing that seed colonizers, such as *Pseudomonas putida* or *Acidovorax citrulli*, require QS and biofilm formation for seed to seedlings transmission (Espinosa-Urgel & Ramos, 2004; Johnson & Walcott, 2013; Tian et al., 2015). Biofilm formation could also account for the microbial resistance to various antimicrobial compounds produced by the plant during germination or by microbial competitors. QS is also involved in the production of a number of antimicrobial compounds such as phenazines or 2,4-diacetylphloroglucinol (Nelson, 2004) and in the release of bacterial toxins through the Type VI secretion system (Tian et al., 2015).

Microorganisms associated with seeds represent the first populations to colonize seedlings and ultimately roots.

3. PLANT COMMUNICATION WITH ASSOCIATED MICROORGANISMS IN THE RHIZOSPHERE

3.1 Evidence for the Rhizosphere Effect

3.1.1 Description of the Rhizosphere Effect

Roots are indeed densely colonized at their surface by a myriad of microorganisms as the result of major release of organic compounds by the plant (i.e., rhizodeposits) which may represent as much as 5–21% of all fixed carbon (Bais, Weir, Perry, Gilroy, & Vivanco, 2006). Rhizodeposits include a wide variety of substances originating from sloughed-off root cells and tissues, mucilages, volatiles, and soluble lysates and exudates that are released from damaged and intact cells, respectively (Baetz & Martinoia, 2014; Curl &

Truelove, 1986; Dakora & Phillips, 2002; van Dam & Bouwmeester, 2016; Uren, 2001). The corresponding massive transfer of nutrients in the zone of soil surrounding roots, while soils are mostly mesotrophic/oligotrophic, induces a major impact on soil microbial communities, which are mostly heterotrophic. This zone of soil, termed rhizosphere (Hiltner, 1904), encompasses together plant roots (including endophytic microorganisms) plus environment surrounding or attached to and influenced by the roots (Hartmann, Rothballer, & Schmid, 2008). Rhizodeposits promote the abundance and activities of microorganisms in the rhizosphere by providing nutrient sources that support their growth, persistence and physiology (Fig. 1). However, not all the populations of the soil community are favoured in the rhizosphere, as indicated by a lower biodiversity in the rhizosphere than in the corresponding bulk soil (García-Salamanca et al., 2013; Lemanceau et al., 1995; Marilley, Vogt, Blanc, & Aragno, 1998; Semenov, van Bruggen, & Zelenev, 1999). Only the most competitive populations develop in the rhizosphere. Microbial competitiveness is greatly related to their carbon and energetic metabolism (Latour, Delorme, Mirleau, & Lemanceau, 2003); the populations having the enzymatic activities enabling to value the organic compounds from the rhizodeposits are favoured in the rhizosphere. Similarly, the most competitive populations of pseudomonads, a major beneficial group for the plants (Lemanceau, 1992), share the ability to use efficiently two different types of electron acceptors (iron and nitrogen oxides) (Ghirardi et al., 2012). Besides their carbon and energetic metabolism, they also share the ability to produce antibiotic compounds and N-acylhomoserine lactones (N-AHSL) mediating QS (Ghirardi et al., 2012). This is in agreement with the higher frequency of N-AHSL-producer pseudomonads in the rhizosphere than in bulk soil (Elasri et al., 2001). Preferential associations between plants and microorganisms are indeed not only regulated via trophic interactions but also via molecular signalization (1) from the plant, with the most documented Nod factors involved in the establishment of nitrogen-fixing symbiosis (see chapter: Commonalities in Symbiotic Plant-Microbe Signalling), (2) among microorganisms, with as example, the density regulation of bacterial activities and phenotypes (i.e., quorum-sensing) (Fuqua, Parsek, & Greenberg, 2001) and (3) through the so-called quorum quenching, an interference of the plant with this microbial communication (Grandclément, Tannières, Moréra, Dessaux, & Faure, 2016).

Plant communication leading to modifications of microbial diversity and activities in the soil surrounding roots and within roots (endophytes), collectively part of the so-called rhizosphere effect, will be reviewed herein.

3.1.2 Impact of the Plant on the Rhizosphere Effect

As expected from variations of rhizodeposits released by the host plant, microbial communities and populations significantly vary among plant species (Bulgarelli et al., 2015; Grayston, Wang, Campbell, & Edwards, 1998; Latour, Corberand, Laguerre, Allard, & Lemanceau, 1996; Pivato et al., 2007) and even genotypes of the same plant species (Inceoglu, Falcao Salles, van Overbeek, & van Elsas, 2010; Zancarini, Mougel, Terrat, Salon, & Munier-Jolain, 2013). They also differ between root zones (Baudoin, Benizri, & Guckert, 2002; Liljeroth, Burgers, & van Veen, 1991; Marschner, Crowley, & Rengel, 2011; Yang & Crowley, 2000) and during plant development (Chaparro et al., 2013; De-la-Pena et al., 2010; Mougel et al., 2006). The specificity of the rhizosphere effect remains strongly impacted by the soil communities and even physicochemical properties. Indeed, plants recruit populations from the soil reservoir whose composition is likely to vary according to the soil type (Ranjard et al., 2013). Latour, Philippot, Corberand, and Lemanceau (1999) have shown that the rhizosphere effect of a given host plant may vary from one soil type to the other, even when the sterilized soils were inoculated with the same pseudomonad community.

3.1.3 Impact of Plant–Fungal Associations (Symbiotic and Pathogenic) on the Rhizosphere Effect

Most of the plants from terrestrial ecosystems establish symbiotic associations with mycorrhizal fungi (for a review, see Smith & Read, 2008; see also chapter: Commonalities in Symbiotic Plant-Microbe Signalling). These associations are widespread and ancient, suggesting that they have evolved together with soilborne bacterial communities. This has been illustrated by the identification of bacterial groups (Comamonadaceae, Oxalobacteraceae) preferentially associated with mycorrhizal roots of *Medicago truncatula* (Offre et al., 2007, 2008; Pivato et al., 2009). In turn, bacteria preferentially associated not only with arbuscular mycorrhiza but also ectomycorrhiza (Frey, Frey-Klett, Garbaye, Berge, & Heulin, 1997) were showed to promote fungal symbioses (Frey-Klett, Garbaye, & Tarkka, 2007; Pivato et al., 2009). These bacteria were named mycorrhiza helper bacteria (MHB – Garbaye, 1994). These synergistic bacteria–mycorrhiza associations promote plant growth (Cusano et al., 2011; Gamalero et al., 2004; Pivato et al., 2009; Viollet et al., 2016). Plant roots are not only colonized by symbiotic fungi but may also be by pathogens which impact bacterial communities. A nice illustration is the take-all decline process during which wheat roots infected with the phytopathogenic fungus *Gaeumannomyces graminis* var. *tritici* select specific populations of fluorescent pseudomonads producing the 2,4-diacetylphloroglucinol antibiotic that inhibit the pathogen and therefore suppress the corresponding disease (Raaijmakers, Weller, & Thomashow, 1997; Raaijmakers & Weller, 1998).

3.1.4 Core Microbiome and Metagenomic Approaches to the Rhizosphere

Because of the importance of the associated microbiota, in terms of abundance, diversity and beneficial effects for the host plant, plants can no longer be considered as stand-alone entities and a more holistic perception is needed (Hacquard & Schadt, 2015; Theis et al., 2016; Vandenkoornhuyse et al., 2015). Plant fitness indeed results not only from plant traits but also from those of associated microorganisms, which with their plant host collectively form the holobiont. The core microbiome is defined as microbial phylotypes that are universally associated with a plant genotype or a plant species (Lundberg et al., 2012). For instance, Lundberg et al. (2012) identified a consistent presence of OTUs affiliated to Streptomycetaceae (Actinobacteria) in the *Arabidopsis thaliana* root endophytic compartment. However, composition of soil communities may to some extent impact the taxonomic composition of the core microbiome, as showed with *A. thaliana* (Bulgarelli et al., 2012; Lundberg et al., 2012). Therefore, a functional definition of the core microbiome has recently been proposed (Lakshmanan, Selvaraj, & Bais, 2014). It can be delineated using either a targeted approach (e.g., selection of antibiotic producers distributed in different genomic background of fluorescent pseudomonads according to the soils, see above) or an untargeted approach based on metagenomic or metatranscriptomic analyses to identify microbial functions (Mendes et al., 2011). The core microbiome is then defined by the shared microbial functions distributed in different microbial taxa.

3.2 Plant Traits Involved in the Rhizosphere Effect

Considering the importance of the rhizosphere effect for plant growth and health (Philippot, Raaijmakers, Lemanceau, & van der Putten, 2013), there is a major effort made by the scientific community in identifying plant traits involved in the selection of specific microbial populations and activities to ultimately monitor them through plant breeding. Among these traits a special attention is given to trophic mediated and signal-mediated communications between plant roots and microorganisms.

3.2.1 Trophic Mediated Communication

Carbon forms present in rhizodeposits are essential for plant–microbe trophic interactions. Chemically recalcitrant to degradation, polymeric carbon forms like cellulose or lignin and more labile sugars are present in the rhizosphere in different proportions; different chemical forms can be present and differ with plant genotypes and environmental factors (Neumann et al., 2014; Nguyen, 2003). Soil microbial taxa respond differently to various carbon sources (Eichorst & Kuske, 2012; Goldfarb et al., 2011; Shi et al., 2011). Only few soil bacterial taxa, such as Burkholderiales and Pseudomonales, are favoured by both labile and chemically recalcitrant substrates (Goldfarb et al., 2011). These groups appear to be commonly associated with plants (e.g., Bakker, Berendsen, Doornbos, Wintermans, & Pieterse, 2013; Bodenhausen, Horton, & Bergelson, 2013; Mendes, Garbeva, & Raaijmakers, 2013), suggesting that the presence of both chemically recalcitrant polymeric carbon forms and more labile sugars in the plant rhizosphere would favour these bacterial groups. The types of sugar molecules released by the plant may also impact the microbial diversity. Even if fluorescent pseudomonads are known for their ability to use a wide range of sugars, there is a high metabolism diversity among these populations. As examples, only some of these pseudomanad populations are able to use trehalose (Latour & Lemanceau, 1997; Raaijmakers et al., 2001) and as a result are favoured in the tomato rhizosphere (Latour et al., 1996) expected to produce the corresponding sugar; furthermore trehalose users were indeed shown to be efficient root colonizers of tomato rhizosphere (Ghirardi et al., 2012). Amino acids are the second most abundant class of compounds released in the rhizosphere behind sugars. They impact rhizosphere microorganisms both as nutrient sources and signal molecules modulating microbial activities such as biofilm formation (Moe, 2013). The most important trophic source for its impact on microbial communities seems to be the organic acids (Shi et al., 2011). Phenolic compounds may also impact the composition of the rhizosphere microbiota by either promoting or depressing a range of microbial taxa depending on their chemical structure (Badri, Chaparro, Zhang, Shen, & Vivanco, 2013). The interaction between plants and the pathogen *Agrobacterium* spp. represents a well-documented trophic mediated communication. These pathogenic bacteria cause crown gall tumours induced after transfer and insertion of bacterial genes (T-DNA) into plant genome (Platt, Morton, Barton, Bever, & Fuqua, 2014). As a result tumour cells produce xenotopic compounds (opines), usually not found in noninfected plant cells, that favour specific microorganisms able to use opines in the infected plant rhizosphere. These low molecular weight molecules (e.g., ordinary secondary amine derivatives derived from the condensation of amino acid and sugar) provide both C and N to these microorganisms. T-DNA found in the genome of tobacco ancestors that probably results from ancient bacterial transfer encodes synthesis of protein involved in opine production (Chen et al., 2016).

Plants not only impact rhizosphere microbiota via the release of nutrient but also via scavenging nutrients. As an example, the hyperaccumulation of iron by a transgenic tobacco deregulated in the ferritin synthesis led both to a depletion of that element in the rhizosphere and to the selection of pseudomonads populations highly competitive for this element, thanks to the synthesis of specific siderophores (Robin et al., 2006, 2007).

3.2.2 Signalling-Mediated Communication

Plant molecules, even present in small amount, such as hormones and hormone-like compounds, impact the composition of rhizosphere microbiome, microbial activities and behaviour. Some phytochemicals can indeed be key signals in well-known molecular communication described in symbiotic or parasitic interactions such as flavonoids involved in the establishment of the nitrogen-fixing symbiosis between legumes and rhizobia (see chapter: Commonalities in Symbiotic Plant-Microbe Signalling) and strigolactones essential in mycorrhizal symbiosis (see chapter: Commonalities in Symbiotic Plant-Microbe Signalling). Phytoalexins and plant phenolics such as acetosyringone are also known to be involved in parasitic interactions between plants and phytopathogenic bacteria (Ahuja, Kissen, & Bones, 2012; Baker et al., 2005; Jeandet, Clément, Courot, & Cordelier, 2013; Subramoni, Nathoo, Klimov, & Yuan, 2014).

Plants may also interfere with the QS pathway among bacteria by degrading acylhomoserine lactones (AHL) via lactonase synthesis that consequently disturb the QS resulting in a pathogenicity decrease of phytopathogenic bacteria (Kalia, 2013). Phytohormones, such as auxin, cytokinins, or gibberellins are involved in the regulation of parasitic and symbiotic plant—microbial interactions (Foo, McAdam, Weller, & Reid, 2016; Foo, Ross, Jones, & Reid, 2013; Groβkinsky et al., 2011; Miri, Janakirama, Held, Ross, & Szczyglowski, 2016; Ng, Perrine-Walker, Wasson, & Mathesius, 2015).

Hormones involved in plant immunity also shape the rhizosphere microbiome. A key hormone mediating plant defence reactions, salicylic acid (SA), impacts the root microbiome of *A. thaliana*, and the enrichment of specific taxa was further ascribed to their ability to use SA (Lebeis et al., 2015). Organic acids, such as citric acid and/or malic acid, also act as signals by inducing bacterial chemotaxis towards plants as showed with various species of plant-growth promoting rhizobacteria (Ling, Raza, Ma, Huang, & Shen, 2011; Rudrappa, Czymmek, Paré, & Bais, 2008; de Weert et al., 2002; Zhang et al., 2014; see also chapter: Chatting With a Tiny Belowground Member of the Holobiome: Communication Between Plants and Growth Promoting Rhizobacteria by Sharifi & Ryu, 2017). Organic acids also promote biofilm formation and therefore root colonization (Rudrappa et al., 2008; Zhang et al., 2014). Biofilm formation is indeed known to favour root bacterial colonization by creating an ecological niche that protects bacteria against biotic and abiotic stresses (Bogino, de las Mercedes Oliva, Sorroche, & Giordano, 2013). The production of QS molecules is also required for the production of mature biofilm, and consequently modulation of QS signalling may affect the ability to this production. Finally, not only these acids but also phenolics regulate expression of bacterial genes such as those encoding synthesis of antibiotics active against fungal phytopathogens (Pérez-Jaramillo, Mendes, & Raaijmakers, 2016).

Other plant compounds may interfere with microbial activities central for soil functioning. As an example, plants may inhibit nitrification (biological inhibition of nitrification, BNI) via suppressors consisting in nonesterified fatty acids, corresponding methyl esters and a cyclic diterpene (Philippot & Hallin, 2011; Skiba, George, Baggs, & Daniell, 2011; Subbarao et al., 2009, 2012). These plant compounds inhibit nitrification by suppressing microbial enzymatic pathways which catalyze the essential reactions of ammonia oxidation (Subbarao et al., 2012).

3.3 Microbial Traits Favoured in the Rhizosphere

As indicated above, elements or molecules, released or scavenged by plants, impact microbial communities in the rhizosphere. Microbial populations favoured by the plant exhibit the ability to (1) feed on specific nutrients and/or signal molecules released by the plant, (2) adapt to deprivation of nutrients scavenged by the plants, (3) respond by positive chemotaxis and (4) establish biofilms. Many studies relied on colonization studies of introduced strains (Pérez-Jaramillo et al., 2016). As an example, Ghirardi et al. (2012) compared the colonization of tomato rhizosphere in a natural soil of a collection of *Pseudomonas* strains representative of a wide diversity (plants, soils). These best colonizers share in common their ability to use a wide range of organic compounds, supporting the results of Goldfarb et al. (2011) following a community approach. These efficient colonizers were also able to efficiently use iron and nitrogen oxides as electron acceptors

(Ghirardi et al., 2012); this efficiency was related to the synthesis of specific siderophores and of nitrogen oxide reductases, respectively. These results confirmed the importance of the ability to adapt to iron starvation shown by Robin et al. (2007) and are in agreement with reported increased level of denitrification rates in the rhizosphere (Philippot et al., 2013).

Taken together, these studies underline the importance of the carbon and energetic metabolism in the adaptation of soilborne bacteria in the rhizosphere (Latour & Lemanceau, 1997), which is in agreement with the properties of that specific environment: (1) increased availability and forms in organic compounds (electron donors) and (2) decreased availability of oxygen and ferric iron as electron acceptors (Philippot et al., 2013; Robin et al., 2008).

Knowledge of plant–microbiota communication in the rhizosphere mostly relies on targeted studies using specific bacterial strains and considering specific traits hypothesized to play a major role. Over the last decade, progress in molecular methodologies has allowed the analysis of changes in microbial communities (Maron, Ranjard, Mougel, & Lemanceau, 2007) associated with different rhizosphere conditions (e.g., plant genotype, mycorrhizal root or not). Overall, these studies pinpointed taxa that are favoured or depressed in different environments (recently reviewed in Lakshmanan et al., 2014) without information on possible bacterial traits involved in the community shift with some exceptions such as the recent report of Lebeis et al. (2015) showing the enhancement of populations using SA. Reciprocally, plant traits accounting for community changes upon variations of the rhizosphere environment are rarely understood. As an example, rationale for the increased occurrence of pseudomonads harbouring type three secretion systems (T3SS) in the rhizosphere (Mazurier, Lemunier, Siblot, Mougel, & Lemanceau, 2004) as related to mycorrhization (Viollet et al., 2011) is not understood despite the beneficial effect of these bacterial populations on the fungal symbiosis and plant growth (Viollet et al., 2016).

3.4 Prospects for Monitoring Rhizosphere Microbiota via the Host Plant

Demonstration that the composition and activities of the microbial communities in the rhizosphere are regulated by the host plant through communication processes opens stimulating prospect for monitoring rhizosphere microbiota to promote populations the most favourable for plant growth and health. This relies on the knowledge of plant traits involved in the corresponding communication as described in

Sections 2.2 and 2.3. Knowledge of these traits may result from targeted or untargeted approaches.

3.4.1 Targeted Approaches

These approaches consist in engineering root exudation (Ryan, Dessaux, Thomashow, & Weller, 2009; Savka, Dessaux, Oger, & Rossbach, 2002) to make the host plant releasing molecules either interfering with bacterial signalling and resulting activities or promoting specific beneficial microbial populations.

Interference with bacterial signalling. Manipulating plants to express AHL-lactonase lead to a decreased pathogenicity of phytopathogenic bacteria requiring QS to express their virulence. This interference of the host plant with bacterial communication therefore decreases disease severity (Helman & Chernin, 2015; Kalia, 2013).

Promotion of specific populations. Agrobacterium—plant interaction represent a well-studied manipulation of the host plant by the bacteria to make it produce compounds (opines) that favour specific bacterial populations. This manipulation relies on the transfer of bacterial T-DNA to induce opine synthesis by the modified host plant giving to the pathogenic bacteria a competitive advantage with the creation of a specific niche (Dessaux, Petit, Farrand, & Murphy, 1998). The 'Opine concept' proposed by these authors stipulates that opines favour the pathogen growth with strain specific pattern as these molecules differ upon Agrobacteria strain and more specifically upon the nature of bacterial carrying tumour-inducing plasmid (pTi) transferred in the host plant. To discriminate the specific role of various opines on microbiota, a range of transgenic plants producing these different opines was generated by transfer to the plant of T-DNA from *Agrobacterium* and opine-degrading populations were characterized (Oger, Petit, & Dessaux, 1997), with different soils and plant species (Mansouri, Petit, Oger, & Dessaux, 2002). An increasing opine carbon bias was tested in artificial exudation systems (with artificial root exudate with increasing amount of octopine) and genetically modified plant rhizospheres that exuded octopine (Mondy et al., 2014). *Ensifer* and *Pseudomonas* spp. were significantly promoted in soil amended with these artificial exudates and in the rhizosphere of octopine-producers transgenic lines. However, the density of *Pseudomonas* was increased as a response to carbon supplementation while that of *Ensifer* only correlated with octopine concentrations suggesting two opposed colonization strategies for rhizosphere bacteria. *Pseudomonas* would be representative of bacteria using the copiotrophic strategy and *Ensifer* of bacteria using the oligotrophic one. Moreover 16S metabarcoding analysis showed that *Ensifer* became dominant in soil supplemented with the highest amount of octopine, indicating a strong reshape of bacterial community in response to octopine supply.

These two examples represent nice proofs of concept of the possibility to monitor microbial populations and activities via the host plant (i.e., engineering the rhizosphere, Dessaux, Grandclément, & Faure, 2016).

3.4.2 Untargeted Approaches

In their pioneer study, Smith, Handelsman, and Goodman (1999) evidenced that ability of inbred lines of tomato to promote the survival of biocontrol bacterial strain and their efficiency to suppress damping-off induced by *Pythium* differed, suggesting that plant traits may influence survival and activities of beneficial bacterial strains. More recently, plant quantitative loci from *A. thaliana* involved in the beneficial effects (root architecture and shoot growth) of a PGPR bacterial strain, *Pseudomonas simiae* WCS417r, were identified by comparing the response of 302 *Arabidopsis* accession to the bacterial effects (Wintermans, Bakker, & Pieterse, 2016). Several of the underlying candidate genes described by the authors have been assigned to important plant growth-related processes.

Attempts are also being made to identify plant traits that select beneficial microbial populations from soilborne communities and/or functional microbial genes encoding beneficial activities by comparing the rhizosphere metagenome associated to plant genotypes chosen for their genetic distances and or for their ecophysological dissimilarities (Lambers, Mougel, Jaillard, & Hinsinger, 2009; Moreau et al., 2015; Zancarini et al., 2013). Root traits mediating plant nutrition appear to be involved in plant—microbial community interaction, as supported by (1) the significant differences of bacterial communities in the rhizosphere of plants with ecophysiological profiles corresponding to different nutritional strategies (Zancarini et al., 2013) and (2) the promotion of nitrate-reducing microbial community in the rhizosphere of plants with a slow nitrogen uptake rate (Moreau et al., 2015).

4. PLANT COMMUNICATION WITH ASSOCIATED MICROORGANISMS IN THE PHYLLOSPHERE

The above-ground parts of plants also sustain an active microbial life (Vorholt, 2012). These microbial communities, which are at the interface between the plant and the surrounding atmosphere, have not fully been integrated into the global change research agenda yet, although they can mediate plant responses to climate change (Vacher et al., 2016). The phyllosphere can be considered as the above-ground homolog of the rhizosphere. It can be defined as '*the microenvironment extending from the leaf surface outward to the outer edge of the boundary layer surrounding the leaf and inward into the leaf tissues*' (Morris, 2002). According to this definition, the phyllosphere microbiota comprises all microorganisms living on the lower and upper leaf surface (i.e., leaf epiphytes) and inside the leaf tissues (i.e., leaf endophytes). The phyllosphere microbiota is composed of bacteria, fungi, Oomycetes and, less frequently, Archaea and Nematodes (Lindow & Brandl, 2003; Vorholt, 2012). These microorganisms interact with the host plant (Baker et al., 2010; Melotto, Underwood, & He, 2008) and with each other (Agler et al., 2016; Jakuschkin et al., 2016; Kemen, 2014). They also interact with viruses, such as mycoviruses (Marzano & Domier, 2015) and bacteriophages (Koskella, 2013; Koskella & Parr, 2015), and with herbivorous arthropods (Crawford, Land, & Rudgers, 2010; Humphrey, Nguyen, Villalobos, & Whiteman, 2014).

4.1 Acquisition of the Phyllosphere Microbiota by the Plant

The colonization of the phyllosphere by microorganisms starts at seedling emergence and starts over every year at leaf flush in the case of perennial plant species. Most phyllosphere microorganisms are acquired horizontally, from the environment. Some of them can also be transmitted vertically, from maternal plants onto offspring via seeds (Fig. 1).

4.1.1 Horizontal Transmission

Phyllosphere microorganisms can originate from various environmental sources, including soil and litter of the germination environment (Barret et al., 2015; Copeland, Yuan, Layeghifard, Wang, & Guttman, 2015) and bioaerosols (Bulgarelli, Schlaeppi, Spaepen, Ver Loren van Themaat, & Schulze-Lefert, 2013) (Fig. 1). Bioaerosols themselves originate from various sources, including aquatic environments, soil, animals (Bulgarelli et al., 2013) and of course neighbouring plants (Wilson, Carroll, Roy, & Blaisdell, 2014). Raindrops, irrigation water (Morris, 2002) and leaf-dwelling insects (Osono, 2014) can also bring some microorganisms onto plant leaves. In the case of perennial plants, the phyllosphere can also be colonized at bud burst by the microorganisms associated to the perennial organs of the plant, such as twigs (Osono, 2014).

4.1.2 Vertical Transmission

In addition, some phyllosphere microorganisms are transmitted from maternal plants onto offspring via seeds (Fig. 1). The transmission can occur via the seed tissues or the exterior of seed coats. Some beneficial bacteria inoculated on maize seeds have for instance been retrieved from leaves, after having colonized the rhizosphere, the roots and the stems (Fig. 1) and have been shown to significantly increase maize tolerance to drought (Naveed, Mitter, Reichenauer, Wieczorek, & Sessitsch, 2014). Such vertical transmission is also common in the endophytic fungal species protecting plants against herbivory (Hodgson et al., 2014; Rodriguez, White, Arnold, & Redman, 2009). Some species of *Epichloë*, for instance, are transmitted via the seeds of temperate grasses. The mycelium first colonizes the stems, rhizomes and leaf tissues of the mother plant. It may also be present on the leaf surface. Then it colonizes the inflorescence, grows into ovules and colonizes the embryo within the seed (Rodriguez et al., 2009). Some foliar endophytes of forbs have also been found in and on pollen grains, suggesting that the transmission to seeds can also occur via the pollen tube (Hodgson et al., 2014).

4.2 Evidence for the Phyllosphere Effect

Phyllosphere microbial communities are significantly different from airborne microbial communities, although they are in direct contact with the atmosphere surrounding the plant (Vokou et al., 2012). Each leaf indeed creates a specific environment in which only some microorganisms can thrive (Vacher et al., 2016; Vorholt, 2012). This selective effect of leaves on microbial communities can be termed the phyllosphere effect, by analogy to the rhizosphere effect (See Section 2.2).

4.2.1 Differentiation Between Foliar, Airborne and Soil Microbial Communities

A first line of evidence of the phyllosphere effect comes from the seasonal dynamics of phyllosphere microbial communities. At the beginning of the vegetative season, phyllosphere microbial communities are strongly influenced by the communities of surrounding environmental sources (soil and air, in particular). Then, they gradually acquire a higher proportion of leaf-specific taxa (Copeland et al., 2015; Maignien, DeForce, Chafee, Eren, & Simmons, 2014). The environmental source influencing the initial assemblage depends on the plant growing conditions. For instance, Maignien et al. (2014) found that phyllosphere microbial communities of

A. thaliana plants growing in sterile soil initially mirrored airborne communities. In contrast, Copeland et al. (2015) showed that phyllosphere microbial communities of several annual crop species growing in field conditions were initially influenced by soil microbial communities. In both cases, phyllosphere microbial communities subsequently differentiated from those of environmental sources, indicating that the leaf selects for specific microbial species.

4.2.2 Variations in Phyllosphere Microbial Communities Among Plant Species and Genotypes

A second line of evidence of the phyllosphere effect comes from the influence of the host plant genotype on phyllosphere microbial communities. Plant species growing in the same location, and therefore submitted to the same inoculum, develop distinct phyllosphere microbial communities (Arfi, Buée, Marchand, Levasseur, & Record, 2012; Dees, Lysøe, Nordskog, & Brurberg, 2015; Inácio, Ludwig, Spencer-Martins, & Fonseca, 2010; Kembel & Mueller, 2014; Kembel, Connor, Arnold, Hubbell, & Wright, 2014; Kim, Singh, Lai-hoe, Chun, & Adams, 2012; Laforest-Lapointe, Messier, & Kembel, 2016; Lambais, Lucheta, & Crowley, 2014; Redford, Bowers, Knight, Linhart, & Fierer, 2010; Sapkota, Knorr, Jørgensen, O'Hanlon, & Nicolaisen, 2015). Similarly, within the same plant species, the composition of phyllosphere microbial communities differs among plant genotypes growing in the same location (Bálint et al., 2013; Balint-Kurti, Simmons, Blum, Ballaré, & Stapleton, 2010; Bodenhausen, Bortfeld-Miller, Ackermann, & Vorholt, 2014; Cordier, Robin, Capdevielle, Desprez-Loustau, & Vacher, 2012; Horton et al., 2014; Hunter, Hand, Pink, Whipps, & Bending, 2010; Hunter, Pink, & Bending, 2015). These variations in phyllosphere microbial communities among plant species and genotypes can be explained by variations in foliar functional traits among plant genotypes; each plant species or genotype has its own foliar features, which selects for specific foliar microbial communities.

4.3 Functional Traits and Signalling Molecules Underlying the Phyllosphere Effect

Phyllosphere microbial communities differ from soil and airborne microbial communities, and among plant species and genotypes, because the host plant exerts a selective pressure on microbial communities (Vacher et al., 2016). This selective pressure (i.e., the phyllosphere effect) depends both on plant and microbial traits. The plant traits define the features of the microbial habitat, whereas the microbial traits underlie the ability of microorganisms to thrive in that habitat.

4.3.1 Plant Traits Shaping Phyllosphere Microbial Communities

Leaf anatomy and chemistry. At the beginning of the vegetative season, many microbial cells and spores reach the leaf via airborne dispersal. Only microorganisms able to attach to the leaf cuticle then have a chance to survive and multiply in the phyllosphere. The structure and composition of the cuticle is thus expected to have a major influence on phyllosphere microbial communities. Mutant lines of *A. thaliana* were used to test this hypothesis; the results confirmed the significant effect of cuticular chemistry on the composition of phyllosphere bacterial communities (Bodenhausen et al., 2014; Reisberg, Hildebrandt, Riederer, & Hentschel, 2013). Several other foliar morphological traits have been studied (Hunter et al., 2010; Yadav, Karamanoli, & Vokou, 2005), but surprisingly they have little influence on phyllosphere microbial communities. For instance, the presence of trichomes does not influence the composition and diversity of phyllosphere bacterial communities (Reisberg, Hildebrandt, Riederer, & Hentschel, 2012), whereas leaf size does not influence the richness of phyllosphere fungal communities (Kinkel, Andrews, Berbee, & Nordheim, 1987). Artificial leaf surfaces, that mimic the leaf surface topography down to the nanometre resolution, constitute a promising tool for better understanding the selective pressures exerted by leaf anatomy on phyllosphere microbial communities (Doan & Leveau, 2015).

Leaf primary metabolism. Among the microorganisms able to attach to the leaf cuticle, only those able to use the nutrients available in the phyllosphere have a chance to thrive and reproduce. These nutrients include the carbohydrates produced by photosynthesis and exported to the leaf surface through leaching and guttation and some volatile organic compounds such as the methanol emitted during leaf growth (Vacher et al., 2016). The nature and abundance of the foliar nutrients is expected to have a major influence on phyllosphere microbial communities. Accordingly, several studies showed that the leaf nitrogen content, which is related to the maximum photosynthetic rate (Cornelissen et al., 2003), correlates with the structure of phyllosphere microbial communities across plant species (Kembel & Mueller, 2014; Kembel et al., 2014; Laforest-Lapointe et al., 2016). In addition, the higher abundance and activity of the genus *Methylobacterium* in the phyllosphere than in the rhizosphere indicates that the foliar production of methanol shapes the composition of phyllosphere bacterial communities (Iguchi, Yurimoto, & Sakai, 2015; Knief et al., 2012).

Leaf secondary metabolism. Finally, only the microorganisms able to cope with the harsh microclimate of the leaf surface and with the foliar defences can thrive and reproduce in the phyllosphere. The genetic and molecular pathways involved in the foliar recognition of pathogens, the foliar production of antimicrobial compounds, the regulation of stomatal aperture and the maintenance of cell wall integrity are expected to have a major influence on phyllosphere microbial communities. Accordingly, Horton et al. (2014) showed that plant loci responsible for defence and cell wall integrity are the main drivers of both phyllosphere fungal and bacterial communities in *A. thaliana*. For instance, a locus involved in the production of callose, a polysaccharide used by the plant for sealing wounds during fungal infection, was found to drive variations in phyllosphere fungal communities (Horton et al., 2014). The structure of phyllosphere microbial communities is also controlled by the production of ethylene (ET) and GABA, two signalling molecules known for mediating the interactions between plants and a variety of organisms (Balint-Kurti et al., 2010; Bodenhausen et al., 2014). Recently, Vogel, Bodenhausen, Gruissem, and Vorholt (2016) showed that the protective ability of some phyllosphere commensals against foliar pathogens is related to the induction of defence-related genes. The colonization of *A. thaliana* by the commensal *Sphingomonas melonis*, which confers protection against the pathogen *Pseudomonas syringae* (Innerebner, Knief, & Vorholt, 2011), changed the expression of almost 400 plant genes, including genes encoding pathogenesis-related proteins and antimicrobial proteins. Among them were genes involved in the salicylic acid (SA)- and ET-dependent defence-signalling pathways. Interestingly, Vogel et al. (2016) showed that *A. thaliana* mutants impaired in the recognition of microbe-associated molecular patterns were less protected by the commensal strain, indicating that the protective ability of *S. melonis* depends on its molecular recognition by the plant. Other commensals did not trigger the same changes in the leaf trancriptome, suggesting that the structure of phyllosphere microbial communities and their impact on plant performance are regulated by fine-tune molecular interactions between the leaf and its microbiota.

4.3.2 Microbial Adaptations to the Phyllosphere

Microbial adaptations to the phyllosphere have been extensively reviewed elsewhere (Vacher et al., 2016; Vorholt, 2012; Whipps, Hand, Pink, &

Bending, 2008). They include microbial traits favouring (1) adhesion to the leaf cuticle, (2) nutrient acquisition and (3) tolerance to biotic and abiotic stresses. The production of hygroscopic biosurfactants, for instance, benefits phyllosphere bacteria by increasing cuticle permeability, thus increasing the diffusion of nutrients to the leaf surface (Burch, Zeisler, Yokota, Schreiber, & Lindow, 2014; Schreiber et al., 2005). Some phyllosphere bacteria also produce multiple transporters, such as TonB-dependent transporters, to scavenge carbohydrates at low concentrations (Delmotte et al., 2009). The discovery of microbial rhodopsins in the phyllosphere suggests that some phyllosphere microorganisms have alternatively evolved a phototrophic lifestyle to cope with the scarcity of nutrients on the leaf surface (Atamna-Ismaeel et al., 2012). In addition, many microorganisms have evolved the ability to enter into the apoplast, which is richer in nutrients than the leaf surface. Some strains of the pathogenic bacteria *P. syringae*, for instance, synthesize phytotoxins regulating stomatal aperture. Stomata close upon molecular recognition of the pathogen by the guard cells but the phytotoxins reopen them, thus allowing the pathogen to colonize the apoplast (Baker et al., 2010; Melotto et al., 2008). Again, these findings show that the phyllosphere microbiota is shaped by the molecular communication between the plant and the microorganisms. Future research needs to better decipher these molecular interactions and to investigate whether the plant can control them to adjust its phyllosphere microbiota to environmental conditions (Vandenkoornhuyse et al., 2015).

5. CONCLUSIONS AND PROSPECTS

Plants are surrounded by microorganisms at the seed, root and shoot levels. There is a continuum between these three compartments: (1) microorganisms in the spermosphere originate both from the flowers and the soilborne microbial community, (2) microorganisms in the rhizosphere originate both from seedborne and soilborne microbial community, (3) microorganisms in the phyllosphere partly originate from the soil and the rhizosphere, as suggested by the overlap between phyllosphere and rhizosphere microbial communities (Bai et al., 2015; Bodenhausen et al., 2013), and from airborne communities (Fig. 1). The recruitment of microbial communities in these different spheres (i.e., spermosphere, rhizosphere, phyllosphere) relies on a subtle communication network between plant and microorganisms. This communication plays a major role in the plant selection of specific microbial populations, and there is growing evidence that the host plant is able to disturb communications among microorganisms to its advantage. There is also increasing awareness that the plant is not an autonomous entity and that it depends on microorganisms, at least some of them, for its growth, nutrition and health (Partida-Martínez et al., 2011; Vandenkoornhuyse et al., 2015). According to this new vision, the biomolecular network formed by the plant and associated microorganisms, termed the holobiont, has been proposed to be a unit of selection in evolution (Bordenstein & Theis, 2015; Zilber-Rosenberg & Rosenberg, 2008).

The properties of these spheres differ significantly, especially in their trophic state. Indeed while the spermosphere and rhizosphere are habitats with relatively abundant nutrients, the phyllosphere is a low-nutrient habitat. Therefore selection of microbial populations in the spermosphere and rhizosphere relies mostly on efficient resources uses, while selection of population in the phyllosphere is dependent on attachment and adaptation to biotic and abiotic stresses. In all cases, molecular signalization plays a major role in the plant selection of microbial populations. The stability of plant–microbe associations requires reciprocal benefices for the partners, the plant providing niches and nutrients to the microorganisms, mostly heterotrophic, and in return microorganisms, at least some of them, enhancing plant growth, nutrition and health. The association of all these bionts (plant and microorganisms) has been proposed to be called holobiont, this holobiont leading to a so-called extended phenotype since plant growth and development are significantly impacted by the associated microbiota (Partida-Martínez et al., 2011). According to this new vision, the biomolecular network formed by the plant and its microorganisms, i.e., holobiont, can be considered as a unit of selection in evolution (Bordenstein & Theis, 2015; Zilber-Rosenberg & Rosenberg, 2008). This network relies on molecular signalization mediated by phytochemicals, some of them being known but probably many remaining to be identified. Progress in metabolomics is expected to allow us to discover new signal molecules and to better detangle the complex communication between plants and associated microorganisms (van Dam & Bouwmeester, 2016).

Development of molecular approaches relying on nucleic acids extraction and analysis has also boosted research on plant–microbe interactions and open stimulating prospects in agroecology. Research on plant–microbe association as a whole (holobiont) is expected to give more insight to the impact of rhizosphere communities on those of the phyllosphere and reciprocally, with potential applications in plant health management (Lemanceau, Expert, Gaymard, Bakker, & Briat, 2009). A better knowledge of plant traits involved in the selection of beneficial microbial populations and its application to plant breeding is expected to allow us to decrease synthetic inputs by growing cultivars better valuing environmental biotic resources (Wissuwa, Mazzola, & Picard, 2009). This represents a major paradigm change since over the last decades agriculture has relied less on biodiversity and more on inputs (Philippot et al., 2013). Moreover, during that period, plant breeding may have led to the counterselection of plant traits involved in the selection of beneficial microbial populations (Bulgarelli et al., 2015; Pérez-Jaramillo et al., 2016; Schmidt, Bowles, & Gaudin, 2016; Wissuwa et al., 2009). Future research in plant breeding will have to reverse this trend, by fully taking the plant microbiota into account, and this will require a better knowledge of the communication network between plants and associated microorganisms.

REFERENCES

Agler, M. T., Ruhe, J., Kroll, S., Morhenn, C., Kim, S.-T., Weigel, D., & Kemen, E. M. (2016). Microbial hub taxa link host and abiotic factors to plant microbiome variation. *PLoS Biology, 14*, e1002352. http://dx.doi.org/10.1371/journal.pbio.1002352.

Ahuja, I., Kissen, R., & Bones, A. M. (2012). Phytoalexins in defense against pathogens. *Trends in Plant Science, 17*, 73–90.

Arfi, Y., Buée, M., Marchand, C., Levasseur, A., & Record, E. (2012). Multiple markers pyrosequencing reveals highly diverse and host-specific fungal communities on the mangrove trees *Avicennia marina* and *Rhizophora stylosa*. *FEMS Microbiology Ecology, 79*, 433–444.

Atamna-Ismaeel, N., Finkel, O. M., Glaser, F., Sharon, I., Schneider, R., Post, A. F., … Belkin, S. (2012). Microbial rhodopsins on leaf surfaces of terrestrial plants. *Environmental Microbiology, 14*, 140–146.

Badri, D. V., Chaparro, J. M., Zhang, R., Shen, Q., & Vivanco, J. M. (2013). Application of natural blends of phytochemicals derived from the root exudates of *Arabidopsis* to the soil reveal that phenolic-related compounds predominantly modulate the soil microbiome. *Journal of Biological Chemistry, 288*, 4502–4512.

Baetz, U., & Martinoia, E. (2014). Root exudates: the hidden part of plant defense. *Trends in Plant Science, 19*, 90–98.

Bai, Y., Müller, D. B., Srinivas, G., Garrido-Oter, R., Potthoff, E., Rott, M., … Schulze-Lefert, P. (2015). Functional overlap of the *Arabidopsis* leaf and root microbiota. *Nature, 528*, 364–369.

Bais, H., Weir, T., Perry, L., Gilroy, S., & Vivanco, J. (2006). The role of root exudates in rhizosphere interactions with plants and other organisms. *Annual Review of Plant Biology, 57*, 233–266.

Baker, C. J., Mock, N. M., Whitaker, B. D., Roberts, D. P., Rice, C. P., Deahl, K. L., & Aver'yanov, A. A. (2005). Involvement of acetosyringone in plant-pathogen recognition. *Biochemical and Biophysical Research Communications, 4*, 130–136.

Baker, C. M., Chitrakar, R., Obulareddy, N., Panchal, S., Williams, P., & Melotto, M. (2010). Molecular battles between plant and pathogenic bacteria in the phyllosphere. *Brazilian Journal of Medical and Biological Research, 43*, 698–704.
Bakker, P. A. H. M., Berendsen, R. L., Doornbos, R. F., Wintermans, P. C. A., & Pieterse, C. M. J. (2013). The rhizosphere revisited: root microbiomics. *Frontiers in Plant Science, 4*(165). http://dx.doi.org/10.3389/fpls.2013.00165.
Balint-Kurti, P., Simmons, S. J., Blum, J. E., Ballaré, C. L., & Stapleton, A. E. (2010). Maize leaf epiphytic bacteria diversity patterns are genetically correlated with resistance to fungal pathogen infection. *Molecular Plant-Microbe Interactions, 23*, 473–484.
Bálint, M., Tiffin, P., Hallström, B., O'Hara, R. B., Olson, M. S., Fankhauser, J. D., … Schmitt, I. (2013). Host genotype shapes the foliar fungal microbiome of balsam poplar (*Populus balsamifera*). *PLoS One, 8*, e53987. http://dx.doi.org/10.1371/journal.pone.0053987.
Barret, M., Briand, M., Bonneau, S., Préveaux, A., Valière, S., Bouchez, O., … Jacques, M.-A. (2015). Emergence shapes the structure of the seed-microbiota. *Applied and Environmental Microbiology, 81*, 1257–1266.
Barret, M., Guimbaud, J.-F., Darrasse, A., & Jacques, M.-A. (2016). Plant microbiota affects seed transmission of phytopathogenic microorganisms. *Molecular Plant Pathology, 17*, 791–795.
Baudoin, E., Benizri, E., & Guckert, A. (2002). Impact of growth stages on the bacterial community structure along maize roots by metabolic and genetic fingerprinting. *Applied Soil Ecology, 19*, 135–145.
Bodenhausen, N., Bortfeld-Miller, M., Ackermann, M., & Vorholt, J. A. (2014). A synthetic community approach reveals plant genotypes affecting the phyllosphere microbiota. *PLoS Genetics, 10*, e1004283. http://dx.doi.org/10.1371/journal.pgen.1004283.
Bodenhausen, N., Horton, M. W., & Bergelson, J. (2013). Bacterial communities associated with the leaves and the roots of *Arabidopsis thaliana*. *PLoS One, 8*(2), e56329. http://dx.doi.org/10.1371/journal.pone.0056329.
Bogino, P. C., de las Mercedes Oliva, L., Sorroche, F. G., & Giordano, W. (2013). The role of bacterial biofilms and surface components in plant-bacterial associations. *International Journal of Molecular Sciences, 14*, 15838–15859.
Bordenstein, S. R., & Theis, K. R. (2015). Host biology in light of the microbiome: ten principles of holobionts and hologenomes. *PLoS Biology, 13*, 1–23.
Bulgarelli, D., Garrido-Oter, R., Münch, P. C., Weiman, A., Dröge, J., Pan, Y., … Schulze-Lefert, P. (2015). Structure and function of the bacterial root microbiota in wild and domesticated barley. *Cell Host & Microbe, 17*, 392–403.
Bulgarelli, D., Rott, M., Schlaeppi, K., van Themaat, E. V. L., Ahmadinejad, N., Assenza, F., … Schulze-Lefert, P. (2012). Revealing structure and assembly cues for *Arabidopsis* root-inhabiting bacterial microbiota. *Nature, 488*, 91–95.
Bulgarelli, D., Schlaeppi, K., Spaepen, S., Ver Loren van Themaat, E., & Schulze-Lefert, P. (2013). Structure and functions of the bacterial microbiota of plants. *Annual Reviews of Plant Biology, 64*, 807–838.
Burch, A. Y., Zeisler, V., Yokota, K., Schreiber, L., & Lindow, S. E. (2014). The hygroscopic biosurfactant syringafactin produced by *Pseudomonas syringae* enhances fitness on leaf surfaces during fluctuating humidity. *Environmental Microbiology, 16*, 2086–2098.
Chaparro, J. M., Badri, D. V., Bakker, M. G., Sugiyama, A., Manter, D. K., & Vivanco, J. M. (2013). Root exudation of phytochemicals in *Arabidopsis* follows specific patterns that are developmentally programmed and correlate with soil microbial functions. *PLoS One, 8*, e55731.
Chen, K., de Borne, F. D., Julio, E., Obszynski, J., Pale, P., & Otten, L. (2016). Root-specific expression of opine genes and opine accumulation in some cultivars of the naturally

occurring genetically modified organism *Nicotiana tabacum*. *The Plant Journal*. http://dx.doi.org/10.1111/tpj.13196.
Copeland, J. K., Yuan, L., Layeghifard, M., Wang, P. W., & Guttman, D. S. (2015). Seasonal community succession of the phyllosphere microbiome. *Molecular Plant-Microbe Interactions, 28*, 274–285.
Cordier, T., Robin, C., Capdevielle, X., Desprez-Loustau, M.-L., & Vacher, C. (2012). Spatial variability of phyllosphere fungal assemblages: genetic distance predominates over geographic distance in a European beech stand (*Fagus sylvatica*). *Fungal Ecology, 5*, 509–520.
Cornelissen, J. H. C., Lavorel, S., Garnier, E., Díaz, S., Buchmann, N., Gurvich, D. E., … Poorter, H. (2003). A handbook of protocols for standardised and easy measurement of plant functional traits worldwide. *Australian Journal of Botany, 51*, 335–380.
Crawford, K. M., Land, J. M., & Rudgers, J. A. (2010). Fungal endophytes of native grasses decrease insect herbivore preference and performance. *Oecologia, 164*, 431–444.
Curl, E. A., & Truelove, B. (1986). The rhizosphere. In *Advanced Series in Agricultural Sciences* (Vol. 15). Berlin-Heidelberg-New York-Tokyo: Springer-Verlag.
Cusano, A. M., Burlinson, P., Deveau, A., Vion, P., Uroz, S., Preston, G. M., & Frey-Klett, P. (2011). *Pseudomonas fluorescens* BBc6R8 type III secretion mutants no longer promote ectomycorrhizal symbiosis. *Environmental Microbiology Reports, 3*, 203–210.
Dakora, F. D., & Phillips, D. A. (2002). Root exudates as mediators of mineral acquisition in low-nutrient environments. *Plant and Soil, 245*, 35–47.
van Dam, N. M., & Bouwmeester, H. J. (2016). Metabolomics in the rhizosphere: tapping into belowground chemical communication. *Trends in Plant Science, 21*, 256–265.
Darrasse, A., Darsonval, A., Boureau, T., Brisset, M.-N., Durand, K., & Jacques, M.-A. (2010). Transmission of plant-pathogenic bacteria by nonhost seeds without induction of an associated defense reaction at emergence. *Applied and Environmental Microbiology, 76*, 6787–6796.
Darsonval, A., Darrasse, A., Durand, K., Bureau, C., Cesbron, S., & Jacques, M.-A. (2009). Adhesion and fitness in the bean phyllosphere and transmission to seed of *Xanthomonas fuscans* subsp. *fuscans*. *Molecular Plant-Microbe Interactions, 22*, 747–757.
De-la-Pena, C., Badri, D. V., Lei, Z., Watson, B. S., Brandao, M. M., Silva-Filho, M. C., … Vivanco, J. M. (2010). Root secretion of defense-related proteins is development-dependent and correlated with flowering time. *Journal of Biological Chemistry, 285*, 30654–30665.
Dees, M. W., Lysøe, E., Nordskog, B., & Brurberg, M. B. (2015). Bacterial communities associated with surfaces of leafy greens: shift in composition and decrease in richness over time. *Applied and Environmental Microbiology, 81*, 1530–1539.
Deflaun, M. F., Marshall, B. M., Kulle, E. P., & Levy, S. B. (1994). Tn5 insertion mutants of *Pseudomonas fluorescens* defective in adhesion to soil and seeds. *Applied and Environmental Microbiology, 60*, 2637–2642.
Delmotte, N., Knief, C., Chaffron, S., Innerebner, G., Roschitzki, B., Schlapbach, R., … Vorholt, J. A. (2009). Community proteogenomics reveals insights into the physiology of phyllosphere bacteria. *Proceedings of the National Academy of Sciences of the United States of America, 106*, 16428–16433.
Dessaux, Y., Grandclément, C., & Faure, D. (2016). Engineering the rhizosphere. *Trends in Plant Science, 21*, 266–278.
Dessaux, Y., Petit, A., Farrand, S. K., & Murphy, P. J. (1998). Opines and opine-like molecules involved in plant–Rhizobiaceae interactions. In H. P. Spaink, et al. (Eds.), *The Rhizobiaceae* (pp. 173–197). The Netherlands: Springer.
Doan, H. K., & Leveau, J. H. J. (2015). Artificial surfaces in phyllosphere microbiology. *Phytopathology, 105*, 1036–1042.

Eichorst, S. A., & Kuske, C. R. (2012). Identification of cellulose-responsive bacterial and fungal communities in geographically and edaphically different soils by using stable isotope probing. *Applied and Environmental Microbiology, 78*, 2316–2327.

Elasri, M., Delorme, S., Lemanceau, P., Stewart, G., Laue, B., Glickmann, E., … Dessaux, Y. (2001). Acyl-homoserine lactone production is more common amongst plant-associated than soil-borne *Pseudomonas* spp. *Applied and Environmental Microbiology, 67*, 1198–1209.

Espinosa-Urgel, M., & Ramos, J. L. (2004). Cell density-dependent gene contributes to efficient seed colonization by *Pseudomonas putida* KT2440. *Applied and Environmental Microbiology, 70*, 5190–5198.

Espinosa-Urgel, M., Salido, A., & Ramos, J. L. (2000). Genetic analysis of functions involved in adhesion of *Pseudomonas putida* to seeds. *Journal of Bacteriology, 182*, 2363–2369.

Foo, E., McAdam, E. L., Weller, J. L., & Reid, J. B. (2016). Interactions between ethylene, gibberellins, and brassinosteroids in the development of rhizobial and mycorrhizal symbioses of pea. *Journal of Experimental Botany*. http://dx.doi.org/10.1093/jxb/erw047.

Foo, E., Ross, J. J., Jones, W. T., & Reid, J. B. (2013). Plant hormones in arbuscular mycorrhizal symbioses: an emerging role for gibberellins. *Annals of Botany, 111*, 769–779.

Frey, P., Frey-Klett, P., Garbaye, J., Berge, O., & Heulin, T. (1997). Metabolic and genotypic fingerprinting of fluorescent pseudomonads associated with the Douglas fir *Laccaria bicolor* mycorrhizosphere. *Applied and Environmental Microbiology, 63*, 1852–1860.

Frey-Klett, P., Garbaye, J., & Tarkka, M. (2007). The mycorrhiza helper bacteria revisited. *New Phytologist, 176*, 22–36.

Fuqua, C., Parsek, M. R., & Greenberg, E. P. (2001). Regulation of gene expression by cell-to-cell communication: acyl-homoserine lactone quorum sensing. *Annual Review of Genetics, 35*, 439–468.

Gamalero, E., Trotta, A., Massa, N., Copetta, A., Martinotti, M. G., & Berta, G. (2004). Impact of two fluorescent pseudomonads and an arbuscular mycorrhizal fungus on tomato plant growth, root architecture and P acquisition. *Mycorrhiza, 14*, 185–192.

Gamliel, A., & Katan, J. (1992). Chemotaxis of fluorescent pseudomonads towards seed exudates and germinating seeds in solarized soil. *Phytopathology, 82*, 320–327.

Garbaye, J. (1994). Helper bacteria: a new dimension to the mycorrhizal symbiosis. *New Phytologist, 128*, 197–210.

García-Salamanca, A., Molina-Henares, M. A., van Dillewijn, P., Solano, J., Pizarro-Tobías, P., Roca, A., … Ramos, J. L. (2013). Bacterial diversity in the rhizosphere of maize and the surrounding carbonate-rich bulk soil. *Microbial Biotechnology, 6*, 36–44.

Ghirardi, S., Dessaint, F., Mazurier, S., Corberand, T., Raaijmakers, J. M., Meyer, J.-M., … Lemanceau, P. (2012). Identification of traits shared by rhizosphere-competent strains of fluorescent pseudomonads. *Microbial Ecology, 64*, 725–737.

Goldfarb, K. C., Karaoz, U., Hanson, C. A., Santee, C. A., Bradford, M. A., Treseder, K. K., … Brodie, E. L. (2011). Differential growth responses of soil bacterial taxa to carbon substrates of varying chemical recalcitrance. *Frontiers in Microbiology, 2*, 94. http://dx.doi.org/10.3389/fmicb.2011.00094.

Grandclément, C., Tannières, M., Moréra, S., Dessaux, Y., & Faure, D. (2016). Quorum quenching: role in nature and applied developments. *FEMS Microbiology Reviews, 40*, 86–116.

Grayston, S. J., Wang, S., Campbell, C. D., & Edwards, A. C. (1998). Selective influence of plant species on microbial diversity in the rhizosphere. *Soil Biology and Biochemistry, 30*, 369–378.

Großkinsky, D. K., Naseem, M., Abdelmohsen, U. R., Plickert, N., Engelke, T., Griebel, T., … Roitsch, T. (2011). Cytokinins mediate resistance against *Pseudomonas syringae* in tobacco through increased antimicrobial phytoalexin synthesis independent of salicylic acid signaling. *Plant Physiology, 157*, 815–830.

Hacquard, S., & Schadt, C. W. (2015). Towards a holistic understanding of the beneficial interactions across the *Populus* microbiome. *New Phytologist, 205*, 1424–1430.
Hartmann, A., Rothballer, M., & Schmid, M. (2008). Lorenz Hiltner, a pioneer in rhizosphere microbial ecology and soil bacteriology research. *Plant and Soil, 312*, 7.
Helman, Y., & Chernin, L. (2015). Silencing the mob: disrupting quorum sensing as a means to fight plant disease. *Molecular Plant Pathology, 16*, 316–329.
Hiltner, L. (1904). Uber neuere erfahrungen und probleme auf dem gebiete der bodenbakteriologie unter besonderden berucksichtigung und Brache. *Arbeiten der Deutschen Landwirtschaftlichen Gesellschaft, 98*, 59–78.
Hodgson, S., de Cates, C., Hodgson, J., Morley, N. J., Sutton, B. C., & Gange, A. C. (2014). Vertical transmission of fungal endophytes is widespread in forbs. *Ecology and Evolution, 4*, 1199–1208.
Horton, M. W., Bodenhausen, N., Beilsmith, K., Meng, D., Muegge, B. D., Subramanian, S., … Bergelson, J. (2014). Genome-wide association study of *Arabidopsis thaliana* leaf microbial community. *Nature Communications, 5*, 5320. http://dx.doi.org/10.1038/ncomm.
Humphrey, P. T., Nguyen, T. T., Villalobos, M. M., & Whiteman, N. K. (2014). Diversity and abundance of phyllosphere bacteria are linked to insect herbivory. *Molecular Ecology, 23*, 1497–1515.
Hunter, P. J., Hand, P., Pink, D., Whipps, J. M., & Bending, G. D. (2010). Both leaf properties and microbe-microbe interactions influence within-species variation in bacterial population diversity and structure in the lettuce (*Lactuca* species) phyllosphere. *Applied and Environmental Microbiology, 76*, 8117–8125.
Hunter, P. J., Pink, D. A. C., & Bending, G. D. (2015). Cultivar-level genotype differences influence diversity and composition of lettuce (*Lactuca* sp.) phyllosphere fungal communities. *Fungal Ecology, 17*, 183–186.
Iguchi, H., Yurimoto, H., & Sakai, Y. (2015). Interactions of methylotrophs with plants and other heterotrophic bacteria. *Microorganisms, 3*, 137–151.
Inácio, J., Ludwig, W., Spencer-Martins, I., & Fonseca, A. (2010). Assessment of phylloplane yeasts on selected Mediterranean plants by FISH with group- and species-specific oligonucleotide probes. *FEMS Microbiology Ecology, 71*, 61–72.
Inceoglu, O., Falcao Salles, J., van Overbeek, L., & van Elsas, J. D. (2010). Effect of plant genotype and growth stage on the Betaproteobacterial communities associated with different potato cultivars in two fields. *Applied and Environmental Microbiology, 11*, 3675–3684.
Innerebner, G., Knief, C., & Vorholt, J. A. (2011). Protection of *Arabidopsis thaliana* against leaf-pathogenic *Pseudomonas syringae* by *Sphingomonas* strains in a controlled model system. *Applied and Environmental Microbiology,* 77, 3202–3210.
Jakuschkin, B., Fievet, V., Schwaller, L., Ouadah, S., Robin, S., Robin, C., & Vacher, C. (2016). Deciphering the pathobiome: intra-and inter-kingdom interactions involving the pathogen *Erysiphe alphitoides*. *Microbial Ecology*. http://dx.doi.org/10.1007/s00248-016-0777-x.
Jeandet, P., Clément, C., Courot, E., & Cordelier, S. (2013). Modulation of phytoalexin biosynthesis in engineered plants for disease resistance. *International Journal of Molecular Sciences, 14*, 14136–14170.
Johnson, K. L., & Walcott, R. R. (2013). Quorum sensing contributes to seed-to-seedling transmission of *Acidovorax citrulli* on watermelon. *Journal of Phytopathology, 161*, 562–573.
Kalia, V. C. (2013). Quorum sensing inhibitors: an overview. *Biotechnology Advances, 31*, 224–245.
Kembel, S. W., Connor, T. K. O., Arnold, H. K., Hubbell, S. P., & Wright, S. J. (2014). Relationships between phyllosphere bacterial communities and plant functional traits

in a neotropical forest. *Proceedings of the National Academy of Sciences of the United States of America, 111*, 13715–13720.

Kembel, S. W., & Mueller, R. C. (2014). Plant traits and taxonomy drive host associations in tropical phyllosphere fungal communities. *Botany, 92*, 303–311.

Kemen, E. (2014). Microbe-microbe interactions determine oomycete and fungal host colonization. *Current Opinion in Plant Biology, 20*, 75–81.

Kim, M., Singh, D., Lai-hoe, A., Chun, J., & Adams, J. M. (2012). Distinctive phyllosphere bacterial communities in tropical trees. *Microbial Ecology, 63*, 674–681.

Kinkel, L. L., Andrews, J. H., Berbee, F. M., & Nordheim, E. V. (1987). Leaves as islands for microbes. *Oecologia, 71*, 405–408.

Knief, C., Delmotte, N., Chaffron, S., Stark, M., Innerebner, G., Wassmann, R., … Vorholt, J. A. (2012). Metaproteogenomic analysis of microbial communities in the phyllosphere and rhizosphere of rice. *ISME Journal, 6*, 1378–1390.

Koskella, B. (2013). Phage-mediated selection on microbiota of a long-lived host. *Current Biology, 23*, 1256–1260.

Koskella, B., & Parr, N. (2015). The evolution of bacterial resistance against bacteriophages in the horse chestnut phyllosphere is general across both space and time. *Philosophical transactions of the Royal Society of London. Series B, 370*, 20140297. http://dx.doi.org/10.1098/rstb.2014.0297.

Kwan, G., Pisithkul, T., Amador-Noguez, D., & Barak, J. (2015). De novo amino acid biosynthesis contributes to *Salmonella enterica* growth in alfalfa seedling exudates. *Applied and Environmental Microbiology, 81*, 861–873.

Laforest-Lapointe, I., Messier, C., & Kembel, S. W. (2016). Host species identity, site and time drive temperate tree phyllosphere bacterial community structure. *Microbiome, 4*, 27. http://dx.doi.org/10.1186/s40168-016-0174-1.

Lakshmanan, V., Selvaraj, G., & Bais, H. P. (2014). Functional soil microbiome: belowground solutions to an aboveground problem. *Plant Physiology, 166*, 689–700.

Lambais, M. R., Lucheta, A. R., & Crowley, D. E. (2014). Bacterial community assemblages associated with the phyllosphere, dermosphere, and rhizosphere of tree species of the Atlantic forest are host taxon dependent. *Microbial Ecology, 68*, 567–574.

Lambers, H., Mougel, C., Jaillard, B., & Hinsinger, P. (2009). Plant-microbe-soil interactions in the rhizosphere: an evolutionary perspective. *Plant and Soil, 321*, 83–115.

Latour, X., Corberand, T., Laguerre, G., Allard, F., & Lemanceau, P. (1996). The composition of fluorescent pseudomonad population associated with roots is influenced by plant and soil type. *Applied and Environmental Microbiology, 62*, 2449–2556.

Latour, X., Delorme, S., Mirleau, P., & Lemanceau, P. (2003). Identification of traits implicated in the rhizosphere competence of fluorescent pseudomonads: description of a strategy based on population and model strain studies. *Agronomie, 23*, 397–405.

Latour, X., & Lemanceau, P. (1997). Carbon and energy metabolism of oxidase-positive saprophytic fluorescent *Pseudomonas* spp. *Agronomie, 17*, 427–443.

Latour, X., Philippot, L., Corberand, T., & Lemanceau, P. (1999). The establishment of an introduced community of fluorescent pseudomonads is affected both by the soil-type and the rhizosphere. *FEMS Microbiology Ecology, 30*, 163–170.

Lebeis, S. L., Paredes, S. H., Lundberg, D. S., Breakfield, N., Gehring, J., McDonald, M., … Dangl, J. L. (2015). Salicylic acid modulates colonization of the root microbiome by specific bacterial taxa. *Science, 349*, 860–864.

Lemanceau, P. (1992). Effets bénéfiques de rhizobactéries sur les plantes : exemple des *Pseudomonas* spp. fluorescents. *Agronomie, 12*, 413–437.

Lemanceau, P., Corberand, T., Gardan, L., Latour, X., Laguerre, G., Boeufgras, J.-M., & Alabouvette, C. (1995). Effect of two plant species, flax (*Linum usitatissimum* L.) and tomato (*Lycopersicum esculentum* Mill.), on the diversity of soilborne populations of fluorescent pseudomonads. *Applied and Environmental Microbiology, 61*, 1004–1012.

Lemanceau, P., Expert, D., Gaymard, F., Bakker, P. A. H. M., & Briat, J.-F. (2009). Role of iron in plant-microbe interactions. *Advances in Botanical Research, 51*, 491–549.
Liljeroth, E., Burgers, S. L. G. E., & van Veen, J. A. (1991). Changes in bacterial populations along roots of wheat (*Triticum aestivum* L.) seedlings. *Biology and Fertility of Soils, 10*, 276–280.
Lindow, S. E., & Brandl, M. T. (2003). Microbiology of the phyllosphere. *Applied and Environmental Microbiology, 69*, 1875–1883.
Ling, N., Raza, W., Ma, J., Huang, Q., & Shen, Q. (2011). Identification and role of organic acids in watermelon root exudates for recruiting *Paenibacillus polymyxa* SQR-21 in the rhizosphere. *European Journal of Soil Biology, 47*, 374–379.
Lundberg, D. S., Lebeis, S. L., Paredes, S. H., Yourstone, S., Gehring, J., Malfatti, S., … Dangl, J. L. (2012). Defining the core *Arabidopsis thaliana* root microbiome. *Nature, 488*, 86–90.
Maignien, L., DeForce, E. A., Chafee, M. E., Eren, A. M., & Simmons, S. L. (2014). Ecological succession and stochastic variation in the assembly of *Arabidopsis thaliana* phyllosphere communities. *mBio, 5*, e00682-13. http://dx.doi.org/10.1128/mBio.00682-13.
Mansouri, H., Petit, A., Oger, P., & Dessaux, Y. (2002). Engineered rhizosphere: the trophic bias generated by opine-producing plants is independent of the opine type, the soil origin, and the plant species. *Applied and Environmental Microbiology, 68*, 2562–2566.
Marilley, L., Vogt, G., Blanc, M., & Aragno, M. (1998). Bacterial diversity in the bulk soil and rhizosphere fractions of *Lolium perenne* and *Trifolium repens* as revealed by PCR restriction analysis of 16S rDNA. *Plant and Soil, 198*, 219–224.
Maron, P.-A., Ranjard, L., Mougel, C., & Lemanceau, P. (2007). Metaproteomics: a new approach for studying functional microbial ecology. *Microbial Ecology, 53*, 486–493.
Marschner, P., Crowley, D., & Rengel, Z. (2011). Rhizosphere interactions between microorganisms and plants govern iron and phosphorus acquisition along the root axis–model and research methods. *Soil Biology & Biochemistry, 43*, 883–894.
Marzano, S.-Y. L., & Domier, L. (2015). Novel mycoviruses discovered from metatranscriptomics survey of soybean phyllosphere phytobiomes. *Virus Research, 213*, 332–342.
Maude, R. B. (1996). *Seedborne diseases and their control: Principles and practice*. Wallinford, Oxon, UK: CAB International.
Mazurier, S., Lemunier, M., Siblot, S., Mougel, C., & Lemanceau, P. (2004). Distribution and diversity of type III secretion system-like genes in saprophytic and phytopathogenic fluorescent pseudomonads. *FEMS Micobiology Ecology, 49*, 455–467.
Melotto, M., Underwood, W., & He, S. Y. (2008). Role of stomata in plant innate immunity and foliar bacterial diseases. *Annual Review of Phytopathology, 46*, 101–122.
Mendes, R., Garbeva, P., & Raaijmakers, J. M. (2013). The rhizosphere microbiome: significance of plant beneficial, plant pathogenic, and human pathogenic microorganisms. *FEMS Microbiology Reviews, 37*, 634–663.
Mendes, R., Kruijt, M., de Bruijn, I., Dekkers, E., van der Voort, M., Schneider, J. H., … Raaijmakers, J. M. (2011). Deciphering the rhizosphere microbiome for disease-suppressive bacteria. *Science, 332*, 1097–1100.
Miri, M., Janakirama, P., Held, M., Ross, L., & Szczyglowski, K. (2016). Into the root: how cytokinin controls rhizobial infection. *Trends in Plant Science, 21*, 178–186.
Moe, L. A. (2013). Amino acids in the rhizosphere: from plants to microbes. *American Journal of Botany, 100*, 1692–1705.
Mondy, S., Lenglet, A., Beury-Cirou, A., Libanga, C., Ratet, P., Faure, D., & Dessaux, Y. (2014). An increasing opine carbon bias in artificial exudation systems and genetically modified plant rhizospheres leads to an increasing reshaping of bacterial populations. *Molecular Ecology, 23*, 4846–4861.

Moreau, D., Pivato, B., Bru, D., Busset, H., Deau, F., Faivre, C., ... Mougel, C. (2015). Plant traits related to nitrogen uptake influence plant-microbe competition. *Ecology, 96*, 2300–2310.

Morris, C. E. (2002). *Phyllosphere* (Vol. 26). eLS. http://dx.doi.org/10.1007/978-81-322-1575-2.

Mougel, C., Offre, P., Ranjard, L., Corberand, T., Gamalero, E., Robin, C., & Lemanceau, P. (2006). Dynamic of the genetic structure of bacterial and fungal communities at different developmental stages of *Medicago truncatula* Gaertn. cv. Jemalong line J5. *New Phytologist, 170*, 165–175.

Naveed, M., Mitter, B., Reichenauer, T. G., Wieczorek, K., & Sessitsch, A. (2014). Increased drought stress resilience of maize through endophytic colonization by *Burkholderia phytofirmans* PsJN and *Enterobacter* sp. FD17. *Environmental Experimental Botany, 97*, 30–39.

Nelson, E. B. (2004). Microbial dynamics and interactions in the spermosphere. *Annual Review of Phytopathology, 42*, 271–309.

Neumann, G., Bott, S., Ohler, M., Mock, H.-P., Lippmann, R., Grosch, R., & Smalla, K. (2014). Root exudation and root development of lettuce (*Lactuca sativa* L. cv. Tizian) as affected by different soils. *Frontiers in Microbiology, 5*, 2. http://dx.doi.org/10.3389/fmicb.2014.00002.

Ng, J. L. P., Perrine-Walker, F., Wasson, A. P., & Mathesius, U. (2015). The control of auxin transport in parasitic and symbiotic root–microbe interactions. *Plants, 4*, 606–643.

Nguyen, C. (2003). Rhizodeposition of organic C by plants: mechanisms and controls. *Agronomie, 23*, 375–396.

Offre, P., Pivato, B., Mazurier, S., Siblot, S., Berta, G., Lemanceau, P., & Mougel, C. (2008). Microdiversity of Burkholderiales associated with mycorrhizal and non-mycorrhizal roots of *Medicago truncatula. FEMS Microbiology Ecology, 65*, 180–192.

Offre, P., Pivato, B., Siblot, S., Gamalero, E., Corberand, T., Lemanceau, P., & Mougel, C. (2007). Identification of bacterial groups preferentially associated with mycorrhizal roots of *Medicago truncatula. Applied and Environmental Microbiology, 73*, 913–921.

Oger, P., Petit, A., & Dessaux, Y. (1997). Genetically engineered plants producing opines alter their biological environment. *Nature Biotechnology, 15*, 369–372.

Osono, T. (2014). Diversity and ecology of endophytic and epiphytic fungi of tree leaves in Japan: a Review. In V. C. Verma, & A. C. Gange (Eds.), *Advances in endophytic research* (pp. 3–26). New Delhi: Springer India.

Partida-Martínez, L. P., & Heil, M. (2011). The microbe-free plant: fact or artifact? *Frontiers in Plant Science, 2*, 100. http://dx.doi.org/10.3389/fpls.2011.00100.

Pérez-Jaramillo, J. E., Mendes, R., & Raaijmakers, J. M. (2016). Impact of plant domestication on rhizosphere microbiome assembly and functions. *Plant Molecular Biology, 90*, 635–644.

Philippot, L., & Hallin, S. (2011). Towards food, feed and energy crops mitigating climate change. *Trends in Plant Science, 16*, 476–480.

Philippot, L., Raaijmakers, J. M., Lemanceau, P., & van der Putten, W. H. (2013). Going back to the roots: the microbial ecology of the rhizosphere. *Nature Reviews Microbiology, 11*, 789–799.

Pivato, B., Mazurier, S., Lemanceau, P., Siblot, S., Berta, G., Mougel, C., & van Tuinen, D. (2007). *Medicago* species affect the community composition of arbuscular mycorrhizal fungi associated with roots. *New Phytologist, 176*, 197–210.

Pivato, B., Offre, P., Marchelli, S., Barbonaglia, B., Mougel, C., Lemanceau, P., & Berta, G. (2009). Bacterial effects on arbuscular mycorrhizal fungi and mycorrhiza development as influenced by the bacteria, fungi, and host plant. *Mycorrhiza, 19*, 81–90.

Platt, T. G., Morton, E. R., Barton, I. S., Bever, J. D., & Fuqua, C. (2014). Ecological dynamics and complex interactions of *Agrobacterium* megaplasmids. *Frontiers in Plant Science, 14*, 635. http://dx.doi.org/10.3389/fpls.2014.00635.

Quesada-Moraga, E., López-Díaz, C., & Landa, B. B. (2014). The hidden habit of the entomopathogenic fungus *Beauveria bassiana*: first demonstration of vertical plant transmission. *PLoS One, 9*(2), e89278.

Raaijmakers, J. M., & Weller, D. M. (1998). Natural plant protection by 2,4-diacetylphloroglucinol-producing *Pseudomonas* spp. in take-all decline soils. *Molecular Plant-Microbe Interactions, 11*, 144–152.

Raaijmakers, J. M., & Weller, D. M. (2001). Exploiting genotypic diversity of 2,4-diacetylphloroglucinol-producing Pseudomonas spp.: characterization of superior root-colonizing P. fluorescens strain Q8r1-96. *Applied and Environmental Microbiology, 67*, 2545–2554.

Raaijmakers, J. M., Weller, D. M., & Thomashow, L. S. (1997). Frequency of antibiotic-producing *Pseudomonas* spp. in natural environments. *Applied and Environmental Microbiology, 63*, 881–887.

Ranjard, L., Dequiedt, S., Chemidlin Prévost-Bouré, N., Thioulouse, J., Saby, N. P. A., Lelievre, M., … Lemanceau, P. (2013). Turnover of soil bacterial diversity driven by wide-scale environmental heterogeneity. *Nature Communications, 4*, 1434.

Redecker, D., Kodner, R., & Graham, L. E. (2000). Glomalean fungi from the Ordovician. *Science, 289*, 1920–1921.

Redford, A. J., Bowers, R. M., Knight, R., Linhart, Y., & Fierer, N. (2010). The ecology of the phyllosphere: geographic and phylogenetic variability in the distribution of bacteria on tree leaves. *Environmental Microbiology, 12*, 2885–2893.

Reisberg, E. E., Hildebrandt, U., Riederer, M., & Hentschel, U. (2012). Phyllosphere bacterial communities of trichome-bearing and trichomeless *Arabidopsis thaliana* leaves. *Antonie Van Leeuwenhoek, 101*, 551–560.

Reisberg, E. E., Hildebrandt, U., Riederer, M., & Hentschel, U. (2013). Distinct phyllosphere bacterial communities on *Arabidopsis* wax mutant leaves. *PLoS One, 8*, e78613. http://dx.doi.org/10.1371/journal.pone.0078613.

Roberts, D. P., Dery, P. D., Yucel, I., & Buyer, J. S. (2000). Importance of *pfkA* for rapid growth of *Enterobacter cloacae* during colonization of crop seeds. *Applied and Environmental Microbiology, 66*, 87–91.

Robin, A., Mazurier, S., Mougel, C., Vansuyt, G., Corberand, T., Meyer, J.-M., & Lemanceau, P. (2007). Diversity of root-associated fluorescent pseudomonads as affected by ferritin over-expression in tobacco. *Environmental Microbiology, 9*, 1724–1737.

Robin, A., Mougel, C., Siblot, S., Vansuyt, G., Mazurier, S., & Lemanceau, P. (2006). Effect of ferritin over-expression in tobacco on the structure of bacterial and pseudomonad communities associated with the roots. *FEMS Micobiology Ecology, 58*, 492–502.

Robin, A., Vansuyt, G., Hinsinger, P., Meyer, J.-M., Briat, J.-F., & Lemanceau, P. (2008). Iron dynamics in the rhizosphere: consequences for plant health and nutrition. *Advances in Agronomy, 99*, 183–225.

Rodriguez, R. J., White, J. F., Arnold, A. E., & Redman, R. S. (2009). Fungal endophytes: diversity and functional roles. *New Phytologist, 182*, 314–330.

Rudrappa, T., Czymmek, K. J., Paré, P. W., & Bais, H. P. (2008). Root-secreted malic acid recruits beneficial soil bacteria. *Plant Physiology, 148*, 1547–1556.

Ryan, P. R., Dessaux, Y., Thomashow, L. S., & Weller, D. M. (2009). Rhizosphere engineering and management for sustainable agriculture. *Plant and Soil, 321*, 363–383.

Sapkota, R., Knorr, K., Jørgensen, L. N., O'Hanlon, K. A., & Nicolaisen, M. (2015). Host genotype is an important determinant of the cereal phyllosphere mycobiome. *New Phytologist, 207*, 1134–1144.

Savka, M. A., Dessaux, Y., Oger, P., & Rossbach, S. (2002). Engineering bacterial competitiveness and persistence in the phytosphere. *Molecular Plant-Microbe Interactions, 15*, 866–874.

Scarafoni, A., Ronchi, A., Prinsi, B., Espen, L., Assante, G., Venturini, G., & Duranti, M. (2013). The proteome of exudates from germinating *Lupinus albus* seeds is secreted through a selective dual-step process and contains proteins involved in plant defence. *FEBS Journal, 280*, 1443–1459.

Schiltz, S., Gaillard, I., Pawlicki-Jullian, N., Thiombiano, B., Mesnard, F., & Gontier, E. (2015). A review: what is the spermosphere and how can it be studied? *Journal of Applied Microbiology, 119*, 1467–1481.

Schmidt, J. E., Bowles, T. M., & Gaudin, A. C. M. (2016). Using ancient traits to convert soil health into crop yield: Impact of selection on maize root and rhizosphere function. *Frontiers in Plant Science, 7*, 373. http://dx.doi.org/10.3389/fpls.2016.00373.

Schreiber, L., Krimm, U., Knoll, D., Sayed, M., Auling, G., & Kroppenstedt, R. M. (2005). Plant-microbe interactions: identification of epiphytic bacteria and their ability to alter leaf surface permeability. *New Phytologist, 166*, 589–594.

Semenov, A. M., van Bruggen, A. H. C., & Zelenev, V. V. (1999). Moving waves of bacterial populations and total organic carbon along roots of wheat. *Microbiology Ecology, 37*, 116–128.

Sharifi, R., & Ryu, C.-M. (2017). Chatting with a tiny belowground member of the holobiome: communication between plants and growth-promoting rhizobacteria. In G. Becard (Ed.), *How plants communicate with their biotic environment* (Vol. 82, pp. 135–160).

Shi, S., Richardson, A. E., O'Callaghan, M., DeAngelis, K. M., Jones, E. E., Stewart, A., … Condron, L. M. (2011). Effects of selected root exudate components on soil bacterial communities. *FEMS Microbiology Ecology, 77*, 600–610.

Singh, D., & Mathur, S. B. (2004). Location of fungal hyphae in seeds. In D. Singh, & S. B. Mathur (Eds.), *Histopathology of seed-borne infections* (pp. 101–168). Boca Raton, FL, USA: CRC Press.

Skiba, M. W., George, T. S., Baggs, E. M., & Daniell, T. J. (2011). Plant influence on nitrification. *Biochemical Society Transactions, 39*, 851–856.

Smith, K. P., Handelsman, J., & Goodman, R. M. (1999). Genetic basis in plants for interactions with disease-suppressive bacteria. *Proceedings of the National Academy of Sciences of the United States of America, 96*, 4786–4790.

Smith, S. E., & Read, D. (2008). *Mycorrhizal symbiosis* (3rd ed.). New York: Academic Press.

Subbarao, G. V., Kishii, M., Nakahara, K., Ishikawa, T., Ban, T., Tsujimoto, H., … Ito, O. (2009). Biological nitrification inhibition (BNI)—is there potential for genetic interventions in the *Triticeae*? *Breeding Science, 59*, 529–545.

Subbarao, G. V., Sahrawat, K. L., Nakahara, K., Rao, I. M., Ishitani, M., Hash, C. T., … Lata, J. C. (2012). A paradigm shift towards low-nitrifying production systems: the role of biological nitrification inhibition (BNI). *Annals of Botany, 112*, 297–316.

Subramoni, S., Nathoo, N., Klimov, E., & Yuan, Z.-C. (2014). *Agrobacterium tumefaciens* responses to plant-derived signaling molecules. *Frontiers in Plant Science, 5*(322). http://dx.doi.org/10.3389/fpls.2014.00322.

Terras, F. R. G., Eggermont, K., Kovaleva, V., Raikhel, N. V., Osborn, R. W., Kester, A., et al. (1995). Small cysteine-rich antifungal proteins from radish – their role in host-defense. *Plant Cell, 7*, 573–588.

Theis, K. R., Dheilly, N. M., Klassen, J. L., Brucker, R. M., Baines, J. F., Bosch, T. C. G., … Bordenstein, S. R. (2016). Getting the hologenome concept right: an eco-evolutionary framework for hosts and their microbiomes. *mSystems, 1*, e00028-16.

Tian, Y., Zhao, Y., Wu, X., Liu, F., Hu, B., & Walcott, R. R. (2015). The type VI protein secretion system contributes to biofilm formation and seed-to-seedling transmission of *Acidovorax citrulli* on melon. *Molecular Plant Pathology, 16*, 38–47.

Truyens, S., Weyens, N., Cuypers, A., & Vangronsveld, J. (2015). Bacterial seed endophytes: genera, vertical transmission and interaction with plants. *Environmental Microbiology Reports, 7*, 40–50.

Uren, N. C. (2001). Types, amounts and possible functions of compounds released into the rhizosphere by soil-grown plants. In R. Pinton, Z. Varanini, & P. Nannipieri (Eds.), *The rhizosphere. Biochemistry and organic substances at the soil-plant interface* (pp. 19–40). New York, USA: Marcel Dekker.

Vacher, C., Hampe, A., Porté, A. J., Sauer, U., Compant, S., & Morris, C. E. (2016). The phyllosphere: microbial jungle at the plant-climate interface. *Annual Review of Ecology Evolution and Systematics, 47*, 1–24.

Vandenkoornhuyse, P., Quaiser, A., Duhamel, M., Le Van, A., & Dufresne, A. (2015). The importance of the microbiome of the plant holobiont. *New Phytologist, 206*, 1196–1206.

Verona, O. (1958). La spermosphère. *Annales de l'Institut Pasteur, 95*, 795–798.

Viollet, A., Corberand, T., Mougel, C., Robin, A., Lemanceau, P., & Mazurier, S. (2011). Fluorescent pseudomonads harboring type III secretion genes are enriched in the mycorrhizosphere of *Medicago truncatula*. *FEMS Micobiology Ecology, 75*, 457–467.

Viollet, A., Pivato, B., Mougel, C., Cleyet-Marel, J.-C., Gubry-Rangin, C., Lemanceau, P., & Mazurier, S. (2016). *Pseudomonas fluorescens* C7R12 type III secretion system impacts mycorrhization of *Medicago truncatula* and associated microbial communities. *Mycorrhiza*. http://dx.doi.org/10.1007/s00572-016-0730-3.

Vogel, C., Bodenhausen, N., Gruissem, W., & Vorholt, J. A. (2016). The Arabidopsis leaf transcriptome reveals distinct but also overlapping responses to colonization by phyllosphere commensals and pathogen infection with impact on plant health. *New Phytologist, 212*.

Vokou, D., Vareli, K., Zarali, E., Karamanoli, K., Constantinidou, H.-I. A., Monokrousos, N., ... Sainis, I. (2012). Exploring biodiversity in the bacterial community of the Mediterranean phyllosphere and its relationship with airborne bacteria. *Microbial Ecology, 64*, 714–724.

Vorholt, J. A. (2012). Microbial life in the phyllosphere. *Nature Reviews Microbiology, 10*, 828–840. http://dx.doi.org/10.1038/nrmicro2910.

de Weert, S., Vermeiren, H., Mulders, I. H., Kuiper, I., Hendrickx, N., Bloemberg, G. V., ... Lugtenberg, B. J. (2002). Flagella-driven chemotaxis towards exudate components is an important trait for tomato root colonization by *Pseudomonas fluorescens*. *Molecular Plant-Microbe Interactions, 15*, 1173–1180.

Whipps, J. M., Hand, P., Pink, D., & Bending, G. D. (2008). Phyllosphere microbiology with special reference to diversity and plant genotype. *Journal of Applied Microbioliogy, 105*, 1744–1755.

Wilson, H. E., Carroll, G. C., Roy, B. A., & Blaisdell, G. K. (2014). Tall fescue is a potential spillover reservoir host for *Alternaria* species. *Mycologia, 106*, 22–31.

Wintermans, P. C. A., Bakker, P. A. H. M., & Pieterse, C. M. J. (2016). Natural genetic variation in Arabidopsis for responsiveness to plant growth-promoting rhizobacteria. *Plant Molecular Biology, 90*, 623–634.

Wissuwa, M., Mazzola, M., & Picard, C. (2009). Novel approaches in plant breeding for rhizosphere-related traits. *Plant and Soil, 321*, 409–430.

Yadav, R. K. P., Karamanoli, K., & Vokou, D. (2005). Bacterial colonization of the phyllosphere of mediterranean perennial species as influenced by leaf structural and chemical Features. *Microbial Ecology, 50*(2), 185–196.

Yang, C. H., & Crowley, D. E. (2000). Rhizosphere microbial community structure in relation to root location and plant iron nutritional status. *Applied and Environmental Microbiology, 66*, 345–351.

Zancarini, A., Mougel, C., Terrat, S., Salon, C., & Munier-Jolain, N. (2013). Combining ecophysiological and microbial ecological approaches to study the relationship between *Medicago truncatula* genotypes and their associated rhizosphere bacterial communities. *Plant and Soil, 365*, 183–199.

Zhang, N., Wang, D., Liu, Y., Li, S., Shen, Q., & Zhang, R. (2014). Effects of different plant root exudates and their organic acid components on chemotaxis, biofilm formation and colonization by beneficial rhizosphere-associated bacterial strains. *Plant and Soil, 374*, 689–700.

Zheng, X. Y., & Sinclair, J. B. (1996). Chemotactic response of *Bacillus megaterium* strain B153-2-2 to soybean root and seed exudates. *Physiological and Molecular Plant Pathology, 48*, 21–35.

Zilber-Rosenberg, I., & Rosenberg, E. (2008). Role of microorganisms in the evolution of animals and plants: the hologenome theory of evolution. *FEMS Microbiology Reviews, 32*, 723–735.

CHAPTER SIX

Chatting With a Tiny Belowground Member of the Holobiome: Communication Between Plants and Growth-Promoting Rhizobacteria

R. Sharifi*, C.-M. Ryu[§,¶,1]
*Razi University, Kermanshah, Iran
[§]Korea Research Institute of Bioscience and Biotechnology, Daejeon, South Korea
[¶]University of Science and Technology, Daejeon, South Korea
[1]Corresponding author: E-mail: cmryu@kribb.re.kr

Contents

Abstract

As sessile organisms, plants have facilitated to actively protect themselves against biotic and abiotic stresses. For this, plants keep communicating with other organisms including insect and microbes inside and outside plant surface including root surface. In the area around the root referred to as the rhizosphere, diverse root-associated bacteria interact with plants with both positive and negative effects. In the past three decades, a group of rhizosphere bacteria known as plant growth-promoting rhizobacteria (PGPR) have been intensively studied for possible use in improving plant health. Here, we provide an overview of the current knowledge on the nature of PGPR, their signatures and their

Advances in Botanical Research, Volume 82
ISSN 0065-2296
http://dx.doi.org/10.1016/bs.abr.2016.09.002

roles in plant growth promotion and biological control against plant pathogens, as revealed by classic physiological and morphological observations and recent cutting-edge technology. We also discuss new information on insect–plant–PGPR tritrophic interactions, as well as technology transfer from the laboratory to the field. We use the new concept of the holobiome to help elucidate plant–PGPR interactions. Understanding plant root–bacterial communications in this novel framework may facilitate the improvement of plant health in agricultural production systems.

1. BELOWGROUND ZERO

The term 'rhizosphere' was first coined by Lorenz Hiltner in 1904 (Hartmann, Rothballer, & Schmid, 2007). This term refers to the narrow zone of soil around plant roots that is under the influence of root exudates chemically, physically and biologically. Plants deposit approximately 6–12% of their fixed carbon into the rhizosphere, mainly in the form of sugars and organic acids. Moreover, total nitrogen exudation comprises 10–16% of total plant nitrogen (Bais, Weir, Perry, Gilroy, & Vivanco, 2006; Dennis, Miller, & Hirsch, 2010). The amount of root exudate varies along the root, with the highest level found below the root tip and elongation zone. The volume of the rhizosphere, which is affected by soil type, plant species and phenological growth stage, ranges from less than 1 mm to several centimetres around the root system.

Compared to the surrounding (bulk) soil, the rhizosphere is a niche full of nutrients, which makes it attractive to soil microorganisms. The microbial population is 10–1000 times higher in the rhizosphere than in bulk soil (Haas & Defago, 2005). Microbial population levels are highest near the root, with the largest population (1×10^{12} CFU/g of soil) occurring within 50 μm of the root (Miransari, 2011; Pinton, Varanini, & Nannipieri, 2007). Bacterial populations in the rhizosphere are large, but since they belong to distinct genera and families of bacteria, their diversity is low in the rhizosphere (Lakshmanan, 2015; Lundberg et al., 2012), because the rhizosphere environment is highly competitive. Bacteria must have competitive and ecological advantages to thrive in this zone (Bulgarelli, Schlaeppi, Spaepen, Ver Loren van Themaat, & Schulze-Lefert, 2013). Several reports revealed that Proteobacteria, especially Pseudomonadaceae, comprise the highest bacterial population in the rhizosphere. High populations of Gram-positive Firmicutes and Actinobacteria are also present in the rhizosphere (Lareen, Burton, & Schafer, 2016; Lundberg et al., 2012). In one study, the largest populations in the rhizosphere included Proteobacteria (46% of the total), followed by Firmicutes (18%), Actinobacteria (11%), Bacteroidetes (7%) and Acidobacteria (3%) (Weinert et al., 2011). Only small groups of these bacteria can enter the root tissues. These microorganisms are commonly known as root endophytes or endorhizosphere microbes. Populations of endophytic bacteria make up less than 10^4 CFU/g of root tissues (Bulgarelli et al., 2013). Actinobacteria, especially members of Streptomycetaceae, are the top colonizers of the endorhizosphere (Lundberg et al., 2012).

Single-plant species have specific root exudate profiles which allow them to support distinctive groups of microbes. Indeed, the diversity of the root microbiome (the total microbial community) differs when different plant species are cultivated in the same soil (Berg & Smalla, 2009). Even different cultivars of the same plant species have different microbial diversity when grown in the same soil (Raaijmakers & Mazzola, 2012). Three different wheat cultivars exhibited different abilities to recruit pseudomonads for biocontrol (Mazzola, Funnell, & Raaijmakers, 2004). Interestingly, plants under biotic stress attract specific microorganisms. Plants infected with the plant pathogen *Verticillium* spp. attracted 2,4-diacetylphloroglucinol (DAPG) producing fluorescent pseudomonads to control this fungal pathogen (Hartmann, Schmid, Tuinen, & Berg, 2008). DAPG is a phenolic antibiotic produced by various *Pseudomonas* species which exhibits antifungal activity against many fungal plant pathogens. Plants secrete different groups of root exudates with various effects on root microbiome diversity. In-depth studies showed that low molecular weight carbon exudates, especially organic acids, play a critical role in altering microbiome diversity in the rhizosphere. Citric acid treatment, 240 mg/g soil as standard concentration, increased the sizes of populations of beta and gamma Proteobacteria, especially members of Burkholderiales, and treatment with the same concentration of amino acid glycine increased the population sizes of Actinobacteria and Bradyrhizobiales (Eilers, Lauber, Knight, & Fierer, 2010). Malic acid was attractive to *Bacillus*, and quinolic acid increased Proteobacteria populations (Shi et al., 2011).

As mentioned earlier, plants release high amounts of carbon and nitrogen into the rhizosphere, which alters the rhizosphere microbiome. Does the plant benefit from this investment, and do rhizobacteria respond to this benevolence? The relationship between plants and rhizobacteria is more than just a cost–benefit relationship. Indeed, the plant–microbiome is considered to resemble an individual superorganism. This superorganism is described by the term holobiont, i.e., a superorganism of a host and its interacting microbes in the microbiome (Partida-Martinez & Heil, 2011).

2. PLANT HOLOBIOME: DO PLANT-ASSOCIATED MICROBES AND THE HOST PLANT REPRESENT A SUPERORGANISM?

Symbiosis of plants or animals with microbes is thought to be the norm rather than the exception (Zilber-Rosenberg & Rosenberg, 2008). Indeed, if we consider all of the genes from all of the microbes (microbiome) that help plants acquire mineral nutrients, modulate growth and increase resistance against biotic and abiotic stresses, we find that a network of thousands of microbial genes work for plant fitness and vice versa. The hologenome refers to the sum of the plant and microbiome genomes. Van Der Heijden, Bardgett, and Van Straalen (2008) suggested that 20,000 plant species have obligatory beneficial relationships with microbes. Approximately 80% of plant nitrogen and 75% of phosphate are provided by such microorganisms. Plants in the Orchidaceae family do not have enough nutrients and energy for germination, making their relationship with specific symbiotic fungi called mycorrhizal fungi indispensable for germination (McCormick et al., 2012). Sometimes, bacteria from other holobionts can act as friends or foes of plants. The helper bacteria help mycorrhizal fungi to grow and colonize both the plant surface and its interior (Bonfante & Anca, 2009). *Rhizobium radiobacter* is an endosymbiont of the beneficial fungus *Piriformospora indica*. This bacterium exhibits plant growth-promoting and biocontrol activities in *P. indica* (Bonfante & Anca, 2009). In another example, microbial endosymbionts of aphids make them more attractive to aphid parasitoids (aphids' natural enemy). The bacterium also alters the host ranges of herbivorous insects (Sugio, Dubreuil, Giron, & Simon, 2014).

In the holobiont, the host plant and mutual rhizobacteria share their genetic capacities to reach their highest ecological competence and to undergo rapid evolution. These beneficial rhizobacteria are known as plant growth-promoting rhizobacteria (PGPR), a term coined by Kloepper, Schroth, and Miller (1980) to describe groups of rhizobacteria that colonize plants and improve their growth and resistance to biotic and abiotic stresses. Well-known genera of PGPR include the genera of *Pseudomonas*, *Bacillus*, *Azospirillum*, *Azotobacter*, *Streptomyces* and Rhizobia genera. PGPR improve plant growth and health by providing mineral nutrients, stimulating plant growth via the production of phytohormones, facilitate mycorrhizal formation, functioning in biological control of plant pathogens, increasing tolerance to abiotic stresses and functioning in the bioremediation of toxic compounds (Bonfante & Anca, 2009; Ruzzi & Aroca, 2015).

3. TINY ORGANISM WITH BIG ROLES IN PLANT HEALTH

PGPR play critical roles in providing nutrient elements for the host plant and enriching plant yields and the contents of mineral elements required for human. PGPR fix atmospheric nitrogen, solubilize phosphorous and produce iron and zinc chelates (known as siderophores). The application of PGPR might represent an alternative to chemical fertilization or at least reduce their application dose (Adesemoye, Torbert, & Kloepper, 2009; Miransari, 2011). Chemical fertilizers represent 30–35% of total agricultural input in some countries (Rose et al., 2014), whose less than 50% and 20% of the added nitrogen and phosphate are absorbed by the plant, respectively (Miransari, 2011). Excess phosphate adds to the inorganic phase of sediment, while excess nitrogen leaches out of the soil into the groundwater or emitted as nitrogen oxides (Eichner, 1990). The application of PGPR *biostimulants* represents a promising technology for reducing the negative effects of chemical fertilizers. In one study, the application of PGPR reduced the need for nitrogen fertilizers by about 50% of recommended amount (Rose et al., 2014), and, in another study, PGPR reduced the need for nitrogen and phosphate by approximately 20–30% in tomato (Adesemoye, Torbert, & Kloepper, 2008).

PGPR also help plants take up microelements such as Fe, Zn and Cu. These elements are abundant in soil, but the levels of soluble microelements are too low in normal agricultural soils with good aeration. The concentration of Fe is 10^{-10} to 10^{-18} in normal soils, but the proper concentration for easy absorption by plants and microbes should be 10^{-6} (Lemanceau, Expert, Gaymard, Bakker, & Briat, 2009). Bacteria produce siderophores with high affinity for iron such as pyoverdine, which solubilize iron and deliver it to producer microorganisms and host plants (Sharifi, Ahmadzadeh, Sharifi-Tehrani, & Talebi-Jahromi, 2010). PGPR also increase iron acquisition efficiency in plants. Volatiles of *Bacillus subtilis* GB03 increased the expression of plant genes encoding Fe-deficiency-induced transcription factor 1 and iron transporter1. They also decreased the rhizosphere pH by increasing proton release into the rhizosphere. Altogether, these mechanisms increased iron uptake and transport in *Arabidopsis*, which subsequently improved photosynthesis in PGPR-treated plants (Zhang et al., 2009).

PGPR use a chemical language to communicate with plants and to improve their growth and tolerance against biotic and abiotic stresses. In this chapter, we describe recent findings on this topic. First, we describe how PGPR directly produce phytohormones or modulate plant hormone signalling to enhance root and shoot growth leading to plant growth promotion. Secondly, we discuss PGPR-elicited induced systemic resistance and induced systemic tolerance against biotic and abiotic stresses. We also describe bacterial determinants that activate plant defences against pests and pathogens, as well as plant defence mechanisms and signalling. Third, we introduce new data on insect-modulated plant root exudation resulting in rhizosphere bacteria that reengineer plant defence responses against insect herbivory. Finally, we evaluate technology transfer from scientists to farmers and provide examples. Since similar reviews focussing on Gram-negative PGPR, primarily *Pseudomonas* spp., have been published (Bloemberg & Lugtenberg, 2001; Van Loon, 2007), we will focus on the Gram-positive genus *Bacillus*. *Bacillus*-based products represent more than 70% of total microbial biopesticide sales (Ongena & Jacques, 2008). The capacity for endospore production makes *Bacillus* spp. easy to formulate, increases their shelf life and makes them durable (Schisler, Slininger, Behle, & Jackson, 2004). Furthermore, the multifaceted biocontrol and plant growth-promoting features of these bacteria made them good candidates for in-depth molecular research.

4. PHYTOHORMONE PRODUCTION AND MODULATION BY RHIZOBACTERIA FOR DIRECT IMPROVEMENT OF PLANT GROWTH

PGPR, as part of a holobiont, improve plant fitness by producing plant hormones or modulating hormonal sensing and signalling in plants. PGPR can synthesize most critical hormones such as auxins, cytokinins, gibberellic acid, abscisic acid (ABA) and salicylic acid (SA) (Ortíz-Castro, Contreras-Cornejo, Macías-Rodríguez, & López-Bucio, 2009; Ruzzi & Aroca, 2015). Among these hormones, auxin production appears to be widespread in rhizobacteria. Approximately 80% of bacteria isolated from the rhizosphere can produce auxin (Glick, Cheng, Czarny, & Duan, 2007). Abbas-Zadeh et al. (2009) tested 40 strains of fluorescent pseudomonads for auxin production and found that all of them produced auxin *in vitro*.

Auxin: The role of auxin in plant—PGPR communications can be investigated by either examining the production of auxin in bacteria or by examining how PGPR modulate auxin signalling in plants. The role of PGPR-produced auxin in plant growth can be assessed by altering auxin production in rhizobacteria through mutation. *Pseudomonas putida*

GR12-2 produces high levels of auxin. The application of the wild-type strain increased plant growth by 35—50% more than that of the auxin mutant (Xie, Pasternak, & Glick, 1996). An auxin deficient mutant of *Bacillus amyloliquefaciens* failed to increase plant growth. *Pseudomonas thivervalensis* MLG45 increased *Arabidopsis* growth by up to 70%, but a mutation in an auxin biosynthesis gene compromised its plant growth-promoting activity (Dodd, Zinovkina, Safronova, & Belimov, 2010). However, some bacteria are unable to produce auxin but they can increase auxin biosynthesis in the host plant. For example, *Phyllobacterium brassicacearum* increased lateral root formation by up to 50% by modulating auxin biosynthesis in *Arabidopsis* (Contesto et al., 2010).

PGPR increase shoot growth but reduce primary root elongation by modulating auxin signalling (Shi, Park, Lee, Ryu, & Ryu, 2010). However, secondary root and hair root proliferation increase in plants treated with auxin-producing or auxin-modulating rhizobacteria. This increase in root surface area increases the absorption of minerals. In-depth investigations have also shown that rhizobacteria determinants such as volatile organic compounds (VOCs), acyl homoserine lactones and DAPG affect auxin signalling and transport in *Arabidopsis* and tomato, respectively (Bai, Todd, Desikan, Yang, & Hu, 2012; Brazelton, Pfeufer, Sweat, Gardener, & Coenen, 2008; Zhang et al., 2007).

Bacterial VOCs increase basipetal transport of auxin from shoot to root. Naphthylphthalamic acid, an inhibitor of auxin transport, abolishes the plant response to *B. subtilis* GB03 VOCs. Low concentrations of auxin are sufficient for improving the growth of the aerial parts of plants, but high concentrations can be suppressive. By contrast, root growth increases in response to higher concentrations of auxin. Therefore, basipetal transports of auxin favours either shoot or root growth. Search for specific volatile compound showed that indole plays an important role in the activation of auxin signalling. The greatest lateral root formation in *Arabidopsis* occurred in the presence of 600 μg of indole (Bailly et al., 2014).

Cytokinins: PGPR belonging to the genera *Rhizobium*, *Bacillus*, *Pseudomonas*, *Pantoea* and *Paenibacillus* produce cytokinin (Bishnoi, 2015). This hormone increases cell division, cell differentiation, stomatal opening, shoot elongation and root hair proliferation, but decreases primary and lateral root formation (Liu, Xing, Ma, Du, & Ma, 2013; Miri, Janakirama, Held, Ross, & Szczyglowski, 2016; Ortíz-Castro, Valencia-Cantero, & López-Bucio, 2008). Cytokinin can reduce root growth via activating the ethylene signalling. However, the holobiont has a strategy for bypassing the negative effects of its cytokinin production on root growth. Treating plants with bacterial cytokinin (Zeatin) leads to increased cytokinin biosynthesis in the aerial parts of plants, but cytokinin production in roots is not affected by this treatment (Liu et al., 2013). Therefore, both root and shoot biomass increase in response to rhizobacteria. *Bacillus megaterium* has an effect on cytokinin perception. Single or multiple mutations in plant cytokinin receptors reduce the plant growth-promoting activity of this bacterium (Ortíz-Castro et al., 2008). In addition, plants with a mutation in *cre*, encoding a cytokinin receptor, are insensitive to *B. subtilis* GB03 volatiles indicating that bacterial volatiles increase plant growth through the cytokinin signalling pathway (Ryu, Farag, et al., 2003).

ABA: ABA increases drought tolerance in plants by promoting stomatal closure, which reduces evaporation through the stomata (Dodd et al., 2010). Furthermore, ABA increases primary root elongation to help the plant reach new sources of water. This hormone also reduces lateral root formation (Ruzzi & Aroca, 2015). PGPR increase ABA biosynthesis in plants under stress conditions, thereby reducing water evaporation in the stomata. Application of *Azospirillum brasilense* sp245 and *Burkholderia* increased ABA biosynthesis in *Arabidopsis* in drought stress conditions (Cohen, Bottini, & Piccoli, 2008; Ruzzi & Aroca, 2015). Treatment of grapevines with Bacillus *licheniformis* and Pseudomonas *fluorescens* under water stress increased ABA production by 70% and 40%, respectively. Plants inoculated with these strains lost 10% and 4% less water, respectively (Salomon et al., 2014). In one case, the application of *Burkholderia* LD-11 increased drought tolerance in plants but reduced ABA levels. Further work revealed that rhizobacteria increase the sensitivity of the ABA receptor, allowing plants to respond to low concentrations of ABA (Fan et al., 2015). In addition to its role in drought stress, ABA functions in the plant growth-promoting activity of rhizobacteria. *B. megaterium* treatment fails to increase plant growth in ABA mutants indicating that ABA is required for the activity of this bacterium. Indeed, bacteria treatment increases ethylene biosynthesis in ABA mutant plants (Porcel, Zamarreño, García-Mina, & Aroca, 2014).

Treatments with volatiles of *B. subtilis* GB03 were found to increase the chlorophyll content, the photosynthesis efficiency and the amount of hexose sugars (60%) in *Arabidopsis*. As treatment with GB03 volatiles also reduced ABA levels in plant shoots by 20–50%, it was proposed that this low level of ABA repressed sugar sensing, therefore preventing the feedback inhibition of photosynthesis by elevated photosynthate level (Zhang et al., 2008). The application of exogenous ABA also removed this ability of *B. subtilis* GB03 volatiles.

Ethylene (ET): Ethylene is involved in senescence and abscission processes. In addition this phytohormone plays regulatory roles in seed germination, flowering, fruit ripening, defoliation, and it inhibits shoot and root growth (Glick, 2015). The role of PGPR in modulating plant ethylene levels was first described by Glick, Jacobson, Schwarze, and Pasternak (1994) who discovered the production of 1-aminocyclopropane-1-carboxylate (ACC) deaminase by rhizobacteria. Rhizobacterial ACC deaminase converts ACC, an immediate precursor of ET, to α-ketobutyrate, a carbon source for bacterial nutrition. ACC deaminase is widespread in both Gram-positive and Gram-negative bacteria; 18 of 20 species of *Burkholderia* can produce ACC deaminase (Onofre-Lemus, Hernández-Lucas, Girard, & Caballero-Mellado, 2009). Treatment with rhizobacteria alleviates the inhibitory effect of ET by repressing ethylene production in roots. ACC deaminase not only increases plant root biomass, but it also increases plant tolerance to abiotic stresses. Treatment of 55-day-old tomato plants with the ACC deaminase producer *P. putida* increased plant resistance to flooding. Transgenic tomato expressing a bacterial ACC deaminase gene showed increased tolerance to flooding (Grichko & Glick, 2001).

5. RELIEVING PLANT STRESS I: RHIZOBACTERIA ACTIVATE ABIOTIC TOLERANCE

Plants suffer much more from abiotic stress than from biotic challenges. The proportion of yield loss caused by abiotic and biotic stress was found to be 70% and 30%, respectively (Jewell, Campbell, & Godwin, 2010). The major abiotic stresses include high or low temperature, flooding or drought, deficiency or toxicity of minerals and soil salinity.

PGPR alleviates abiotic stress via various mechanisms, such as increasing micro- and macronutrient uptake, root volume and extracellular polysaccharide (EPS) production as well as modulating plant stress hormone levels. Ethylene and ABA are the main hormones that function in abiotic stress tolerance (Fig. 1). The roles of rhizobacterial hormones in stress tolerance are described earlier.

Cyanobacteria increased tolerance to salt stress by reducing soil pH and salt concentrations. They decreased pH from 5.0 to 8.8 and Na^{+} level from 0.78 to 0.60 ppm (Singh, Pandey, & Singh, 2011). PGPR produce EPS, binding agents for the production of soil micro- and macroaggregates. These

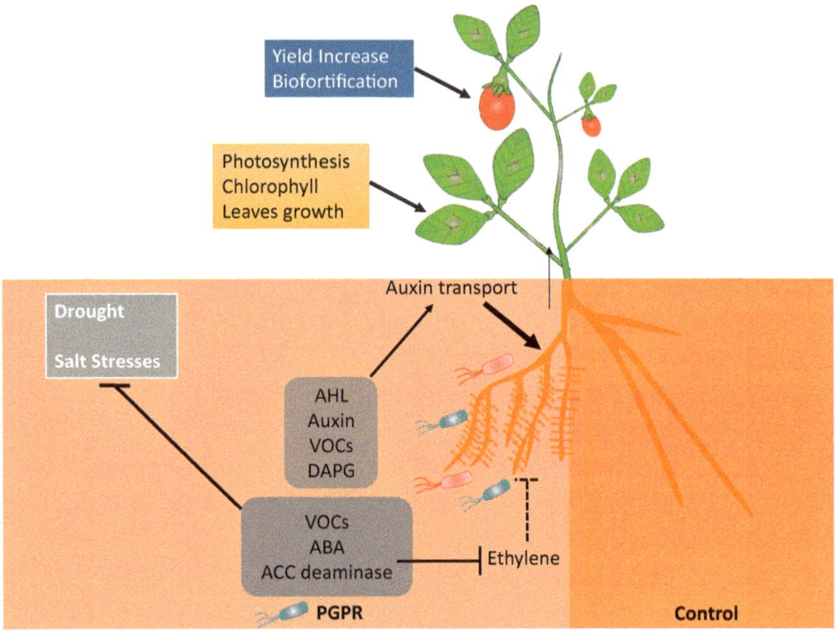

Figure 1 Plant growth-promoting rhizobacteria (PGPR) promote plant growth by providing nutrients, modulating hormone levels and alleviating abiotic stresses. Rhizobacteria produce volatile organic compounds (VOCs), acyl homoserine lactones (AHL) and the antibiotic diacetylphloroglucinol (DAPG) for modulating auxin homeostasis in plants. Auxin increases shoot growth, lateral root and root hair proliferation. Bacterial volatiles also increase chlorophyll content and photosynthesis efficiency by improving iron uptake and removing the negative impact of sugar accumulation on photosynthesis. The bacterial enzyme aminocyclopropane-1-carboxylate (ACC) deaminase represses the negative effect of ethylene on root growth by degrading the ethylene precursor ACC. Rhizobacteria increase plant tolerance against abiotic stresses such as drought, flooding and salt by modulating the levels of the hormones abscisic acid (ABA) and ethylene or diminishing Na^+ uptake in plants.

particles improve soil aeration and water capacity. In addition, EPS absorb Na^+ cations, making them unavailable for plants (Paul & Lade, 2014; Upadhyay, Singh, & Singh, 2011). *Pseudomonas aeruginosa*, with high EPS production, has high tolerance to salt compared to other strains. This strain increases plant tolerance to salt stress (Tewari & Arora, 2014). An EPS mutant of this bacterium failed to increase plant salt stress tolerance.

The production of metabolites such as the osmoprotectant trehalose-6-phosphate by rhizobacteria confers abiotic stress resistance to plants (Rodrigues, Bonifacio, De Araujo, Lira, & Figueiredo, 2015). Overexpressing trehalose-6-phosphate in *Rhizobium etli* increased drought tolerance in common bean (Suárez et al., 2008). Plants inoculated by an overexpression mutant exhibited increased nodule formation and high nitrogenase activity compared to the control, but a mutation in trehalose-6-phosphate had a negative effect on these two factors. Transformation of *Azospirillum brasilense* with a trehalose-6-phosphate gene increased its tolerance to high salt levels compared to the wild type. Corn plants inoculated with this mutant exhibited increased drought tolerance of more than 85% compared to noninoculated plants. Application of a wild-type strain increased drought tolerance by only 55% compared to the noninoculated control (Rodríguez-Salazar, Suárez, Caballero-Mellado, & Iturriaga, 2009).

6. RELIEVING PLANT STRESS II: RHIZOBACTERIA ACTIVATE BIOTIC RESISTANCE

Plants and animals have been invaded by pests and pathogens throughout their evolutionary history and have therefore developed efficient strategies for defence against these enemies. Animals have innate and adaptive immunity based on mobile lymphocyte cells (Flajnik & Kasahara, 2010). Plants use pattern recognition receptors for non-self-detection of microbial molecular patterns in places of attack (Newman, Sundelin, Nielsen, & Erbs, 2013). Same as animals, plants recognize molecular determinants on the surfaces of pathogens. These patterns are known as microbe-associated molecular patterns (MAMPs). Well-studied MAMPs include molecules such as flagellin, chitin, peptidoglycan, lipopolysaccharide, siderophores and VOCs (Newman et al., 2013; Pieterse et al., 2014). After detecting MAMPs, plants use mobile signals such as SA, azelaic acid, pipecolic acid and methyl-Jasmonate (Me-JA) to spread the defence mechanism throughout the plant (Shah & Zeier, 2014).

In addition to using mobile signals, plants use defence hormones to regulate the expression of specific sets of defence genes. Two important pathways are involved in the regulation of defence gene: the SA and jasmonic acid (JA)/ET pathways (Pieterse, Van Der Does, Zamioudis, Leon-Reyes, & Van Wees, 2012). In the SA pathway, defence inducers such as SA, benzothiadiazole (BTH), necrotizing pathogens and some rhizobacteria increase SA biosynthesis in the cytoplasm. Accumulation of SA alters cytoplasmic redox and cleaves disulfide bonds in nonexpressor of PR genes1 (NPR1) oligomers. NPR1 monomers are transported to the nucleus, where they act as cotranscription factors with TGA transcription factors. This complex activates genes such as *wrky70*, which subsequently activates a set of SA-dependent defence genes such as those encoding pathogenesis-related proteins (PRs). PR-1 is a marker for the activation of this pathway (Fu & Dong, 2013).

In the JA pathway, defence inducers such as JA, β-aminobutyric acid (BABA), and some rhizobacteria are perceived by plant cells, leading to an increase in JA biosynthesis. JAR1 convert JA to the highly active form, JA-isoleucine (Ile). JA-Ile mediates recognition and degradation of jasmonate ZIM-domain (JAZ) by the 26s proteasome. JAZ is a transcriptional repressor of JA-dependent genes (Katsir, Chung, Koo, & Howe, 2008). Thus, degradation of JAZ releases its repressive activity and activates JA transcription factors such as MYC2 and ERF1 for the expression of JA-dependent genes such as *pdf1.2* and *vsp2* (Pieterse et al., 2012).

The SA pathway is more effective against biotrophic pathogens, whereas the JA pathway is effective against necrotrophic pathogens and herbivorous insects (Glazebrook, 2005). Plants have the ability to fine-tune these two pathways. Most often, the activation of one of the pathways suppresses the other pathway. For example, the application of exogenous SA represses JA-dependent genes or the production of the JA-Ile mimic coronatine produced by *Pseudomonas syringae*, thereby repressing SA-dependent genes and sensitizing the plant to biotrophic pathogens (Brooks, Bender, & Kunkel, 2005). However, there are examples of synergistic activation of both pathways. Low concentrations of exogenous SA or JA activate both *PR-1* and *PDF1.2*. In addition, some rhizobacteria activate both pathways (Sharifi & Ryu, 2016), which is common in the genus *Bacillus* (Table 1).

High efficiency systemic defence against an array of pathogens has inspired chemical companies to release chemical inducers into the market. SA has negative effects on some plants, but its derivative BTH is more compatible with plants. Commercial formulations of BTH (BION and Actigard), Messenger and Laminarin have been introduced into the market as promising products for the control of plant disease (Leadbeater & Staub, 2007). However, it was later revealed that treating plants with these compounds has a negative impact on plant growth and yield (Goellner & Conrath, 2008). This finding indicates that plants have limited sources of nutrients, and activating thousands of defence genes diverts these resources away from plant growth and fitness. Constitutive activation of the SA pathway in the *cpr1* mutant of *Arabidopsis* reduces rosette volume and seed production (Heidel, Clarke, Antonovics, & Dong, 2004).

Table 1 Signalling Pathways Involved in the Induction of Systemic Resistance by Gram-Positive Rhizobacteria

Strains	Plants	Pathogen	Signalling Pathway: Salicylic Acid	Signalling Pathway: Jasmonic Acid	Signalling Pathway: Ethylene	References
Bacillus pumilus SE34	Tobacco	*Peronospora tabacina*[b]	+	nd	nd	Zhang, Moyne, Reddy, and Kloepper (2002)
B. pumilus SE34	Tomato	*Phytophthora infestans*[n]	0	+	+	Yan et al. (2002)
B. pumilus SE34	*Arabidopsis*	*Pseudomonas syringae* pv. *tomato*[h]	++	+	0	Ryu, Hu, Reddy, and Kloepper (2003)
Paenibacillus alvei K165	*Arabidopsis*	*Verticillium dahliae*[h]	+	0	0	Tjamos, Flemetakis, Paplomatas, and Katinakis (2005)
Paenibacillus polymyxa E681	*Arabidopsis*	*Pseudomonas syringae* pv. *maculicola*[h]	+	+	+	Lee, Farag, et al. (2012)
Streptomyces sp. EN28	*Arabidopsis*	*Fusarium oxysporum*[h]	++	+	+	Conn, Walker, and Franco (2008)
Streptomyces sp. EN28	*Arabidopsis*	*Erwinia carotovora*[n]	0	+	+	Conn et al. (2008)
Bacillus cereus AR156	*Arabidopsis*	*P. syringae* pv. *tomato*[h]	+	+	+	Niu et al. (2011)
B. cereus AR156 Leaf Infiltration	*Arabidopsis*	*P. syringae* pv. *tomato*[h]	+	+	+	Niu et al. (2016)
Bacillus amyloliquefaciens KPS46	*Arabidopsis*	*P. syringae* pv. *tomato*[h]	+	+	+	Buensanteai, Yuen, and Prathuangwong (2009)
B. amyloliquefaciens KPS46	Soya bean	*Xanthomonas axonopodis* pv. *glycines*[h]	+	+	nd	Buensanteai et al. (2009)
Acetoin (Volatile From *Bacillus subtilis* FB17)	*Arabidopsis*	*P. syringae* pv. *tomato*[h]	+	+	+	Rudrappa et al. (2010)

(*Continued*)

Table 1 Signalling Pathways Involved in the Induction of Systemic Resistance by Gram-Positive Rhizobacteria—cont'd

Strains	Plants	Pathogen	Signalling Pathway: Salicylic Acid	Signalling Pathway: Jasmonic Acid	Signalling Pathway: Ethylene	References
3-Pentanol (Volatile From *Bacillus amyloliquefaciens* IN937a)	Pepper	*Xanthomonas axonopodis pv. vesicatoria*[h]	+	+	0	Choi et al. (2014)
3-Pentanol	*Arabidopsis*	*P. syringae* pv. *tomato*[h]	+	+	0	Song, Choi, and Ryu (2015)
Bacillus vallismortis EXTN-1	Tobacco	*Pepper mild mottle virus*[b]	+	+	+	Ahn, Park, and Kim (2002)
B. vallismortis EXTN-1	Potato	Potato virus Y[b]	+	+	nd	Park, Ryu, and Kim (2006)
Bacillus cereus UMAF8564	Cucurbits	*Powdery mildew*[b]	+	0	nd	Garcia-Gutierrez et al. (2013)
Bacillus amyloliquefaciens LJ02	Cucurbits	*Powdery mildew*[b]	+	nd	nd	Li et al. (2015)
Mycosubtilin From *Bacillus* sp.	Grapevine	*Botrytis cinerea*[n]	+	+	0	Farace et al. (2015)
Bacillus subtilis GB03	*Arabidopsis*	*B. cinerea*[n]	+	+	0	Sharifi and Ryu (2016)
B. subtilis GB03	*Arabidopsis*	*E. carotovora*[n]	0	0	+	Ryu et al. (2004)
C13 hexadecane	*Arabidopsis*	*E. carotovora*[n]	0	+	0	Park, Lee, Kloepper, and Ryu (2013)
Bacillus amyloliquefaciens	Tomato	*Ralstonia solanacearum*[n]	0	++	+	Tan et al. (2013)
Bacillus sp. JS	Tobacco	*Rhizoctonia solani*[n]	0	+	nd	Kim et al. (2015)
Surfactin from *Bacillus* FZB42	Lettuce	*R. solani*[n]	0	+	0	Chowdhury et al. (2015)

(+) positive effect, (−) negative effect, (0) no effect, (nd) no data. Superscript bold letters 'b' indicates biotroph, 'h' indicates hemibiotroph and 'n' indicates necrotroph.

Constitutive expression of the JA pathway via constitutive expression of *VSP1* (*cev1*) also has a negative effect on plant growth (Ellis & Turner, 2001).

Interestingly, rhizobacteria activate plant defence responses against several invaders. Not only do these bacteria have no negative effect on plant growth, but they actually improve plant growth and yield significantly. Indeed, plants treated with PGPR do not exhibit noticeable changes in defence gene expression, but, after pathogen challenge, they respond more rapidly and strongly than nontreated plants. This phenomenon is known as priming of defence (Conrath et al., 2006). Priming makes the plant ready for battle. Recent studies shed light on some molecular aspects of priming. Treating plants with priming inducers increased the expression of mitogen-activated protein kinases, MPK3 and MPK6, key signalling molecules in plant defence, in *Arabidopsis* (Conrath, Beckers, Langenbach, & Jaskiewicz, 2015). However, these molecules were in the nonphosphorylated, inactive form. After pathogen attack, plants do not need to spend time synthesizing these signalling molecules; they are present in the proper amounts. Epigenetic regulation is also important in defence priming. Methylation of siRNA, histone modification and chromatin modification are involved in the priming of resistance; see Conrath et al. (2015) for a more comprehensive review of the molecular mechanisms of defence priming. Examples of priming of resistance by PGPR or their metabolites are given in Table 1. As shown in this table, rhizobacteria utilize different signalling pathways with the type of signalling pathway depending on the type of pathosystem. Resistance against biotrophs is commonly mediated by modulation of the SA pathway. Biotrophic pathogens include powdery and downy mildews, viruses and rusts, as well as hemibiotrophs such as *P. syringae*, *Xanthomonas* and *Colletotrichum*. However, modulation of plant defence is not limited to rhizobacteria, pathogens also modulate plant defence as virulence mechanism. Some strains of *P. syringae* produce coronatine, which activates the JA pathway and represses the SA pathway (Brooks et al., 2005). Rhizobacteria also activate the JA pathway against necrotrophic pathogens such as *Rhizoctonia solani*, *Ralstonia solanacearum*, *Pectobacterium caratovorum* and *Botrytis cinerea*. However, *B. cinerea* can activate the SA pathway by producing an exopolysaccharide as virulence factor (El Oirdi et al., 2011). As shown in Table 1, the activation of both JA and SA signalling is common in Gram-positive bacteria.

7. YOU ARE NOT ALONE! – PLANTS CRY FOR HELP FROM RHIZOBACTERIA UNDER INSECT ATTACK

As mentioned earlier, rhizobacteria are attracted by root exudates in the rhizosphere. Plants can change the microbiome by changing their exudates. Some exudates are attractive and some are repulsive to specific groups of microbes. Carvalhais et al. (2015) showed that mutation in two branch of JA signalling in *Arabidopsis* had different exudate profiles and attracted specific groups of microbes. A mutation in *med25* increased the population sizes of *Bacillus* and *Streptomyces*, and a mutation in *myc2* increased the abundance of members of Enterobacteriaceae. Application of exogenous Me-JA also altered the populations of bacteria and Archaea.

Aphid and whitefly infestation of the aerial parts of pepper increases the abundance of Gram-positive PGPR bacteria and fungi, as shown in Fig. 2. The insects activate SA and JA signalling pathways in aboveground and belowground plant parts (Lee, Lee, & Ryu, 2012; Yang et al., 2011). Sometimes, plants challenged by invaders recruit specific biocontrol agents for defence against the invaders, a phenomenon known as cry for help. Aphid infestation in sweet pepper activates the SA and JA pathways. These plants actively recruit *B. subtilis* GB03 to colonize their roots (Lee, Farag, et al., 2012). Moreover, infection with *Pseudomonas syringae* pv. tomato DC3000 in aerial parts of plants increases the population of *Bacillus subtilis* FB17 in roots. Infected *Arabidopsis* plants secrete malic acid into the rhizosphere; this organic acid specifically attracts *B. subtilis* FB17 to the root (Fig. 2). FB17 induces systemic resistance against *P. syringae* pv. tomato DC3000 (Rudrappa, Czymmek, Paré, & Bais, 2008). Organic acids are the main root exudates that alter bacterial abundance. Malic acid affects chemotaxis in bacteria, and fumaric acid improves biofilm formation (Yuan et al., 2015). Cucumber plants inoculated with *Fusarium oxysporum* f.sp. *cucumerinum* exhibit increased fumaric acid secretion and improved colonization by *Bacillus amyloliquefaciens* SQR9 (Yuan et al., 2015).

8. FROM THE PETRI DISH TO THE FIELD: COMMERCIAL APPLICATIONS OF RHIZOBACTERIA

The demand for organic farming and healthy foods has increased in recent years, prompting researchers to develop safer chemical pesticides or biopesticides. Moreover, the cost of agrochemical discovery and the period of their registration have increased rapidly due to stricter legislation, both of

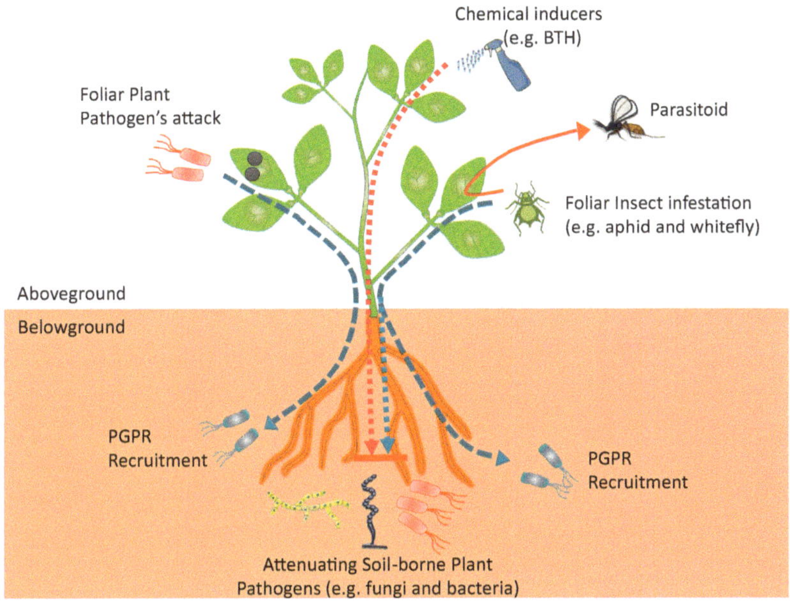

Figure 2 Plants link aboveground and belowground. Infestation with insects such as aphids and whiteflies or even treatment with chemical inducers modulates the levels of defence hormones jasmonic acid and salicylic acid in plants. Whitefly infestation increases the population of Gram-positive plant growth-promoting rhizobacteria (PGPR) [blue colour], and benzothiadiazole (BTH) increases the abundance of Gram-negative PGPR [pink colour] and fungi in the rhizosphere. Aphid damage induces the production of specific volatiles in the plant that function as signals for the attraction of aphid parasitoids. Inoculation of plants with the plant pathogen *Pseudomonas syringae* recruits biocontrol bacterium *Bacillus subtilis* to colonize plant roots and induce resistance against the pathogen.

which favour investment in the production of biopesticides. The annual growth rate of the biopesticide market is more than 15%, but that of chemical pesticides has been declining by 1.5% annually (Thakore, 2006). Biopesticide companies such as Gustafson, Agroquest, Certis, Becker Underwood, BioWorks, ABiTEP GmbH, Marrone Bio Innovation and Prophyta have released effective biopesticides to the market. Agrochemical supercompanies such as Bayer CropScience have also been interested in the biopesticide market. Bayer CropScience acquired Agroquest in 2012 at a price of 425 million US dollars (Kling, 2012). One year later, Bayer also signed an agreement to acquire Prophyta GmbH. BASF, the giant chemical pesticide company, acquired Becker Underwood in 2013.

Nowadays, effective PGPR formulations such as Kodiak, Serenade, Sonata, Yieldshield and Rhizovital are available. Some of these products, such as Serenade and Kodiak, can also induce systemic resistance in plants. Serenade activates JA- and ethylene-dependent genes and increases PAL activity in canola (Lahlali et al., 2013). *B. subtilis* GB03, the active ingredient of Kodiak and Quantum, activates defence against multiple pathogens. Combining two or more PGPR strains is likely to enhance their effects (Raupach & Kloepper, 1998). However, these bacilli strains must have no negative effects on each other and must have different modes of action in the promotion of plant growth or biocontrol of pests and pathogens. BioYield is a mixture of *B. subtilis* GB03 and *Bacillus amyloliquefaciens* IN937a with the carrier chitosan. This formulation induces systemic resistance against two plant pathogens. The type of defence signalling depends on the type of pathogen (Ryu, Murphy, Reddy, & Kloepper, 2007). In addition to combining strains, it is possible to combine PGPR strains with plant activators. The plant activator BTH, a derivative of SA, has a negative effect on plant growth, but treatment of BTH combined with *Bacillus pumilus* INR7 diminishes the negative effect of BTH on plant growth under field conditions (Yi, Yang, & Ryu, 2013).

The increasing information about the molecular biology of plant–PGPR interactions has opened up a new window for improving commercial biocontrol products by genetic engineering. For successful and sustainable application of PGPR, it is important to consider three factors: soil quality, PGPR and plants. The soil should be amended to support physical and biochemical bases for holobiome establishment. Application of compost, vermicompost, sewage sludge, green manure or biochar and good cultural practices will improve soil quality to support the holobiome. Biochar amendment in soil improves soya bean growth either directly or by shifting the microbiome in favour of PGPR (Egamberdieva, Wirth, Behrendt, Abd Allah, & Berg, 2016). On the plant side, root exudation could be engineered. As stated earlier, organic acids such as malic acid and citric acid influence the rhizobacterial population. Overexpressing the pyrophosphatase gene in *Arabidopsis* increased the levels of these two organic acids in exudates. Treating *Arabidopsis* with the toxin coronatine or with flagellin from *Pseudomonas syringae* pv. tomato induced malic acid secretion, which subsequently increased the population size of the biocontrol agent *Bacillus subtilis* (Rudrappa et al., 2008). On the bacterial side, there are several examples of genetic engineering of PGPR to enhance their rhizosphere competence. As noted earlier, overexpressing a trehalose-6-phosphate synthase gene in *R. etli* improved its ability to induce drought stress tolerance in common bean (Suárez et al., 2008). Expressing *Rhizobium leguminosarum*-derived ACC deaminase in *Sinorhizobium meliloti* increased its nodulation in alfalfa (Ma, Charles, & Glick, 2004). Expressing this gene in *Mesorhizobium ciceri* also reduced the negative effects of osmotic stress in chickpea (Brígido, Nascimento, Duan, Glick, & Oliveira, 2013). However, attention should also be paid to the holobiont when applying PGPR or releasing genetically engineered organisms due to the possibility of ecologically disturbing the shaping of each component in the holobiome. To avoid this pitfall, the impact of genetically engineered plants and microbes must be carefully evaluated. Any alteration in a component of the holobiont may have adverse effects on the other components. For example, plants under attack by the insect *Diabrotica virgifera* recruited entomopathogenic nematodes in wild maize, but not in new American hybrid maize lines (Rasmann et al., 2005). Subsequently, the gene corresponding to the determinant of recruitment of entomopathogenic nematodes was introduced into American maize, resulting in protection against insect herbivory. Breeding programmes that take this concept into consideration may greatly improve the control of plant pathogens and insects.

9. PERSPECTIVES

In this review, we explained in detail the process and molecular mechanisms involved in plant–PGPR communications. This advanced knowledge has inspired scientists and has provided a basis for discovery and field application of novel disease-protecting agents. Rhizobacterial determinants are another source of new plant activators. For example, 3-pentanol, a volatile of *Bacillus*, suppresses multiple diseases on pepper in the field (Choi, Song, Yi, & Ryu, 2014). This compound induced the SA and JA signalling pathways. Another bacterial volatile, 2-butanone, induced plant defence against bacteria and aphids under field conditions. This compound also increased the population of beneficial lady bird beetles in the field (Song & Ryu, 2013).

In this new millennium, the strong demand for reducing the use of chemical pesticides and carbon consumption and our concern about environmental side effects must be balanced with the need to produce healthy plants to feed the world's population. In this chapter, we described the many potential applications of a group of rhizosphere bacteria (PGPR), from mechanisms to field applications. We hope that this information broadens our understanding and will be of interest to a general audience. Finally, we again emphasize that we hope we have provided a fresh point of view about the roles of PGPR in plant health based on bacterium–plant communications.

REFERENCES

Abbas-Zadeh, P., Saleh-Rastin, N., Asadi-Rahmani, H., Khavazi, K., Soltani, A., Shoary-Nejati, A. R., & Miransari, M. (2009). Plant growth-promoting activities of fluorescent pseudomonads, isolated from the Iranian soils. *Acta Physiologiae Plantarum, 32*, 281–288.

Adesemoye, A., Torbert, H., & Kloepper, J. (2009). Plant growth-promoting rhizobacteria allow reduced application rates of chemical fertilizers. *Microbial Ecology, 58*, 921–929.

Adesemoye, A. O., Torbert, H. A., & Kloepper, J. W. (2008). Enhanced plant nutrient use efficiency with PGPR and AMF in an integrated nutrient management system. *Canadian Journal of Microbiology, 54*, 876–886.

Ahn, I. P., Park, K., & Kim, C. H. (2002). Rhizobacteria-induced resistance perturbs viral disease progress and triggers defense-related gene expression. *Molecules and Cells, 13*, 302–308.

Bai, X., Todd, C. D., Desikan, R., Yang, Y., & Hu, X. (2012). N-3-oxo-decanoyl-L-homoserine-lactone activates auxin-induced adventitious root formation via hydrogen peroxide-and nitric oxide-dependent cyclic GMP signaling in mung bean. *Plant Physiology, 158*, 725–736.

Bailly, A., Groenhagen, U., Schulz, S., Geisler, M., Eberl, L., & Weisskopf, L. (2014). The inter-kingdom volatile signal indole promotes root development by interfering with auxin signalling. *The Plant Journal, 80*, 758–771.

Bais, H. P., Weir, T. L., Perry, J. L., Gilroy, S., & Vivanco, J. M. (2006). The role of root exudates in rhizosphere interactions with plants and other organisms. *Annual Review of Plant Biology, 57*, 233–266.

Berg, G., & Smalla, K. (2009). Plant species and soil type cooperatively shape the structure and function of microbial communities in the rhizosphere. *FEMS Microbiology Ecology, 68*, 1–13.

Bishnoi, U. (2015). *PGPR Interaction, 75*, 81–113.

Bloemberg, G. V., & Lugtenberg, B. J. (2001). Molecular basis of plant growth promotion and biocontrol by rhizobacteria. *Current Opinion in Plant Biology, 4*, 343–350.

Bonfante, P., & Anca, I.-A. (2009). Plants, mycorrhizal fungi, and bacteria: a network of interactions. *Annual Review of Microbiology, 63*, 363–383.

Brazelton, J. N., Pfeufer, E. E., Sweat, T. A., Gardener, B. B. M., & Coenen, C. (2008). 2, 4-Diacetylphloroglucinol alters plant root development. *Molecular Plant-Microbe Interactions, 21*, 1349–1358.

Brígido, C., Nascimento, F. X., Duan, J., Glick, B. R., & Oliveira, S. (2013). Expression of an exogenous 1-aminocyclopropane-1-carboxylate deaminase gene in *Mesorhizobium* spp. reduces the negative effects of salt stress in chickpea. *FEMS Microbiology Letters, 349*, 46–53.

Brooks, D. M., Bender, C. L., & Kunkel, B. N. (2005). The *Pseudomonas syringae* phytotoxin coronatine promotes virulence by overcoming salicylic acid-dependent defences in *Arabidopsis thaliana*. *Molecular Plant Pathology, 6*, 629–639.

Buensanteai, N., Yuen, G. Y., & Prathuangwong, S. (2009). Priming, signaling, and protein production associated with induced resistance by *Bacillus amyloliquefaciens* KPS46. *World Journal of Microbiology and Biotechnology, 25*, 1275–1286.

Bulgarelli, D., Schlaeppi, K., Spaepen, S., Ver Loren van Themaat, E., & Schulze-Lefert, P. (2013). Structure and functions of the bacterial microbiota of plants. *Annual Review of Plant Biology, 64*, 807–838.

Carvalhais, L. C., Dennis, P. G., Badri, D. V., Kidd, B. N., Vivanco, J. M., & Schenk, P. M. (2015). Linking jasmonic acid signaling, root exudates, and rhizosphere microbiomes. *Molecular Plant-Microbe Interactions, 28*, 1049–1058.

Choi, H. K., Song, G. C., Yi, H. S., & Ryu, C. M. (2014). Field evaluation of the bacterial volatile derivative 3-pentanol in priming for induced resistance in pepper. *Journal of Chemical Ecology, 40*, 882–892.

Chowdhury, S. P., Uhl, J., Grosch, R., Alquéres, S., Pittroff, S., Dietel, K., … Hartmann, A. (2015). Cyclic lipopeptides of *Bacillus amyloliquefaciens* subsp. *plantarum* colonizing the lettuce rhizosphere enhance plant defense responses toward the bottom rot pathogen *Rhizoctonia solani*. *Molecular Plant-Microbe Interactions, 28*, 984–995.

Cohen, A. C., Bottini, R., & Piccoli, P. N. (2008). *Azospirillum brasilense* Sp 245 produces ABA in chemically-defined culture medium and increases ABA content in *Arabidopsis* plants. *Plant Growth Regulation, 54*, 97–103.

Conn, V. M., Walker, A. R., & Franco, C. M. (2008). Endophytic actinobacteria induce defense pathways in *Arabidopsis thaliana*. *Molecular Plant-Microbe Interactions, 21*, 208–218.

Conrath, U., Beckers, G. J., Flors, V., García-Agustín, P., Jakab, G., Mauch, F., … Mauch-Mani, B. (2006). Priming: getting ready for battle. *Molecular Plant-Microbe Interactions, 19*, 1062–1071.

Conrath, U., Beckers, G. J., Langenbach, C. J., & Jaskiewicz, M. R. (2015). Priming for enhanced defense. *Annual Review of Phytopathology, 53*, 97–119.

Contesto, C., Milesi, S., Mantelin, S., Zancarini, A., Desbrosses, G., Varoquaux, F., … Touraine, B. (2010). The auxin-signaling pathway is required for the lateral root response of *Arabidopsis* to the rhizobacterium *Phyllobacterium brassicacearum*. *Planta, 232*, 1455–1470.

Dennis, P. G., Miller, A. J., & Hirsch, P. R. (2010). Are root exudates more important than other sources of rhizo deposits in structuring rhizosphere bacterial communities? *FEMS Microbiology Ecology, 72*, 313–327.

Dodd, I. C., Zinovkina, N. Y., Safronova, V. I., & Belimov, A. A. (2010). Rhizobacterial mediation of plant hormone status. *Annals of Applied Biology, 157*, 361–379.

Egamberdieva, D., Wirth, S., Behrendt, U., Abd Allah, E. F., & Berg, G. (2016). Biochar treatment resulted in a combined effect on soybean growth promotion and a shift in plant growth promoting rhizobacteria. *Frontiers in Microbiology, 7*.

Eichner, M. J. (1990). Nitrous oxide emissions from fertilized soils: summary of available data. *Journal of Environmental Quality, 19*, 272–280.

Eilers, K. G., Lauber, C. L., Knight, R., & Fierer, N. (2010). Shifts in bacterial community structure associated with inputs of low molecular weight carbon compounds to soil. *Soil Biology and Biochemistry, 42*, 896–903.

El Oirdi, M., El Rahman, T. A., Rigano, L., El Hadrami, A., Rodriguez, M. C., Daayf, F., … Bouarab, K. (2011). *Botrytis cinerea* manipulates the antagonistic effects between immune pathways to promote disease development in tomato. *The Plant Cell, 23*, 2405–2421.

Ellis, C., & Turner, J. G. (2001). The *Arabidopsis* mutant cev1 has constitutively active jasmonate and ethylene signal pathways and enhanced resistance to pathogens. *The Plant Cell, 13*, 1025–1033.

Fan, X., Hu, H., Huang, G., Huang, F., Li, Y., & Palta, J. (2015). Soil inoculation with *Burkholderia* sp. LD-11 has positive effect on water-use efficiency in inbred lines of maize. *Plant and Soil, 390*, 337–349.

Farace, G., Fernandez, O., Jacquens, L., Coutte, F., Krier, F., Jacques, P., … Dorey, S. (2015). Cyclic lipopeptides from *Bacillus subtilis* activate distinct patterns of defence responses in grapevine. *Molecular Plant Pathology, 16*, 177–187.

Flajnik, M. F., & Kasahara, M. (2010). Origin and evolution of the adaptive immune system: genetic events and selective pressures. *Nature Reviews Genetics, 11*, 47–59.

Fu, Z. Q., & Dong, X. (2013). Systemic acquired resistance: turning local infection into global defense. *Annual Review of Plant Biology, 64*, 839–863.

Garcia-Gutierrez, L., Zeriouh, H., Romero, D., Cubero, J., De Vicente, A., & Perez-Garcia, A. (2013). The antagonistic strain *Bacillus subtilis* UMAF6639 also confers protection to melon plants against cucurbit powdery mildew by activation of jasmonate- and salicylic acid-dependent defence responses. *Microbial Biotechnology, 6*, 264–274.

Glazebrook, J. (2005). Contrasting mechanisms of defense against biotrophic and necrotrophic pathogens. *Annual Review of Phytopathology, 43*, 205–227.

Glick, B., Cheng, Z., Czarny, J., & Duan, J. (2007). Promotion of plant growth by ACC deaminase-producing soil bacteria. *European Journal of Plant Pathology, 119*, 329–339.

Glick, B. R. (2015). *Beneficial plant-bacterial interactions*. Springer.

Glick, B. R., Jacobson, C. B., Schwarze, M. M., & Pasternak, J. (1994). 1-Aminocyclopropane-1-carboxylic acid deaminase mutants of the plant growth promoting rhizobacterium *Pseudomonas putida* GR12-2 do not stimulate canola root elongation. *Canadian Journal of Microbiology, 40*, 911–915.

Goellner, K., & Conrath, U. (2008). Priming: it's all the world to induced disease resistance. *European Journal of Plant Pathology, 121*, 233–242.

Grichko, V. P., & Glick, B. R. (2001). Flooding tolerance of transgenic tomato plants expressing the bacterial enzyme ACC deaminase controlled by the 35S, rolD or PRB-1b promoter. *Plant Physiology and Biochemistry, 39*, 19–25.

Haas, D., & Defago, G. (2005). Biological control of soil-borne pathogens by fluorescent pseudomonads. *Nature Reviews Microbiology, 3*, 307–319.

Hartmann, A., Rothballer, M., & Schmid, M. (2007). Lorenz Hiltner, a pioneer in rhizosphere microbial ecology and soil bacteriology research. *Plant and Soil, 312*, 7–14.

Hartmann, A., Schmid, M., Tuinen, D. V., & Berg, G. (2008). Plant-driven selection of microbes. *Plant and Soil, 321*, 235–257.

Heidel, A. J., Clarke, J. D., Antonovics, J., & Dong, X. (2004). Fitness costs of mutations affecting the systemic acquired resistance pathway in *Arabidopsis thaliana*. *Genetics, 168*, 2197–2206.

Jewell, M. C., Campbell, B. C., & Godwin, I. D. (2010). Transgenic plants for abiotic stress resistance. In *Transgenic crop plants* (pp. 67–132). Springer.

Katsir, L., Chung, H. S., Koo, A. J., & Howe, G. A. (2008). Jasmonate signaling: a conserved mechanism of hormone sensing. *Current Opinion in Plant Biology, 11*, 428–435.

Kim, J. S., Lee, J., Lee, C. H., Woo, S. Y., Kang, H., Seo, S. G., & Kim, S. H. (2015). Activation of pathogenesis-related genes by the rhizobacterium, *Bacillus* sp. JS, which induces systemic resistance in tobacco plants. *The Plant Pathology Journal, 31*, 195–201.

Kling, J. (2012). Bayer acquisition spotlights biopesticides. *Nature Biotechnology, 30*, 810.

Kloepper, J., Schroth, M., & Miller, T. (1980). Effects of rhizosphere colonization by plant growth-promoting rhizobacteria on potato plant development and yield. *Phytopathology, 70*, 1078–1082.

Lahlali, R., Peng, G., Gossen, B., McGregor, L., Yu, F. Q., Hynes, R. K., … Boyetchko, S. M. (2013). Evidence that the biofungicide Serenade (*Bacillus subtilis*) suppresses clubroot on canola via antibiosis and induced host resistance. *Phytopathology, 103*, 245–254.

Lakshmanan, V. (2015). Root microbiome assemblage is modulated by plant host factors. *Advances in Botanical Research, 75*, 57–79.

Lareen, A., Burton, F., & Schafer, P. (2016). Plant root-microbe communication in shaping root microbiomes. *Plant Molecular Biology, 90*, 575–587.

Leadbeater, A., & Staub, T. (2007). Exploitation of induced resistance: a commercial perspective. *Induced Resistance for Plant Defence: A Sustainable Approach to Crop Protection*, 229–242.
Lee, B., Farag, M. A., Park, H. B., Kloepper, J. W., Lee, S. H., & Ryu, C. M. (2012). Induced resistance by a long-chain bacterial volatile: elicitation of plant systemic defense by a C13 volatile produced by *Paenibacillus polymyxa*. *PLoS One, 7*, e48744.
Lee, B., Lee, S., & Ryu, C. M. (2012). Foliar aphid feeding recruits rhizosphere bacteria and primes plant immunity against pathogenic and non-pathogenic bacteria in pepper. *Annals of Botany, 110*, 281–290.
Lemanceau, P., Expert, D., Gaymard, F., Bakker, P., & Briat, J.-F. (2009). Role of iron in plant–microbe interactions. *Advances in Botanical Research, 51*, 491–549.
Li, Y., Gu, Y., Li, J., Xu, M., Wei, Q., & Wang, Y. (2015). Biocontrol agent *Bacillus amyloliquefaciens* LJ02 induces systemic resistance against cucurbits powdery mildew. *Frontiers in Microbiology, 6*, 883.
Liu, F., Xing, S., Ma, H., Du, Z., & Ma, B. (2013). Cytokinin-producing, plant growth-promoting rhizobacteria that confer resistance to drought stress in *Platycladus orientalis* container seedlings. *Applied Microbiology and Biotechnology, 97*, 9155–9164.
Lundberg, D. S., Lebeis, S. L., Paredes, S. H., Yourstone, S., Gehring, J., Malfatti, S., … Dangl, J. L. (2012). Defining the core *Arabidopsis thaliana* root microbiome. *Nature, 488*, 86–90.
Ma, W., Charles, T. C., & Glick, B. R. (2004). Expression of an exogenous 1-aminocyclopropane-1-carboxylate deaminase gene in *Sinorhizobium meliloti* increases its ability to nodulate alfalfa. *Applied and Environmental Microbiology, 70*, 5891–5897.
Mazzola, M., Funnell, D. L., & Raaijmakers, J. (2004). Wheat cultivar-specific selection of 2, 4-diacetylphloroglucinol-producing fluorescent *Pseudomonas* species from resident soil populations. *Microbial Ecology, 48*, 338–348.
McCormick, M. K., Lee Taylor, D., Juhaszova, K., Burnett, R. K., Whigham, D. F., & O'neill, J. P. (2012). Limitations on orchid recruitment: not a simple picture. *Molecular Ecology, 21*, 1511–1523.
Miransari, M. (2011). Soil microbes and plant fertilization. *Applied Microbiology and Biotechnology, 92*, 875–885.
Miri, M., Janakirama, P., Held, M., Ross, L., & Szczyglowski, K. (2016). Into the root: how cytokinin controls rhizobial infection. *Trends in Plant Science, 21*, 178–186.
Newman, M.-A., Sundelin, T., Nielsen, J. T., & Erbs, G. (2013). MAMP (microbe-associated molecular pattern) triggered immunity in plants. *Frontiers in Plant Science, 4*.
Niu, D., Wang, X., Wang, Y., Song, X., Wang, J., Guo, J., & Zhao, H. (2016). *Bacillus cereus* AR156 activates PAMP-triggered immunity and induces a systemic acquired resistance through a NPR1-and SA-dependent signaling pathway. *Biochemical and Biophysical Research Communications, 469*, 120–125.
Niu, D. D., Liu, H. X., Jiang, C. H., Wang, Y. P., Wang, Q. Y., Jin, H. L., & Guo, J. H. (2011). The plant growth-promoting rhizobacterium *Bacillus cereus* AR156 induces systemic resistance in *Arabidopsis thaliana* by simultaneously activating salicylate- and jasmonate/ethylene-dependent signaling pathways. *Molecular Plant-Microbe Interactions, 24*, 533–542.
Ongena, M., & Jacques, P. (2008). *Bacillus* lipopeptides versatile weapon for plant disease control. *Trends in Microbiology, 16*, 115–125.
Onofre-Lemus, J., Hernández-Lucas, I., Girard, L., & Caballero-Mellado, J. (2009). ACC (1-aminocyclopropane-1-carboxylate) deaminase activity, a widespread trait in *Burkholderia* species, and its growth-promoting effect on tomato plants. *Applied and Environmental Microbiology, 75*, 6581–6590.
Ortíz-Castro, R., Contreras-Cornejo, H. A., Macías-Rodríguez, L., & López-Bucio, J. (2009). The role of microbial signals in plant growth and development. *Plant Signaling and Behavior, 4*, 701–712.

Ortíz-Castro, R., Valencia-Cantero, E., & López-Bucio, J. (2008). Plant growth promotion by *Bacillus megaterium* involves cytokinin signaling. *Plant Signaling and Behavior, 3*, 263–265.

Park, H. B., Lee, B., Kloepper, J. W., & Ryu, C. M. (2013). One shot-two pathogens blocked: exposure of *Arabidopsis* to hexadecane, a long chain volatile organic compound, confers induced resistance against both *Pectobacterium carotovorum* and *Pseudomonas syringae*. *Plant Signaling and Behavior, 8*, e24619.

Park, K. S., Ryu, K. R., & Kim, Y. K. (2006). *Bacillus vallismortis* strain EXTN-1 mediated systemic resistance against potato virus Y and X in the field. *The Plant Pathology Journal, 22*, 360–363.

Partida-Martinez, L. P. P., & Heil, M. (2011). The microbe-free plant: fact or artifact? *Frontiers in Plant Science, 2*, 100.

Paul, D., & Lade, H. (2014). Plant-growth-promoting rhizobacteria to improve crop growth in saline soils: a review. *Agronomy for Sustainable Development, 34*, 737–752.

Pieterse, C. M., Van Der Does, D., Zamioudis, C., Leon-Reyes, A., & Van Wees, S. C. (2012). Hormonal modulation of plant immunity. *Annual Review of Cell and Developmental Biology, 28*, 489–521.

Pieterse, C. M., Zamioudis, C., Berendsen, R. L., Weller, D. M., Van Wees, S. C., & Bakker, P. A. (2014). Induced systemic resistance by beneficial microbes. *Annual Review of Phytopathology, 52*, 347–375.

Pinton, R., Varanini, Z., & Nannipieri, P. (2007). *The rhizosphere: Biochemistry and organic substances at the soil-plant interface*. CRC press.

Porcel, R., Zamarreño, M., García-Mina, J. M., & Aroca, R. (2014). Involvement of plant endogenous ABA in *Bacillus megaterium* PGPR activity in tomato plants. *BMC Plant Biology, 14*, 1.

Raaijmakers, J. M., & Mazzola, M. (2012). Diversity and natural functions of antibiotics produced by beneficial and plant pathogenic bacteria. *Annual Review of Phytopathology, 50*, 403–424.

Rasmann, S., Kollner, T. G., Degenhardt, J., Hiltpold, I., Toepfer, S., Kuhlmann, U., … Turlings, T. C. (2005). Recruitment of entomopathogenic nematodes by insect-damaged maize roots. *Nature, 434*, 732–737.

Raupach, G. S., & Kloepper, J. W. (1998). Mixtures of plant growth-promoting rhizobacteria enhance biological control of multiple cucumber pathogens. *Phytopathology, 88*, 1158–1164.

Rodrigues, A. C., Bonifacio, A., De Araujo, F. F., Lira, M. A., Jr., & Figueiredo, M. D. V. B. (2015). *Azospirillum* sp. as a challenge for agriculture. In *Bacterial metabolites in sustainable agroecosystem* (pp. 29–51). Springer.

Rodríguez-Salazar, J., Suárez, R., Caballero-Mellado, J., & Iturriaga, G. (2009). Trehalose accumulation in *Azospirillum brasilense* improves drought tolerance and biomass in maize plants. *FEMS Microbiology Letters, 296*, 52–59.

Rose, M. T., Phuong, T. L., Nhan, D. K., Cong, P. T., Hien, N. T., & Kennedy, I. R. (2014). Up to 52% N fertilizer replaced by biofertilizer in lowland rice via farmer participatory research. *Agronomy for Sustainable Development, 34*, 857–868.

Rudrappa, T., Biedrzycki, M. L., Kunjeti, S. G., Donofrio, N. M., Czymmek, K. J., Paré, P. W., & Bais, H. P. (2010). The rhizobacterial elicitor acetoin induces systemic resistance in *Arabidopsis thaliana*. *Communicative and Integrative Biology, 3*, 130–138.

Rudrappa, T., Czymmek, K. J., Paré, P. W., & Bais, H. P. (2008). Root-secreted malic acid recruits beneficial soil bacteria. *Plant Physiology, 148*, 1547–1556.

Ruzzi, M., & Aroca, R. (2015). Plant growth-promoting rhizobacteria act as biostimulants in horticulture. *Scientia Horticulturae, 196*, 124–134.

Ryu, C., Murphy, J. F., Reddy, M., & Kloepper, J. W. (2007). A two-strain mixture of rhizobacteria elicits induction of systemic resistance against *Pseudomonas syringae* and

Cucumber mosaic virus coupled to promotion of plant growth on *Arabidopsis thaliana*. *Journal of Microbiology and Biotechnology, 17*, 280.
Ryu, C. M., Farag, M. A., Hu, C. H., Reddy, M. S., Kloepper, J. W., & Pare, P. W. (2004). Bacterial volatiles induce systemic resistance in *Arabidopsis*. *Plant Physiology, 134*, 1017–1026.
Ryu, C. M., Farag, M. A., Hu, C. H., Reddy, M. S., Wei, H. X., Paré, P. W., & Kloepper, J. W. (2003). Bacterial volatiles promote growth in *Arabidopsis*. *Proceedings of the National Academy of Sciences of the United States of America, 100*, 4927–4932.
Ryu, C. M., Hu, C. H., Reddy, M., & Kloepper, J. W. (2003). Different signaling pathways of induced resistance by rhizobacteria in *Arabidopsis thaliana* against two pathovars of *Pseudomonas syringae*. *New Phytologist, 160*, 413–420.
Salomon, M. V., Bottini, R., De Souza Filho, G. A., Cohen, A. C., Moreno, D., Gil, M., & Piccoli, P. (2014). Bacteria isolated from roots and rhizosphere of *Vitis vinifera* retard water losses, induce abscisic acid accumulation and synthesis of defense-related terpenes in in vitro cultured grapevine. *Physiologia Plantarum, 151*, 359–374.
Schisler, D., Slininger, P., Behle, R., & Jackson, M. (2004). Formulation of *Bacillus* spp. for biological control of plant diseases. *Phytopathology, 94*, 1267–1271.
Shah, J., & Zeier, J. (2014). Long-distance communication and signal amplification in systemic acquired resistance. *Induced Plant Responses to Microbes and Insects, 23*.
Sharifi, R., Ahmadzadeh, M., Sharifi-Tehrani, A., & Talebi-Jahromi, K. (2010). Pyoverdine production in *Pseudomonas fluorescens* UTPF5 and its association with suppression of common bean damping off caused by *Rhizoctonia solani* (Kuhn). *Journal of Plant Protection Research, 50*, 72–78.
Sharifi, R., & Ryu, C. M. (2016). Are bacterial volatile compounds poisonous odors to a fungal pathogen *Botrytis cinerea*, alarm signals to *Arabidopsis* seedlings for eliciting induced resistance, or both? *Frontiers in Microbiology*, 7, 196.
Shi, C.-L., Park, H.-B., Lee, J. S., Ryu, S., & Ryu, C.-M. (2010). Inhibition of primary roots and stimulation of lateral root development in *Arabidopsis thaliana* by the rhizobacterium *Serratia marcescens* 90–166 is through both auxin-dependent and-independent signaling pathways. *Molecules and Cells, 29*, 251–258.
Shi, S. J., Richardson, A. E., O'callaghan, M., DeAngelis, K. M., Jones, E. E., Stewart, A., … Condron, L. M. (2011). Effects of selected root exudate components on soil bacterial. *FEMS Microbiology Ecology, 77*, 600–610.
Singh, J. S., Pandey, V. C., & Singh, D. P. (2011). Efficient soil microorganisms: a new dimension for sustainable agriculture and environmental development. *Agriculture, Ecosystems and Environment, 140*, 339–353.
Song, G. C., Choi, H. K., & Ryu, C. M. (2015). Gaseous 3-pentanol primes plant immunity against a bacterial speck pathogen, *Pseudomonas syringae* pv. tomato via salicylic acid and jasmonic acid-dependent signaling pathways in *Arabidopsis*. *Frontiers in Plant Science, 6*, 821.
Song, G. C., & Ryu, C. M. (2013). Two volatile organic compounds trigger plant self-defense against a bacterial pathogen and a sucking insect in cucumber under open field conditions. *International Journal of Molecular Sciences, 14*, 9803–9819.
Suárez, R., Wong, A., Ramírez, M., Barraza, A., Orozco Mdel, C., Cevallos, M. A., … Iturriaga, G. (2008). Improvement of drought tolerance and grain yield in common bean by overexpressing trehalose-6-phosphate synthase in rhizobia. *Molecular Plant-Microbe Interactions, 21*, 958–966.
Sugio, A., Dubreuil, G., Giron, D., & Simon, J.-C. (2014). Plant–insect interactions under bacterial influence: ecological implications and underlying mechanisms. *Journal of Experimental Botany*, eru435.
Tan, S., Dong, Y., Liao, H., Huang, J., Song, S., Xu, Y., & Shen, Q. (2013). Antagonistic bacterium *Bacillus amyloliquefaciens* induces resistance and controls the bacterial wilt of tomato. *Pest Management Science, 69*, 1245–1252.

Tewari, S., & Arora, N. K. (2014). Multifunctional exopolysaccharides from *Pseudomonas aeruginosa* PF23 involved in plant growth stimulation, biocontrol and stress amelioration in sunflower under saline conditions. *Current Microbiology, 69*, 484–494.

Thakore, Y. (2006). The biopesticide market for global agricultural use. *Industrial Biotechnology, 2*, 194–208.

Tjamos, S. E., Flemetakis, E., Paplomatas, E. J., & Katinakis, P. (2005). Induction of resistance to *Verticillium dahliae* in *Arabidopsis thaliana* by the biocontrol agent K-165 and pathogenesis-related proteins gene expression. *Molecular Plant-Microbe Interactions, 18*, 555–561.

Upadhyay, S., Singh, J., & Singh, D. (2011). Exopolysaccharide-producing plant growth-promoting rhizobacteria under salinity condition. *Pedosphere, 21*, 214–222.

Van Der Heijden, M. G., Bardgett, R. D., & Van Straalen, N. M. (2008). The unseen majority: soil microbes as drivers of plant diversity and productivity in terrestrial ecosystems. *Ecology Letters, 11*, 296–310.

Van Loon, L. (2007). Plant responses to plant growth-promoting rhizobacteria. *European Journal of Plant Pathology, 119*, 243–254.

Weinert, N., Piceno, Y., Ding, G.-C., Meincke, R., Heuer, H., Berg, G., … Smalla, K. (2011). PhyloChip hybridization uncovered an enormous bacterial diversity in the rhizosphere of different potato cultivars: many common and few cultivar-dependent taxa. *FEMS Microbiology Ecology, 75*, 497–506.

Xie, H., Pasternak, J., & Glick, B. R. (1996). Isolation and characterization of mutants of the plant growth-promoting rhizobacterium *Pseudomonas putida* GR12-2 that overproduce indoleacetic acid. *Current Microbiology, 32*, 67–71.

Yan, Z., Reddy, M. S., Ryu, C. M., Mcinroy, J. A., Wilson, M., & Kloepper, J. W. (2002). Induced systemic protection against tomato late blight elicited by plant growth-promoting rhizobacteria. *Phytopathology, 92*, 1329–1333.

Yang, J. W., Yi, H.-S., Kim, H., Lee, B., Lee, S., Ghim, S.-Y., & Ryu, C.-M. (2011). Whitefly infestation of pepper plants elicits defence responses against bacterial pathogens in leaves and roots and changes the below-ground microflora. *Journal of Ecology, 99*, 46–56.

Yi, H. S., Yang, J. W., & Ryu, C. M. (2013). ISR meets SAR outside: additive action of the endophyte *Bacillus pumilus* INR7 and the chemical inducer, benzothiadiazole, on induced resistance against bacterial spot in field-grown pepper. *Frontiers in Plant Science, 4*, 122.

Yuan, J., Zhang, N., Huang, Q., Raza, W., Li, R., Vivanco, J. M., & Shen, Q. (2015). Organic acids from root exudates of banana help root colonization of PGPR strain *Bacillus amyloliquefaciens* NJN-6. *Scientific Reports, 5*, 13438.

Zhang, H., Kim, M. S., Krishnamachari, V., Payton, P., Sun, Y., Grimson, M., … Paré, P. W. (2007). Rhizobacterial volatile emissions regulate auxin homeostasis and cell expansion in *Arabidopsis. Planta, 226*, 839–851.

Zhang, H., Sun, Y., Xie, X., Kim, M. S., Dowd, S. E., & Pare, P. W. (2009). A soil bacterium regulates plant acquisition of iron via deficiency-inducible mechanisms. *The Plant Journal, 58*, 568–577.

Zhang, H., Xie, X., Kim, M. S., Kornyeyev, D. A., Holaday, S., & Pare, P. W. (2008). Soil bacteria augment *Arabidopsis* photosynthesis by decreasing glucose sensing and abscisic acid levels in planta. *The Plant Journal, 56*, 264–273.

Zhang, S., Moyne, A.-L., Reddy, M., & Kloepper, J. W. (2002). The role of salicylic acid in induced systemic resistance elicited by plant growth-promoting rhizobacteria against blue mold of tobacco. *Biological Control, 25*, 288–296.

Zilber-Rosenberg, I., & Rosenberg, E. (2008). Role of microorganisms in the evolution of animals and plants: the hologenome theory of evolution. *FEMS Microbiology Reviews, 32*, 723–735.

CHAPTER SEVEN

Effector-Mediated Communication of Filamentous Plant Pathogens With Their Hosts

E. Gaulin

Laboratoire de Recherche en Sciences Végétales, Université de Toulouse, CNRS, UPS, France
E-mail: gaulin@lrsv.ups-tlse.fr

Contents

Abstract

Pathogenic fungi and oomycetes can establish intimate associations with plants. These interactions underlie a molecular dialogue that leads to the successful colonization of host tissues. Major questions driving research in plant pathology these last decades are how pathogenic microorganisms circumvent preformed or induced defences and how pathogens manipulate host physiology to promote virulence. One key actor in this dialogue relies on a class of molecules secreted by pathogens termed effectors. Effectors perturb host processes by targeting a variety of host functions either in the apoplast or in the cytosol of host cells. This chapter focuses on fungal and oomycetal cytoplasmic effectors by reviewing methods to predict and to characterize effectors as well as their activities and role during infection. We provide current knowledge regarding their evolution and their putative role in the shaping of plant-associated microbial communities.

Advances in Botanical Research, Volume 82
ISSN 0065-2296
http://dx.doi.org/10.1016/bs.abr.2016.09.003

1. INTRODUCTION

Plants are exposed to a huge diversity of microorganisms in both the rhizosphere and the phyllosphere (Hacquard et al., 2016; Whipps, Hand, Pink, & Bending, 2008). Not all of these microorganisms can infect plant tissues as plants possess preexistent physical and chemical barriers as well as an inducible immune system that enables them to identify and avoid potential invaders (De Coninck, Timmermans, Vos, Cammue, & Kazan, 2015; Jones & Dangl, 2006). Some fungal and oomycete species have developed ways to circumvent these barriers and to modify host structures and physiology, transforming plant tissues into a suitable niche to obtain carbon sources and ensure the completion of their cycle. They are so successful at this, that diseases caused by these microorganisms stand as the most important on agricultural and natural ecosystems worldwide (Fisher et al., 2012; Gladieux et al., 2015). Representative examples are the oomycete *Phytophthora infestans*, responsible for the potato and tomato blight disease (Fry et al., 2015), or *Magnaporthe oryzae*, a common pathogen of rice recently detected in Asia on wheat and causing up to 90% yield losses in more than 15,000 hectares in Bangladesh (Callaway, 2016).

To infect their host, fungi and oomycetes employ several infection strategies. A key pathogenic strategy resides on the secretion of molecules termed 'effectors' which change host structures and target host functions involved in specific cellular processes, thereby promoting plant susceptibility and disease (Hogenhout, Van der Hoorn, Terauchi, & Kamoun, 2009; Pel & Pieterse, 2013). Various definitions of effectors have been proposed, the simplest being a 'molecule produced by microorganisms exerting an effect on plant cells'. Most described effectors are proteins, but it must be noted that other metabolites, toxins (Amselem et al., 2011; Arias, Theumer, Mary, & Rubinstein, 2012; Collemare et al., 2008) and small RNA (Weiberg et al., 2013), have also been described as effectors. While effectors are also common in plant pathogenic bacteria (Macho, 2016), nematodes (Quentin, Abad, & Favery, 2013) and symbiotic fungi (Kloppholz, Kuhn, & Requena, 2011; Plett et al., 2014), here we focus on proteinaceous effectors from filamentous plant-colonizing pathogens (i.e., fungi and oomycetes).

The initial step of the plant immune responses is generally triggered by molecules essential to the microorganisms referred to Pathogen/Microbe-Associated Molecular Pattern (PAMP/MAMP). These essential components (e.g., flagellin from bacteria, chitin from fungi) are recognized by plant membrane-localized pattern recognition receptors (PRR), which induce the first level of defence called Pathogen-Triggered Immunity (PTI). To successfully facilitate infection pathogens must be able to counteract PTI and thereby secrete numerous effectors, leading to Effector-Triggered Susceptibility (ETS) (Jones & Dangl, 2006). Effectors are delivered either into the apoplast (apoplastic effectors) or inside host cells (cytoplasmic/cytosolic effectors) where they can be addressed to specific cellular compartments and organelles (Lo Presti et al., 2015). While substantial progress has been made in identifying effectors, plant targets and biochemical activities through different experimental methods, the exact mechanism allowing their secretion and uptake into host cells remains a milestone in effector research. The effector repertoire within pathogenic species is constantly shaped as the result of the coevolution with host plants, which impose high selective pressures (Lo Presti et al., 2015). Indeed, some plants can develop mechanisms to detect effector activities and to activate immune responses that lead to resistance or Effector-Triggered Immunity (ETI). In such context the detection of the effector (formerly named avirulence protein) causes the arrest of infection (Hein, Gilroy, Armstrong, & Birch, 2009; Jones & Dangl, 2006). The mechanisms by which effectors evolve might explain how filamentous pathogens surpass plant resistance and the emergence of novel disease outbreaks.

Understanding the phytopathogenic success of filamentous species requires understanding effector biology, an entire research area aimed at identifying their activities in host cells, their host targets and their evolution within genomes. In this chapter we focus on cytoplasmic effectors of fungi and oomycetes in the context of plant susceptibility by reviewing their role in promoting infection. We give insights into their origin and evolution and discuss experimental methods that have been applied to predict candidate effectors in species. We further present current knowledge regarding the mechanisms allowing their delivery inside the plant cell. Finally, we invite to address emerging questions regarding effectors in the shaping of plant-associated microbial communities (Fig. 1).

2. COMPUTATIONAL METHODS TO PREDICT EFFECTORS

The accurate computational identification of effectors is a challenging and evolving exercise. A key criterion in pipelines is the presence of classical

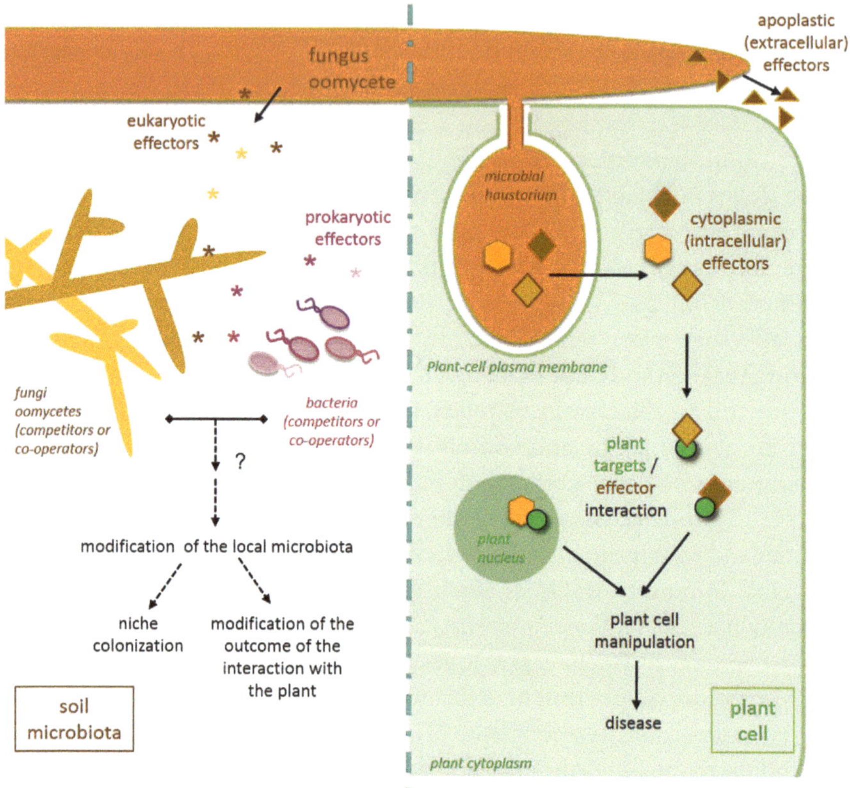

Figure 1 ***How effectors of filamentous plant pathogens influence the outcome of the interaction with the host plant.*** The right part of the diagram represents a plant cell. During host infection numerous filamentous pathogens develop intracellular feeding structures called 'haustoria' into the plant cell that remain separated from the host cytoplasm by an extrahaustorial membrane. To establish a successful infection, intracellular pathogens have evolved cytoplasmic/intracellular effectors that target plant components to manipulate host physiology and facilitate disease development. The left part of the figure illustrates the local biota surrounding plant roots. Effectors produced by pathogenic microorganisms can negatively or positively impact the microbial community leading to a local modification of the microbiome. In this case, effectors are used to combat other microbes (competitors) or to stimulate the other microorganisms that can favour its development (cooperators). Finally these intereukaryotic or eukaryotic–prokaryotic 'dialogues' could facilitate niche colonization by the filamentous pathogen and impact the outcome of the interaction with the host plant.

secretion leaders (signal peptide) and the absence of transmembrane domains as, conceptually, effectors must be secreted to reach the host–microbe interface and exert their activities. Nevertheless, depending on the used software, signal peptides are not always in silico predicted, and their absence does not necessarily indicate that the protein is not secreted (Sperschneider, Williams, Hane, Singh, Taylor, 2015). Frequently, effector proteins are also defined as small secreted proteins (SSPs, arbitrary less than 300 amino acids) (De Carvalho et al., 2016; Gan et al., 2013; Lorrain, Hecker, & Duplessis, 2015; Pellegrin, Morin, Martin, & Veneault-Fourrey, 2015; Sperschneider, Dodds, et al., 2015). Another parameter frequently used to predict an effector is the absence of functional domain or detectable orthologous genes outside the genus, although effectors harbouring putative functional domain or orthologs outside the genus have been found (Dagdas et al., 2016; Ramirez-Garces et al., 2016; Takahara et al., 2016). Based on these different criterions various bioinformatics pipelines have been developed (Sonah, Deshmukh, & Belanger, 2016; Sperschneider et al., 2016).

In silico amino acid sequence inspection of set of secreted proteins has evidenced the presence of commonly occurring amino acid residues or motif for some of them. This was typically the case in oomycete proteins known to be perceived by plant cells (i.e., encoded by avirulence genes) in which a conserved N-terminal RxLR (arginine, any amino acid, leucine, arginine) motif was found few amino acids after the signal peptide (Rehmany et al., 2005). This motif was used as a bait for computational search of putative effectors and for the definition of several RxLR effector families in various oomycetes (Anderson, Deb, Fedkenheuer, & McDowell, 2015; Haas et al., 2009; Jiang, Tripathy, Govers, & Tyler, 2008; Kemen et al., 2011; Pel et al., 2014). Similar to RxLR effectors, the presence of the consensus motif (L/Q/FLAK) led to define the oomycete Crinklers (CRNs) effector family (Gaulin et al., 2008; Haas et al., 2009; Jiang et al., 2013; Stam, Jupe, et al., 2013; Stam, Motion, Boevink, Huitema, 2013; Torto et al., 2003). In addition to these motifs, a conserved N-terminal 'Y/F/WxC' motif was also found after mining proteins encoded by genes expressed in specialized infectious organs as haustoria of powdery mildew of barley, stem rust and leaf rust of wheat (Godfrey et al., 2010; Pedersen et al., 2012). Moreover, seven amino acids ('RSIDELD') at the C-terminus have been detected in candidate effectors of the root endophyte *Piriformospora indica* (Zuccaro et al., 2011).

Other sequence signatures such as the presence of repeated-regions (Mesarich, Bowen, Hamiaux, & Templeton, 2015; Saunders et al., 2012), prediction of the 2D/3D structure (Guyon, Balague, Roby, & Raffaele, 2014; Win et al., 2012), signs of diversifying selection (Pedersen et al., 2012; Sperschneider et al., 2014; Win et al., 2012) could also pinpoint, in a set of secreted proteins, their potential role as effector. In addition genome-wide expression profiling or transcriptomic data from infected samples provided tools to selected subset of candidate effectors that are expressed during host colonization, or during specific stages of the infection (Gaulin et al., 2008; Hacquard et al., 2011; Kleemann et al., 2012; Mogga et al., 2016).

Finally what we learned from those studies is that, generally, 'secretomes' represent 5–10% of the predicted proteome of plant-colonizing fungi and oomycetes (Haas et al., 2009; Lo Presti et al., 2015; Pellegrin et al., 2015; Sonah et al., 2016; Zuccaro et al., 2011), and effector candidates can represent a large part of this secretome. Computational analyses must be completed by transcriptional data (RNA-Seq) to identify effector genes highly induced during pathogenesis.

3. FUNCTIONAL METHODS TO VALIDATE IN SILICO PREDICTION OF EFFECTORS

The computational pipelines usually result in lists of hundreds of putative effectors in a given pathogen, requiring the use of high-throughput screening methods (effectoromics) to validate effector candidates and assess their function.

Transient expression in plants of the microbial effector candidates, using *Agrobacterium tumefaciens*, has become a method of choice widely used either for phenotyping plant cell responses (e.g., cell death, Ma, Lukasik, Gawehns, & Takken, 2012; Petre et al., 2016; Torto et al., 2003) or for subcellular localization of the microbial protein using fluorescent tags (Petre et al., 2015). The delivery of effector candidates directly into the host cytosol using the type-three secretion system (TTSS) of *Pseudomonas syringae* and the assessment of plant cell responses (e.g., callose deposition) constitute another screening method (Fabro et al., 2011). Protoplast transient expression system to check the activity of putative effectors is also an effective method (Chen et al., 2013; Zheng et al., 2014). Proteomics studies on infected tissues can be used to confirm in silico predictions (Gawehns et al., 2015; Gupta et al., 2015). When possible, large-scale gene disruption of predicted secreted proteins that are expressed during host infection can facilitate effector validation (Brefort et al., 2014; Saitoh et al., 2012). However, the major obstacle of this approach is gene redundancy. The CRISPR/Cas9 system (Doudna & Charpentier, 2014) that allows genome editing in various organisms including filamentous fungi and oomycetes (Arazoe et al., 2015; Fang & Tyler, 2016), offers a new opportunity to functionally characterize effector candidates.

Altogether these 'effectoromics' methods can bridge in silico prediction and functional identification of candidates. Nevertheless the bottleneck to accurately assess effector proteins resides in the absence of experimental systems to prioritize and/or screen large sets of candidate effectors in 'natural' host–microorganism interaction. However the continuous development of new techniques will facilitate the characterization of effectors.

4. LOCATION OF EFFECTORS ENCODING GENES WITHIN THE MICROBIAL GENOMES

Genomic plasticity enables organisms to adapt to environmental changes. It is particular relevant in the context of plant–microbe interactions where pathogens evolution may be driven by host selection pressure (Gladieux et al., 2014; Raffaele & Kamoun, 2012). In the genome of filamentous plant pathogens, plastic regions are generally located within the core chromosome/genome or reside in additional dispensable chromosomes (Balesdent et al., 2013; Coleman et al., 2009; Han, Liu, Benny, Kistler, & VanEtten, 2001; Hatta et al., 2002; Ma et al., 2010). For instance, smut fungi harbour genomic clusters of secreted proteins which are dispersed throughout the genome and coregulated during infection (Schirawski et al., 2010). Deletion of several of those clusters in the maize pathogen *Ustilago maydis* led to a reduced virulence on maize (Kämper et al., 2006). By contrast, gene-sparse regions enriched in genes encoding effectors (i.e., RxLR) occur in *Phytophthora* genomes (Haas et al., 2009; Raffaele, Win, Cano, & Kamoun, 2010). Indeed 2000 gene-sparse regions containing less than 10 genes and associated with transposable elements (TEs) have been reported in *P. infestans*. These regions are fast-evolving as compared to the core genome that carries essential genes that encode basal functions (Raffaele, Win, et al., 2010). This observation led to the concept of the 'two-speed genome' model in which pathogen genomes have a bipartite organization to favour pathogen genome evolution while keeping stability of the core genome (Dong, Raffaele, & Kamoun, 2015; Raffaele & Kamoun, 2012). The ascomycete *Leptosphaeria maculans* also harbour a bipartite genome structure that comprises alternating gene-rich GC-isochores and gene-poor AT-isochores (Gout et al., 2006; Rouxel et al., 2011). The AT-rich blocks contain TEs and putative effector gene families that present similar expression patterns (Grandaubert et al., 2014; Rouxel et al., 2011). In *Verticillium dahliae*, effectors are compartmentalized in lineage-specific genomic regions (LS) that correspond to about 5% of the genome and that are highly variable between *V. dahliae* strains (de Jonge et al., 2013). LS regions are enriched in retrotransposons and repetitive sequence elements. Fungal effectors can be located in conditionally dispensable chromosomes. For instance, *Fusarium oxysporum* dispensable genomic regions are enriched in genes important for pathogenicity, some specifically expressed during plant infection such as the SIX (Secreted in Xylem) effector family (Rep et al., 2004). Losses of these supernumerary chromosomes affect fungus pathogenicity (Vlaardingerbroek, Beerens, Schmidt, Cornelissen, & Rep, 2016), and transfer of these regions between strains of *F. oxysporum* may convert a nonpathogenic strain to a pathogenic variant (Ma et al., 2010; Vlaardingerbroek, Beerens, Rose, et al., 2016). In *F. oxysporum* 74% of the TEs are found in these supernumerary chromosomes (Ma et al., 2010). While the wheat pathogen *Zymoseptoria tritici* genome harbours a structure-like *Fusarium* sp. with chromosomes that are dispensable, the eight additional chromosomes do not reveal enrichment in secreted proteins or putative candidate effectors (Schirawski et al., 2010). The functional relevance of these accessory chromosomes is still unknown but they exhibit a higher proportion of repetitive elements and fewer genes as compared to the core chromosomes (Goodwin et al., 2011). All these data sustain the view that compartmentalization of effector genes in the genomes of filamentous pathogens might facilitate adaptation to hosts.

5. EFFECTOR ORIGIN AND EVOLUTION

The observation that numerous genomes of filamentous phytopathogens comprise a core part with essential conserved genes, and lineage-specific effector genes, addresses the question of the origin of the latter (O'Connell et al., 2012; Raffaele & Kamoun, 2012; Schirawski et al., 2010; Spanu et al., 2010). Indeed filamentous plant pathogens are engaged in coevolutionary arms races with their hosts and therefore constantly deploy modified or extended effector repertoires to optimize their virulence (Raffaele & Kamoun, 2012). Thereby these effector genes usually show accelerated mutation rates compared to genes of the core genome not involved in the interaction with a host. Upon genome sequencing, signs of positive selection that increase variability were found in numerous loci and repeat-rich regions of fungal and oomycete effectors (Aguileta et al., 2012; Hacquard et al., 2013; Raffaele, Farrer, et al., 2010; Sharma, Xia, Riess, Bauer, & Thines, 2015; Stergiopoulos et al., 2014; Win & Kamoun, 2008). Moreover as previously mentioned, effectors are usually found in regions enriched in repetitive sequences and TEs. TEs accelerate genome evolution by creating highly dynamic genomics regions that harbour accelerated diversification, and by leading to gene insertions/deletions or duplications (Dutheil et al., 2016; Grandaubert et al., 2014; Iribarren, Pascuan, Soto, & Ayub, 2015; Raffaele & Kamoun, 2012; Seidl & Thomma, 2014). TEs can constitute a large fraction of fungal genome (e.g., 64% of *Blumeria graminis* sp. *hordei* genome, Spanu et al., 2010) and TE-rich regions are frequently associated to highly packed chromatin (heterochromatin) that restricts gene transcription and therefore expression of neighbouring genes like effectors (Qutob, Chapman, & Gijzen, 2013; Soyer et al., 2014). Candidate effectors located in genomic clusters of smut fungi are predicted to evolve by tandem duplication and suspected to originate from an uncharacterized transposable element (Dutheil et al., 2016). Other fungal effectors have also probably evolved from ancestral transposable elements like the EKA family of effectors of the powdery mildew fungus *Blumeria graminis* (with 1350 homologues in the genome) that probably originated from degenerative copies of Class I Line retrotransposons (Amselem et al., 2015). In oomycetes the expanded Crinklers (CRNs) protein family was predicted to arise from bacterial transposons and might retain their transposase-like function thereby facilitating genome diversification (Zhang, Burroughs, Vidal, Iyer, & Aravind, 2016). Thus, rapid evolution of repertoires of effector genes could involve activation of transposition mechanisms.

Acquisition of novel genes by horizontal gene transfer (HGT) can also be a powerful source of adaption. Recently, it has been shown that the secretome of oomycetes is subjected to HGT events (Richards et al., 2011), as exemplified by a genomic comparison of 23 oomycetes genomes that identified secreted proteins, some with plant pathogenicity function, in 69% of the putative HGT families (48 families) (Savory, Leonard, & Richards, 2015). Not only the predicted distribution of some CRN effector proteins (e.g., CRN13 family) and the conservation of DNA-binding capacity of CRN13 of the oomycete *Aphanomyces euteiches* but also of the amphibian pathogen *Batrachochytrium dendrobatidis* is also consistent with HGT (Ramirez-Garces et al., 2016; Sun, Yang, Kosch, Summers, & Huang, 2011). Bacterial and fungal donors of horizontally transferred genes have been identified in oomycetes (Misner, Blouin, Leonard, Richards, & Lane, 2015). In fungi, predicted HGT that concern genes associated with virulence have been described (Friesen et al., 2006; Khaldi & Wolfe, 2011; Soanes & Richards, 2014). Surprisingly HGT events between fungi seem to be more frequent than expected as illustrated by the characterization of more than 90 genes being putatively transferred between *Magnaporthales* and *Colletotrichum*, one-third corresponding to carbohydrate-active enzymes (CAZymes) (Qiu, Cai, Luo, Bhattacharya, & Zhang, 2016).

6. EFFECTOR SECRETION AND TRANSLOCATION INSIDE HOST CELLS

Experimental data show that effector secretion is induced as soon as the pathogens are in contact with their host. For example, the two effectors ChEC6 and ChEC36 of the hemibiotroph *Colletotrichum higginsianum* accumulate inside the appressorial pore before the penetration peg breaks the plant barrier. Then both effectors are detected in the region of the plant cell wall beneath the appressorial pore (Kleemann et al., 2012). The PSE1 effector of *Phytophthora parasitica* is also located at the appressorial surface during the penetration process (Evangelisti et al., 2013). After penetration filamentous pathogens usually develop intercellular invading hyphae and numerous intracellular feeding structures called haustoria (Fig. 1). The haustorium is completely enclosed by the host plasma membrane establishing an interface for molecular exchanges between the microbe and the host (Giraldo & Valent, 2013; Lo Presti et al., 2015). This organization is reminiscent of the double-membrane barrier that separates the intracellular parasite-like *Plasmodium faciparum* from the red blood cytosol (Spielmann, Montagna, Hecht, & Matuschewski, 2012). During the blood stage of infection, numerous parasite-encoded effectors cross the parasite plasma membrane and transfer into erythrocyte cytosol through a *Plasmodium*-encoded translocon machinery (Elsworth et al., 2014; de Koning-Ward et al., 2009; Mesen-Ramirez et al., 2016). In *Phytophthora*, an early study showed that the Nterminus of the RxLR Avr3 effector was required for transfer into potato cells (Whisson et al., 2007) and was functionally interchangeable with the Nterminus 'PEXEL' motif of effector proteins from *Plasmodium* that are transferred into red blood cells (Bhattacharjee et al., 2006). Therefore the conserved Nterminus of RxLR effectors was proposed to play a role in trafficking. The observations that Ntermini of AeCRN5 effector protein from the legume pathogen *A. euteiches* and CRN8 from *P. infestans* favour protein transport into plant cells (Schornack et al., 2010) further sustain the concept that host targeting is mediated by the Nterminal domain of RxLR and CRNs effectors. Numerous studies aimed at demonstrating the 'addressing' function of the RxLR/CRN Ntermini sequences and at establishing whether the host cell entry is pathogen-dependent or not (Boddey et al., 2016; Dou et al., 2008; Kale et al., 2010; Petre & Kamoun, 2014; Schornack et al., 2010; Song et al., 2015; Wawra et al., 2012, 2013). Finally how oomycete effectors target the host cell is still an open question.

In fungi while pioneering works provided direct evidence by immunolocalization of translocation of *Melamspora lini* effector (e.g., AvrM) within the plant cytosol during host infection (Kemen et al., 2005), the precise mechanism of effector entry into host cells also remains challenging. The characterization of 3D structure of AvrM led to the identification of functionally important surface regions (e.g., hydrophobic regions) for cell entry (Ve et al., 2013), suggesting that uptake motif may exist but may not be detectable at the primary amino acid sequence level. During the early steps of host colonization, cell-imaging showed that numerous effectors of the *Colletotrichum orbiculare*, a pathogen that do not form haustoria, are localized in an interface around the biotrophic primary hyphal neck. Nevertheless translocation into the host cell could not be observed (Irieda et al., 2014, Irieda, Ogawa, & Takano, 2016). A focal accumulation of effectors is also observed in *M. oryzae*, which produces apoplastic and cytoplasmic effectors that follow two distinct secretory pathways (Giraldo et al., 2013). For cytoplasmic effectors, the proteins accumulate preferentially in a 'biotrophic interfacial complex' (BIC), which develops at the tip of the primary hyphae that newly invaded host cell, before being transferred within the plant cells (Y. Dong et al., 2015; Khang et al., 2010; Zhang & Xu, 2014). While two cytoplasmic effectors (Tin2 and Cmu1, Djamei et al., 2011; Tanaka et al., 2014) have been functionally characterized in the biotrophic pathogen *U. maydis*, numerous attempts based on distinct experimental approaches did not allow the identification of a mechanism underlying effector uptake into host cells (Tanaka et al., 2015).

In the pathogen a communication between the hyphal tip and the nucleus can also take place to sustain the production and secretion of effectors during infection. In *U. maydis* this communication is mediated by long-distance retrograde motility of early endosomes (EE). Indeed upon perception of the host by the pathogen, penetrating hyphae connect this information to motile EEs that transport the signal to the nucleus through retrograde mechanism and subsequently trigger effector gene expression (Bielska et al., 2014). Surprisingly EEs from *U. maydis* move back and forth along microtubules and this motility is crucial for pathogenicity, suggesting that EEs orchestrate effector production (Bielska et al., 2014; Chen, Ebbole, & Wang, 2015; Higuchi & Steinberg, 2015).

7. EFFECTOR FUNCTIONS

Numerous fungal and oomycetes effectors have been shown to be important for the virulence but our comprehension of effector precise function is still in its infancy. Nevertheless, the combinatorial use of different technologies has made it possible to gain interesting insights into the role of effectors. A major concept when invoking the role in virulence of cytoplasmic effectors is the modulation of plant defences. Several effectors prevent the activation of signalling events following the initial perception of pathogen proteins at the outer surface of the host cell. The RxLR Avr3a of *P. infestans* interacts and stabilizes CMPG1, a U-Box ubiquitin E3 ligase whose degradation positively regulates cell death activated by other effectors (Gilroy et al., 2011). The RxLR Pi02860 of *P. infestans* interacts with the plant susceptibility factor NLR1 (Cullin-3-associated ubiquitin E3 ligase) and suppresses host cell death only when activated by the oomycete INF1 necrosis effector (Yang et al., 2016). The targeting of components of the ubiquitination complex is also a strategy deployed by the fungus *M. oryzae* on rice. This pathogen produces AvrPiz-t to inhibit the host RING E3 ubiquitin ligase APIP6. The loss of APIP6 function is accompanied by the loss of defence gene expression (Park et al., 2012). Effectors can modulate immune responses via their direct action with host transcriptional factors (TFs) as exemplified with the RxLR Pi03192, which associates with NAC TFs in the endoplasmic reticulum blocking their relocalization to the host nucleus (McLellan et al., 2013).

Other cytoplasmic effectors target the metabolism of the host defence-related phytohormones salicylic acid (SA), jasmonic acid (JA) or ethylene (ET). For instance, SA level in maize is modulated by the smut fungus *U. maydis* during host infection. Indeed the pathogen secretes the Cmu1 chorismate mutase effector into maize cells which redirects a metabolic pathway to reduce level of SA (Djamei et al., 2011). Similarly the effector Pslsc1 required for virulence of *Phytophthora sojae* triggered a decrease in the amount of SA (Liu et al., 2014). In addition, RxLR44, a host nuclear-cytoplasmic effector of the powdery mildew *Hyaloperonospora arabidopsidis*, induces expression of JA/ET-responsive genes and suppresses expression of SA genes by degrading the plant MED19a, a subunit of the transcriptional complex mediator (Caillaud et al., 2013). The PSE1 RxLR effector from *P. parasitica* induces aberrant root development probably through modification of host auxin efflux carriers distribution (Evangelisti et al., 2013).

Nucleomodulins effectors known in prokaryotic pathogens as effectors delivered to the host nucleus to subvert host defences by interfering with transcription, chromatin modelling, RNA splicing or DNA replication and repair (Bierne & Cossart, 2012) have also been characterized in oomycetes. Indeed PsCRN108 from *P. sojae* binds regulatory elements of promoters of heat-shock proteins (HSP), probably to displace heat shock TFs, thereby inhibiting HSP gene expression and promoting *P. sojae* virulence (Song et al., 2015). CRN12_997 from *P. capsici* binds to the tomato TF SlTCP14-2 affecting its association with chromatin and possibly the activation of defence genes (Stam, Jupe, et al., 2013; Stam, Motion, et al., 2013). The effector CRN13 from *A. euteiches* (also expressed by various pathogenic oomycetes and by the amphibian pathogen *B. dendrobatidis*) exhibits an HNH-like endonuclease motif interacting with host DNA, resulting in DNA-damages and activation of DNA repair machinery of the host cell (Ramirez-Garces et al., 2016). Targeting of host nuclear machinery is also illustrated by two plant nuclear-localized effectors of *P. sojae* (PSR1 and PSR2) which suppress RNA-silencing, a mechanism known to play a major role in plant defence (Qiao, Shi, Zhai, Hou, & Ma, 2015; Xiong et al., 2014).

Another strategy is the manipulation of the plant secretory pathway. While Avr-Pii from *M. oryzae* interacts with two rice Exo70 proteins probably implicated in plant exocytosis system (Fujisaki et al., 2015), the RxLR effector AVR1 of *P. infestans* interacts with the protein Sec5, an exocytosis component, to affect trafficking of host vesicles and thereby plant immune responses (Du, Mpina, Birch, Bouwmeester, & Govers, 2015). In addition, PexRD54 from *P. infestans* stimulates production of host autophagosomes to benefit the microorganism, by interacting with the autophagy ATG8 protein (Dagdas et al., 2016).

All the findings presented above show how diverse are the functions of effectors of filamentous pathogens involving targeting essential components of host cell machinery.

8. ROLE OF EFFECTORS BEYOND PLANT PATHOGENESIS

Microorganisms live amid extremely diverse and rich microbial environments like, for example, in close proximity of roots (rhizosphere) or on the surface of plant aerial organs (phyllosphere) (Bai et al., 2015; Edwards et al., 2015; Hacquard, 2016; Lundberg et al., 2012; Ploch, Rose, Bass, & Bonkowski, 2016). To survive in these environments one can expect that they produce molecules enabling them to fight against microbial competitors. Recently, it has become apparent that bacteria secrete effectors not only targeted to plant hosts but also against neighbouring prokaryotic cells. These effectors enable niche colonization within a polymicrobial environment (Berg, Grube, Schloter, & Smalla, 2014; Lareen, Burton, & Schafer, 2016) as exemplified with the soil bacterium *A. tumefaciens* that produces an antibacterial DNAse effector to degrade bacterial competitors (Ma, Hachani, Lin, Filloux, & Lai, 2014). Thereby the production of DNAses increases the fitness of the pathogen during host colonization and allows bacteria to thrive in a competitive environment. This example illustrates that pathogens not only have to deal with plant defence responses during host colonization but also have to compete with other microbes. In the recent years, emerging studies revealed that root microbial communities (microbiome) surrounding plants are structured and form interconnected complex (Agler et al., 2016; van der Heijden & Hartmann, 2016). Field experiments and in vitro studies revealed that microbe—microbe interactions (i.e., symbiotic, commensal or parasitic interactions) in addition to abiotic factors and plant genotypes shape the microbiome diversity (Hacquard et al., 2015; Lareen et al., 2016; Vallance et al., 2009). The possibility that effectors from filamentous plant pathogens may have a role in intereukaryotic or eukaryotic—prokaryotic dialogues by targeting microbiome members rather than plant physiology is an emerging question (Fig. 1). One can presume for instance that chitin-binding effectors that protect parasites against plant chitinases (van den Burg, Harrison, Joosten, Vervoort, & de Wit, 2006; van Esse, Bolton, Stergiopoulos, de Wit, & Thomma, 2007; Marshall et al., 2011) may also protect them against microbial enzymes produced by members of the microbiome, such as *Trichoderma* and *Pythium* that produce chitinases (El-Katatny et al., 2001; Naglot et al., 2015; Rao, Raju, & Ravisankar, 2015). The characterization of CRNs effectors that showed features with polymorphic prokaryotic toxins in terms of function and evolution (Zhang et al., 2016) supports this idea. CRN effectors initially detected in oomycetes and fungi are now predicted in a wide range of eukaryotes including free-living eukaryotes (Zhang et al., 2016), suggesting that CRNs may play a function not only during plant colonization but also in microbe—microbe interactions by damaging for instance DNA of microbial competitors (Ramirez-Garces et al., 2016). In this context, future studies regarding microbial effectors should no longer consider these molecules solely in the context of an interaction between one pathogen and one plant, i.e., between two 'individuals'. Effectors should not only be considered as pathogenicity genes, but also as general actors enabling filamentous pathogens to establish in a new environmental niche by eliminating potential competitors.

9. CONCLUDING REMARKS

In summary, here we highlight several research topics that concern the molecular dialogue established between plants and filamentous pathogens with an emphasis on cytoplasmic effectors. Despite numerous advances in the field it remains to clarify how these effectors are addressed into the plant cells. In addition, describing which effector(s) and effector targets pathogens use to reprogram the host cell represent an important perspective for future research. Finally one theme that emerges is that effectors not only play a role during plant infection but also act as weapons to combat other microbes in a polymicrobial environment and facilitate niche colonization.

ACKNOWLEDGEMENTS

We acknowledge funding by the ANR (ANR-JCJC-12-JSV6-0004-01). This work has been done at the LRSV (Laboratoire de Recherche en Sciences Végétales UMR5546), part of the French Laboratory of Excellence 'TULIP' (ANR-10-LABX-41; ANR-11-IDEX-0002-02). We apologize to all colleagues whose original work has not been cited owing to space constraints.

REFERENCES

Agler, M. T., Ruhe, J., Kroll, S., Morhenn, C., Kim, S. T., Weigel, D., & Kemen, E. M. (2016). Microbial hub taxa link host and abiotic factors to plant microbiome variation. *PLoS Biology, 14*, e1002352.

Aguileta, G., Lengelle, J., Chiapello, H., Giraud, T., Viaud, M., Fournier, E., … Gout, L. (2012). Genes under positive selection in a model plant pathogenic fungus, *Botrytis. Infection, Genetics and Evolution*, 987–996 (Amsterdam: Elsevier).

Amselem, J., Cuomo, C. A., van Kan, J. A., Viaud, M., Benito, E. P., Couloux, A., … Dickman, M. (2011). Genomic analysis of the necrotrophic fungal pathogens *Sclerotinia sclerotiorum* and *Botrytis cinerea. PLoS Genetics*, e1002230.

Amselem, J., Vigouroux, M., Oberhaensli, S., Brown, J. K., Bindschedler, L. V., Skamnioti, P., … Sacristan, S. (2015). Evolution of the EKA family of powdery mildew avirulence-effector genes from the ORF 1 of a LINE retrotransposon. *BMC Genomics, 16*, 917–924.

Anderson, R. G., Deb, D., Fedkenheuer, K., & McDowell, J. M. (2015). Recent progress in RxLR effector research. *Molecular Plant-Microbe Interactions, 28*, 1063–1072.

Arazoe, T., Miyoshi, K., Yamato, T., Ogawa, T., Ohsato, S., Arie, T., & Kuwata, S. (2015). Tailor-made CRISPR/Cas system for highly efficient targeted gene replacement in the rice blast fungus. *Biotechnology and Bioengineering, 112*, 2543–2549.

Arias, S. L., Theumer, M. G., Mary, V. S., & Rubinstein, H. R. (2012). Fumonisins: probable role as effectors in the complex interaction of susceptible and resistant maize hybrids and *Fusarium verticillioides. Journal of Agricultural and Food Chemistry, 60*, 5667–5675.

Bai, Y., Muller, D. B., Srinivas, G., Garrido-Oter, R., Potthoff, E., Rott, M., ... Schulze-Lefert, P. (2015). Functional overlap of the *Arabidopsis* leaf and root microbiota. *Nature, 528*, 364–369.

Balesdent, M. H., Fudal, I., Ollivier, B., Bally, P., Grandaubert, J., Eber, F., ... Rouxel, T. (2013). The dispensable chromosome of *Leptosphaeria maculans* shelters an effector gene conferring avirulence towards *Brassica rapa*. *New Phytologist, 198*, 887–898.

Berg, G., Grube, M., Schloter, M., & Smalla, K. (2014). Unraveling the plant microbiome: looking back and future perspectives. *Frontiers in Microbiology, 5*, 148–152.

Bhattacharjee, S., Hiller, N. L., Liolios, K., Win, J., Kanneganti, T. D., Young, C., ... Haldar, K. (2006). The malarial host-targeting signal is conserved in the Irish potato famine pathogen. *PLoS Pathogens, 2*, e50.

Bielska, E., Higuchi, Y., Schuster, M., Steinberg, N., Kilaru, S., Talbot, N. J., & Steinberg, G. (2014). Long-distance endosome trafficking drives fungal effector production during plant infection. *Nature Communications, 5*, 5097.

Bierne, H., & Cossart, P. (2012). When bacteria target the nucleus: the emerging family of nucleomodulins. *Cellular Microbiology, 14*, 622–633.

Boddey, J. A., O'Neill, M. T., Lopaticki, S., Carvalho, T. G., Hodder, A. N., Nebl, T., ... Cowman, A. F. (2016). Export of malaria proteins requires co-translational processing of the PEXEL motif independent of phosphatidylinositol-3-phosphate binding. *Nature Communications, 7*, 10470.

Brefort, T., Tanaka, S., Neidig, N., Doehlemann, G., Vincon, V., & Kahmann, R. (2014). Characterization of the largest effector gene cluster of *Ustilago maydis*. *PLoS Pathogens, 10*, e1003866.

van den Burg, H. A., Harrison, S. J., Joosten, M. H., Vervoort, J., & de Wit, P. J. (2006). *Cladosporium fulvum* Avr4 protects fungal cell walls against hydrolysis by plant chitinases accumulating during infection. *Molecular Plant-Microbe Interactions, 19*, 1420–1430.

Caillaud, M. C., Asai, S., Rallapalli, G., Piquerez, S., Fabro, G., & Jones, J. D. (2013). A downy mildew effector attenuates salicylic acid-triggered immunity in *Arabidopsis* by interacting with the host mediator complex. *PLoS Biology, 11*, e1001732.

Callaway, E. (2016). Devastating wheat fungus appears in Asia for first time. *Nature, 532*, 421–422.

Chen, S., Songkumarn, P., Venu, R. C., Gowda, M., Bellizzi, M., Hu, J., ... Wang, G. L. (2013). Identification and characterization of in planta-expressed secreted effector proteins from *Magnaporthe oryzae* that induce cell death in rice. *Molecular Plant-Microbe Interactions, 26*, 191–202.

Chen, X., Ebbole, D. J., & Wang, Z. (2015). The exocyst complex: delivery hub for morphogenesis and pathogenesis in filamentous fungi. *Current Opinion in Plant Biology, 28*, 48–54.

Coleman, J. J., Rounsley, S. D., Rodriguez-Carres, M., Kuo, A., Wasmann, C. C., Grimwood, J., ... VanEtten, H. D. (2009). The genome of *Nectria haematococca*: contribution of supernumerary chromosomes to gene expansion. *PLoS Genetics, 5*, e1000618.

Collemare, J., Pianfetti, M., Houlle, A. E., Morin, D., Camborde, L., Gagey, M. J., ... Bohnert, H. U. (2008). *Magnaporthe grisea* avirulence gene ACE1 belongs to an infection-specific gene cluster involved in secondary metabolism. *New Phytologist, 179*, 196–208.

Dagdas, Y. F., Belhaj, K., Maqbool, A., Chaparro-Garcia, A., Pandey, P., Petre, B., ... Bozkurt, T. O. (2016). An effector of the Irish potato famine pathogen antagonizes a host autophagy cargo receptor. *eLife, 5*, e10856.

De Carvalho, M. C., Nascimento, L. C., Darben, L. M., Polizel-Podanosqui, A. M., Lopes-Caitar, V. S., Qi, M., ... Marcelino-Guimaraes, F. C. (2016). Prediction of the in planta *P. pachyrhizi* secretome and potential effector families. *Molecular Plant Pathology*. http://dx.doi.org/10.1111/mpp.12405.

De Coninck, B., Timmermans, P., Vos, C., Cammue, B. P., & Kazan, K. (2015). What lies beneath: belowground defense strategies in plants. *Trends in Plant Science, 20*, 91–101.
Djamei, A., Schipper, K., Rabe, F., Ghosh, A., Vincon, V., Kahnt, J., … Kahmann, R. (2011). Metabolic priming by a secreted fungal effector. *Nature, 478*, 395–398.
Dong, S., Raffaele, S., & Kamoun, S. (2015). The two-speed genomes of filamentous pathogens: waltz with plants. *Current Opinion in Genetics & Development, 35*, 57–65.
Dong, Y., Li, Y., Zhao, M., Jing, M., Liu, X., Liu, M., … Zhang, Z. (2015). Global genome and transcriptome analyses of *Magnaporthe oryzae* epidemic isolate 98-06 uncover novel effectors and pathogenicity-related genes, revealing gene gain and lose dynamics in genome evolution. *PLoS Pathogens, 11*, e1004801.
Dou, D., Kale, S. D., Wang, X., Jiang, R. H., Bruce, N. A., Arredondo, F. D., … Tyler, B. M. (2008). RXLR-mediated entry of *Phytophthora sojae* effector Avr1b into soybean cells does not require pathogen-encoded machinery. *Plant Cell, 20*, 1930–1947.
Doudna, J. A., & Charpentier, E. (2014). Genome editing. The new frontier of genome engineering with CRISPR-Cas9. *Science, 346*, 1258096.
Du, Y., Mpina, M. H., Birch, P. R., Bouwmeester, K., & Govers, F. (2015). *Phytophthora infestans* RXLR effector AVR1 interacts with exocyst component Sec5 to manipulate plant immunity. *Plant Physiology, 169*, 1975–1990.
Dutheil, J. Y., Mannhaupt, G., Schweizer, G., Christian, M. K. S., Munsterkotter, M., Guldener, U., … Kahmann, R. (2016). A tale of genome compartmentalization: the evolution of virulence clusters in smut fungi. *Genome Biology and Evolution, 8*, 681–704.
Edwards, J., Johnson, C., Santos-Medellin, C., Lurie, E., Podishetty, N. K., Bhatnagar, S., … Sundaresan, V. (2015). Structure, variation, and assembly of the root-associated microbiomes of rice. *Proceedings of the National Academy of Sciences of the United States of America, 112*, 911–920.
El-Katatny, M. H., Gudelj, M., Robra, K. H., Elnaghy, M. A., & Gubitz, G. M. (2001). Characterization of a chitinase and an endo-beta-1,3-glucanase from *Trichoderma harzianum* Rifai T24 involved in control of the phytopathogen *Sclerotium rolfsii*. *Applied Microbiology and Biotechnology, 56*, 137–143.
Elsworth, B., Matthews, K., Nie, C. Q., Kalanon, M., Charnaud, S. C., Sanders, P. R., … de Koning-Ward, T. F. (2014). PTEX is an essential nexus for protein export in malaria parasites. *Nature, 511*, 587–591.
van Esse, H. P., Bolton, M. D., Stergiopoulos, I., de Wit, P. J., & Thomma, B. P. (2007). The chitin-binding *Cladosporium fulvum* effector protein Avr4 is a virulence factor. *Molecular Plant-Microbe Interactions, 20*, 1092–1101.
Evangelisti, E., Govetto, B., Minet-Kebdani, N., Kuhn, M. L., Attard, A., Ponchet, M., … Gourgues, M. (2013). The *Phytophthora parasitica* RXLR effector penetration-specific effector 1 favours *Arabidopsis thaliana* infection by interfering with auxin physiology. *New Phytologist, 199*, 476–489.
Fabro, G., Steinbrenner, J., Coates, M., Ishaque, N., Baxter, L., Studholme, D. J., … Jones, J. D. (2011). Multiple candidate effectors from the oomycete pathogen *Hyaloperonospora arabidopsidis* suppress host plant immunity. *PLoS Pathogens, 7*, e1002348.
Fang, Y., & Tyler, B. M. (2016). Efficient disruption and replacement of an effector gene in the oomycete *Phytophthora sojae* using CRISPR/Cas9. *Molecular Plant Pathology, 17*, 127–139.
Fisher, M. C., Henk, D. A., Briggs, C. J., Brownstein, J. S., Madoff, L. C., McCraw, S. L., & Gurr, S. J. (2012). Emerging fungal threats to animal, plant and ecosystem health. *Nature, 484*, 186–194.
Friesen, T. L., Stukenbrock, E. H., Liu, Z., Meinhardt, S., Ling, H., Faris, J. D., … Oliver, R. P. (2006). Emergence of a new disease as a result of interspecific virulence gene transfer. *Nature Genetics, 38*, 953–956.

Fry, W. E., Birch, P. R., Judelson, H. S., Grünwald, N. J., Danies, G., Everts, K. L., ... Smart, C. D. (2015). Five reasons to consider *Phytophthora infestans* a reemerging pathogen. *Phytopathology, 105*, 966–981.

Fujisaki, K., Abe, Y., Ito, A., Saitoh, H., Yoshida, K., Kanzaki, H., ... Terauchi, R. (2015). Rice Exo70 interacts with a fungal effector, AVR-Pii, and is required for AVR-Pii-triggered immunity. *The Plant Journal, 83*, 875–887.

Gan, P., Ikeda, K., Irieda, H., Narusaka, M., O'Connell, R. J., Narusaka, Y., ... Shirasu, K. (2013). Comparative genomic and transcriptomic analyses reveal the hemibiotrophic stage shift of *Colletotrichum fungi*. *New Phytologist, 197*, 1236–1249.

Gaulin, E., Madoui, M. A., Bottin, A., Jacquet, C., Mathé, C., Couloux, A., ... Dumas, B. (2008). Transcriptome of *Aphanomyces euteiches*: new oomycete putative pathogenicity factors and metabolic pathways. *PLoS One, 3*, e1723.

Gawehns, F., Ma, L., Bruning, O., Houterman, P. M., Boeren, S., Cornelissen, B. J., ... Takken, F. L. (2015). The effector repertoire of *Fusarium oxysporum* determines the tomato xylem proteome composition following infection. *Frontiers in Plant Science, 6*, 967.

Gilroy, E. M., Taylor, R. M., Hein, I., Boevink, P., Sadanandom, A., & Birch, P. R. (2011). CMPG1-dependent cell death follows perception of diverse pathogen elicitors at the host plasma membrane and is suppressed by *Phytophthora infestans* RXLR effector AVR3a. *New Phytologist, 190*, 653–666.

Giraldo, M. C., Dagdas, Y. F., Gupta, Y. K., Mentlak, T. A., Yi, M., Martinez-Rocha, A. L., ... Valent, B. (2013). Two distinct secretion systems facilitate tissue invasion by the rice blast fungus *Magnaporthe oryzae*. *Nature Communications, 4*, 1996.

Giraldo, M. C., & Valent, B. (2013). Filamentous plant pathogen effectors in action. *Nature Reviews Microbiology, 11*, 800–814.

Gladieux, P., Feurtey, A., Hood, M. E., Snirc, A., Clavel, J., Dutech, C., ... Giraud, T. (2015). The population biology of fungal invasions. *Molecular Ecology, 24*, 1969–1986.

Gladieux, P., Ropars, J., Badouin, H., Branca, A., Aguileta, G., de Vienne, D. M., ... Giraud, T. (2014). Fungal evolutionary genomics provides insight into the mechanisms of adaptive divergence in eukaryotes. *Molecular Ecology, 23*, 753–773.

Godfrey, D., Bohlenius, H., Pedersen, C., Zhang, Z., Emmersen, J., & Thordal-Christensen, H. (2010). Powdery mildew fungal effector candidates share N-terminal Y/F/WxC-motif. *BMC Genomics, 11*, 317.

Goodwin, S. B., M'barek, S. B., Dhillon, B., Wittenberg, A. H., Crane, C. F., Hane, J. K., ... Kema, G. H. J. (2011). Finished genome of the fungal wheat pathogen *Mycosphaerella graminicola* reveals dispensome structure, chromosome plasticity, and stealth pathogenesis. *PLoS Genetics, 7*, e1002070.

Gout, L., Fudal, I., Kuhn, M. L., Blaise, F., Eckert, M., Cattolico, L., ... Rouxel, T. (2006). Lost in the middle of nowhere: the AvrLm1 avirulence gene of the Dothideomycete *Leptosphaeria maculans*. *Molecular Microbiology, 60*, 67–80.

Grandaubert, J., Lowe, R. G., Soyer, J. L., Schoch, C. L., Van de Wouw, A. P., Fudal, I., ... Rouxel, T. (2014). Transposable element-assisted evolution and adaptation to host plant within the *Leptosphaeria maculans-Leptosphaeria biglobosa* species complex of fungal pathogens. *BMC Genomics, 15*, 891.

Gupta, R., Lee, S. E., Agrawal, G. K., Rakwal, R., Park, S., Wang, Y., & Kim, S. T. (2015). Understanding the plant-pathogen interactions in the context of proteomics-generated apoplastic proteins inventory. *Frontiers in Plant Science, 6*, 352.

Guyon, K., Balague, C., Roby, D., & Raffaele, S. (2014). Secretome analysis reveals effector candidates associated with broad host range necrotrophy in the fungal plant pathogen *Sclerotinia sclerotiorum*. *BMC Genomics, 15*, 336.

Haas, B. J., Kamoun, S., Zody, M. C., Jiang, R. H., Handsaker, R. E., Cano, L. M., … Nusbaum, C. (2009). Genome sequence and analysis of the Irish potato famine pathogen *Phytophthora infestans*. *Nature, 461*, 393–398.

Hacquard, S. (2016). Disentangling the factors shaping microbiota composition across the plant holobiont. *New Phytologist, 209*, 454–457.

Hacquard, S., Garrido-Oter, R., Gonzalez, A., Spaepen, S., Ackermann, G., Lebeis, S., … Schulze-Lefert, P. (2015). Microbiota and host nutrition across plant and animal kingdoms. *Cell Host & Microbe, 17*, 603–616.

Hacquard, S., Kracher, B., Hiruma, K., Munch, P. C., Garrido-Oter, R., Thon, M. R., … O'Connell, R. J. (2016). Survival trade-offs in plant roots during colonization by closely related beneficial and pathogenic fungi. *Nature Communications, 7*, 11362.

Hacquard, S., Kracher, B., Maekawa, T., Vernaldi, S., Schulze-Lefert, P., & Ver Loren van Themaat, E. (2013). Mosaic genome structure of the barley powdery mildew pathogen and conservation of transcriptional programs in divergent hosts. *Proceedings of the National Academy of Sciences of the United States of America, 110*, 2219–2228.

Hacquard, S., Petre, B., Frey, P., Hecker, A., Rouhier, N., & Duplessis, S. (2011). The poplar-poplar rust interaction: insights from genomics and transcriptomics. *Journal of Pathogens, 2011*, 716041.

Han, Y., Liu, X., Benny, U., Kistler, H. C., & VanEtten, H. D. (2001). Genes determining pathogenicity to pea are clustered on a supernumerary chromosome in the fungal plant pathogen *Nectria haematococca*. *The Plant Journal, 25*, 305–314.

Hatta, R., Ito, K., Hosaki, Y., Tanaka, T., Tanaka, A., Yamamoto, M., … Tsuge, T. (2002). A conditionally dispensable chromosome controls host-specific pathogenicity in the fungal plant pathogen *Alternaria alternata*. *Genetics, 161*, 59–70.

van der Heijden, M. G., & Hartmann, M. (2016). Networking in the plant microbiome. *PLoS Biology, 14*, e1002378.

Hein, I., Gilroy, E. M., Armstrong, M. R., & Birch, P. R. (2009). The zig-zag-zig in oomycete-plant interactions. *Molecular Plant Pathology, 10*, 547–562.

Higuchi, Y., & Steinberg, G. (2015). Early endosomes motility in filamentous fungi: how and why they move. *Fungal Biology Reviews, 29*, 1–6.

Hogenhout, S. A., Van der Hoorn, R. A., Terauchi, R., & Kamoun, S. (2009). Emerging concepts in effector biology of plant-associated organisms. *Molecular Plant-Microbe Interactions, 22*, 115–122.

Iribarren, M. J., Pascuan, C., Soto, G., & Ayub, N. D. (2015). Genetic analysis of environmental strains of the plant pathogen *Phytophthora capsici* reveals heterogeneous repertoire of effectors and possible effector evolution via genomic island. *FEMS Microbiology Letters, 362*, 22.

Irieda, H., Maeda, H., Akiyama, K., Hagiwara, A., Saitoh, H., Uemura, A., … Takano, Y. (2014). *Colletotrichum orbiculare* secretes virulence effectors to a biotrophic interface at the primary hyphal neck via exocytosis coupled with SEC22-meiated traffic. *Plant Cell, 26*, 2265–2281.

Irieda, H., Ogawa, S., & Takano, Y. (2016). Focal effector accumulation in a biotrophic interface at the primary invasion sites of *Colletotrichum orbiculare* in multiple susceptible plants. *Plant Signaling & Behavior, 11*, e1137407.

Jiang, R. H., de Bruijn, I., Haas, B. J., Belmonte, R., Löbach, L., Christie, J., … van West, P. (2013). Distinctive expansion of potential virulence genes in the genome of the oomycete fish pathogen *Saprolegnia parasitica*. *PLoS Genetics, 9*, e1003272.

Jiang, R. H., Tripathy, S., Govers, F., & Tyler, B. M. (2008). RXLR effector reservoir in two *Phytophthora* species is dominated by a single rapidly evolving superfamily with more than 700 members. *Proceedings of the National Academy of Sciences of the United States of America, 105*, 4874–4879.

Jones, J. D., & Dangl, J. L. (2006). The plant immune system. *Nature, 444*, 323–329.
de Jonge, R., Bolton, M. D., Kombrink, A., van den Berg, G. C., Yadeta, K. A., & Thomma, B. P. (2013). Extensive chromosomal reshuffling drives evolution of virulence in an asexual pathogen. *Genome Research, 23*, 1271–1282.
Kale, S. D., Gu, B., Capelluto, D. G., Dou, D., Feldman, E., Rumore, A., ... Tyler, B. M. (2010). External lipid PI3P mediates entry of eukaryotic pathogen effectors into plant and animal host cells. *Cell, 142*, 284–295.
Kämper, J., Kahmann, R., Bölker, M., Ma, L. J., Brefort, T., Saville, B. J., ... Birren, B. W. (2006). Insights from the genome of the biotrophic fungal plant pathogen *Ustilago maydis*. *Nature, 444*, 97–101.
Kemen, E., Gardiner, A., Schultz-Larsen, T., Kemen, A. C., Balmuth, A. L., Robert-Seilaniantz, A., ... Jones, J. D. (2011). Gene gain and loss during evolution of obligate parasitism in the white rust pathogen of *Arabidopsis thaliana*. *PLoS Biology.*, e1001094.
Kemen, E., Kemen, A. C., Rafiqi, M., Hempel, U., Mendgen, K., Hahn, M., & Voegele, R. T. (2005). Identification of a protein from rust fungi transferred from haustoria into infected plant cells. *Molecular Plant-Microbe Interactions, 18*, 1130–1139.
Khaldi, N., & Wolfe, K. H. (2011). Evolutionary Origins of the fumonisin secondary metabolite gene cluster in *Fusarium verticillioides* and *Aspergillus Niger*. *International Journal of Evolutionary Biology, 2011*, 423821.
Khang, C. H., Berruyer, R., Giraldo, M. C., Kankanala, P., Park, S. Y., Czymmek, K., ... Valent, B. (2010). Translocation of *Magnaporthe oryzae* effectors into rice cells and their subsequent cell-to-cell movement. *Plant Cell, 22*, 1388–1403.
Kleemann, J., Rincon-Rivera, L. J., Takahara, H., Neumann, U., Ver Loren van Themaat, E., van der Does, H. C., ... O'Connell, R. J. (2012). Sequential delivery of host-induced virulence effectors by appressoria and intracellular hyphae of the phytopathogen *Colletotrichum higginsianum*. *PLoS Pathogens, 8*, e1002643.
Kloppholz, S., Kuhn, H., & Requena, N. (2011). A secreted fungal effector of *Glomus intraradices* promotes symbiotic biotrophy. *Current Biology, 21*, 1–6.
de Koning-Ward, T. F., Gilson, P. R., Boddey, J. A., Rug, M., Smith, B. J., Papenfuss, A. T., ... Crabb, B. S. (2009). A newly discovered protein export machine in malaria parasites. *Nature, 459*, 945–949.
Lareen, A., Burton, F., & Schafer, P. (2016). Plant root-microbe communication in shaping root microbiomes. *Plant Molecular Biology, 90*, 575–587.
Liu, T., Song, T., Zhang, X., Yuan, H., Su, L., Li, W., ... Dou, D. (2014). Unconventionally secreted effectors of two filamentous pathogens target plant salicylate biosynthesis. *Nature Communications, 5*, 4686.
Lo Presti, L., Lanver, D., Schweizer, G., Tanaka, S., Liang, L., Tollot, M., ... Kahmann, R. (2015). Fungal effectors and plant susceptibility. *Annual Review of Plant Biology, 66*, 513–545.
Lorrain, C., Hecker, A., & Duplessis, S. (2015). Effector-mining in the poplar rust fungus *Melampsora larici-populina* secretome. *Frontiers in Plant Science, 6*, 1051.
Lundberg, D. S., Lebeis, S. L., Paredes, S. H., Yourstone, S., Gehring, J., Malfatti, S., ... Dangl, J. L. (2012). Defining the core *Arabidopsis thaliana* root microbiome. *Nature, 488*, 86–90.
Ma, L., Lukasik, E., Gawehns, F., & Takken, F. L. (2012). The use of agroinfiltration for transient expression of plant resistance and fungal effector proteins in *Nicotiana benthamiana* leaves. *Methods in Molecular Biology, 835*, 61–74.
Ma, L. J., van der Does, H. C., Borkovich, K. A., Coleman, J. J., Daboussi, M. J., Di Pietro, A., ... Rep, M. (2010). Comparative genomics reveals mobile pathogenicity chromosomes in *Fusarium*. *Nature, 464*, 367–373.

Ma, L. S., Hachani, A., Lin, J. S., Filloux, A., & Lai, E. M. (2014). *Agrobacterium tumefaciens* deploys a superfamily of type VI secretion DNase effectors as weapons for interbacterial competition in planta. *Cell Host & Microbe, 16*, 94–104.

Macho, A. P. (2016). Subversion of plant cellular functions by bacterial type-III effectors: beyond suppression of immunity. *New Phytologist, 210*, 51–57.

Marshall, R., Kombrink, A., Motteram, J., Loza-Reyes, E., Lucas, J., Hammond-Kosack, K. E., … Rudd, J. J. (2011). Analysis of two in planta expressed LysM effector homologs from the fungus *Mycosphaerella graminicola* reveals novel functional properties and varying contributions to virulence on wheat. *Plant Physiology, 156*, 756–769.

McLellan, H., Boevink, P. C., Armstrong, M. R., Pritchard, L., Gomez, S., Morales, J., … Birch, P. R. (2013). An RxLR effector from *Phytophthora infestans* prevents re-localisation of two plant NAC transcription factors from the endoplasmic reticulum to the nucleus. *PLoS Pathogens, 9*, e1003670.

Mesarich, C. H., Bowen, J. K., Hamiaux, C., & Templeton, M. D. (2015). Repeat-containing protein effectors of plant-associated organisms. *Frontiers in Plant Science, 6*, 872.

Mesen-Ramirez, P., Reinsch, F., Blancke Soares, A., Bergmann, B., Ullrich, A. K., Tenzer, S., & Spielmann, T. (2016). Stable translocation intermediates jam global protein export in *Plasmodium falciparum* parasites and link the PTEX component EXP2 with translocation Activity. *PLoS Pathogens, 12*, e1005618.

Misner, I., Blouin, N., Leonard, G., Richards, T. A., & Lane, C. E. (2015). The secreted proteins of *Achlya hypogyna* and *Thraustotheca clavata* identify the ancestral oomycete secretome and reveal gene acquisitions by horizontal gene transfer. *Genome Biology and Evolution, 7*, 120–135.

Mogga, V., Delventhal, R., Weidenbach, D., Langer, S., Bertram, P. M., Andresen, K., … Schaffrath, U. (2016). *Magnaporthe oryzae* effectors MoHEG13 and MoHEG16 interfere with host infection and MoHEG13 counteracts cell death caused by *Magnaporthe*-NLPs in tobacco. *Plant Cell Reports, 35*, 1169–1185.

Naglot, A., Goswami, S., Rahman, I., Shrimali, D. D., Yadav, K. K., Gupta, V. K., … Veer, V. (2015). Antagonistic potential of native *Trichoderma viride* strain against potent tea fungal pathogens in North East India. *The Plant Pathology Journal, 31*, 278–289.

O'Connell, R. J., Thon, M. R., Hacquard, S., Amyotte, S. G., Kleemann, J., Torres, M. F., … Vaillancourt, L. J. (2012). Lifestyle transitions in plant pathogenic *Colletotrichum fungi* deciphered by genome and transcriptome analyses. *Nature Genetics, 44*, 1060–1065.

Park, C. H., Chen, S., Shirsekar, G., Zhou, B., Khang, C. H., Songkumarn, P., … Wang, G. L. (2012). The *Magnaporthe oryzae* effector AvrPiz-t targets the RING E3 ubiquitin ligase APIP6 to suppress pathogen-associated molecular pattern-triggered immunity in rice. *Plant Cell, 24*, 4748–4762.

Pedersen, C., Ver Loren van Themaat, E., McGuffin, L. J., Abbott, J. C., Burgis, T. A., Barton, G., … Spanu, P. D. (2012). Structure and evolution of barley powdery mildew effector candidates. *BMC Genomics, 13*, 694.

Pel, M. J., & Pieterse, C. M. (2013). Microbial recognition and evasion of host immunity. *Journal of Experimental Botany, 64*, 1237–1248.

Pel, M. J., Wintermans, P. C., Cabral, A., Robroek, B. J., Seidl, M. F., Bautor, J., … Pieterse, C. M. (2014). Functional analysis of *Hyaloperonospora arabidopsidis* RXLR effectors. *PLoS One, 9*, e110624.

Pellegrin, C., Morin, E., Martin, F. M., & Veneault-Fourrey, C. (2015). Comparative analysis of secretomes from ectomycorrhizal fungi with an emphasis on Small-Secreted Proteins. *Frontiers in Microbiology, 6*, 1278.

Petre, B., & Kamoun, S. (2014). How do filamentous pathogens deliver effector proteins into plant cells? *PLoS Biology, 12*, e1001801.

Petre, B., Saunders, D. G., Sklenar, J., Lorrain, C., Krasileva, K. V., Win, J., ... Kamoun, S. (2016). Heterologous expression screens in *Nicotiana benthamiana* identify a candidate effector of the wheat yellow rust pathogen that associates with processing bodies. *PLoS One, 11*, e0149035.

Petre, B., Saunders, D. G., Sklenar, J., Lorrain, C., Win, J., Duplessis, S., & Kamoun, S. (2015). Candidate effector proteins of the rust pathogen *Melampsora Larici-Populina* target diverse plant cell compartments. *Molecular Plant-Microbe Interactions, 28*, 689–700.

Plett, J. M., Daguerre, Y., Wittulsky, S., Vayssières, A., Deveau, A., Melton, J., ... Martin, F. (2014). Effector MiSSP7 of the mutualistic fungus *Laccaria bicolor* stabilizes the *Populus* JAZ6 protein and represses jasmonic acid (JA) responsive genes. *Proceedings of the National Academy of Sciences of the United States of America, 111*, 8299–8304.

Ploch, S., Rose, L. E., Bass, D., & Bonkowski, M. (2016). High diversity revealed in leaf-associated protists (*Rhizaria: Cercozoa*) of *Brassicaceae*. *Journal of Eukaryotic Microbiology*. http://dx.doi.org/10.1111/jeu.12314.

Qiao, Y., Shi, J., Zhai, Y., Hou, Y., & Ma, W. (2015). *Phytophthora* effector targets a novel component of small RNA pathway in plants to promote infection. *Proceedings of the National Academy of Sciences of the United States of America, 112*, 5850–5855.

Qiu, H., Cai, G., Luo, J., Bhattacharya, D., & Zhang, N. (2016). Extensive horizontal gene transfers between plant pathogenic fungi. *BMC Biology, 14*, 41.

Quentin, M., Abad, P., & Favery, B. (2013). Plant parasitic nematode effectors target host defense and nuclear functions to establish feeding cells. *Frontiers in Plant Science, 13*, 53.

Qutob, D., Chapman, B. P., & Gijzen, M. (2013). Transgenerational gene silencing causes gain of virulence in a plant pathogen. *Nature Communications, 4*, 1349.

Raffaele, S., Farrer, R. A., Cano, L. M., Studholme, D. J., MacLean, D., Thines, M., ... Kamoun, S. (2010). Genome evolution following host jumps in the Irish potato famine pathogen lineage. *Science, 330*, 1540–1543.

Raffaele, S., & Kamoun, S. (2012). Genome evolution in filamentous plant pathogens: why bigger can be better. *Nature Reviews Microbiology, 10*, 417–430.

Raffaele, S., Win, J., Cano, L. M., & Kamoun, S. (2010). Analyses of genome architecture and gene expression reveal novel candidate virulence factors in the secretome of *Phytophthora infestans*. *BMC Genomics, 11*, 637.

Ramirez-Garces, D., Camborde, L., Pel, M. J., Jauneau, A., Martinez, Y., Neant, I., ... Gaulin, E. (2016). CRN13 candidate effectors from plant and animal eukaryotic pathogens are DNA-binding proteins which trigger host DNA damage response. *New Phytologist, 210*, 602–617.

Rao, K. L., Raju, K. S., & Ravisankar, H. (2015). Antifungal properties of native *Trichoderma* isolates against *Sclerotium rolfsii* and *Pythium aphanidermatum* infecting tobacco. *Journal of Environmental Biology, 36*, 1349–1353.

Rehmany, A. P., Gordon, A., Rose, L. E., Allen, R. L., Armstrong, M. R., Whisson, S. C., ... Beynon, J. L. (2005). Differential recognition of highly divergent downy mildew avirulence gene alleles by RPP1 resistance genes from two *Arabidopsis* lines. *Plant Cell, 17*, 1839–1850.

Rep, M., van der Does, H. C., Meijer, M., van Wijk, R., Houterman, P. M., Dekker, H. L., ... Cornelissen, B. J. (2004). A small, cysteine-rich protein secreted by *Fusarium oxysporum* during colonization of xylem vessels is required for I-3-mediated resistance in tomato. *Molecular Microbiology, 53*, 1373–1383.

Richards, T. A., Soanes, D. M., Jones, M. D., Vasieva, O., Leonard, G., Paszkiewicz, K., ... Talbot, N. J. (2011). Horizontal gene transfer facilitated the evolution of plant parasitic mechanisms in the oomycetes. *Proceedings of the National Academy of Sciences of the United States of America, 108*, 15258–15263.

Rouxel, T., Grandaubert, J., Hane, J. K., Hoede, C., van de Wouw, A. P., Couloux, A., ... Howlett, B. J. (2011). Effector diversification within compartments of the *Leptosphaeria maculans* genome affected by repeat-induced point mutations. *Nature Communications, 2*, 202.

Saitoh, H., Fujisawa, S., Mitsuoka, C., Ito, A., Hirabuchi, A., Ikeda, K., ... Terauchi, R. (2012). Large-scale gene disruption in *Magnaporthe oryzae* identifies MC69, a secreted protein required for infection by monocot and dicot fungal pathogens. *PLoS Pathogens, 8*, e1002711.

Saunders, D. G., Win, J., Cano, L. M., Szabo, L. J., Kamoun, S., & Raffaele, S. (2012). Using hierarchical clustering of secreted protein families to classify and rank candidate effectors of rust fungi. *PLoS One, 7*, e29847.

Savory, F., Leonard, G., & Richards, T. A. (2015). The role of horizontal gene transfer in the evolution of the oomycetes. *PLoS Pathogens, 11*, e1004805.

Schirawski, J., Mannhaupt, G., Münch, K., Brefort, T., Schipper, K., Doehlemann, G., ... Kahmann, R. (2010). Pathogenicity determinants in smut fungi revealed by genome comparison. *Science, 330*, 1546–1548.

Schornack, S., van Damme, M., Bozkurt, T. O., Cano, L. M., Smoker, M., Thines, M., ... Huitema, E. (2010). Ancient class of translocated oomycete effectors targets the host nucleus. *Proceedings of the National Academy of Sciences of the United States of America, 107*, 17421–17426.

Seidl, M. F., & Thomma, B. P. (2014). Sex or no sex: evolutionary adaptation occurs regardless. *Bioessays, 36*, 335–345.

Sharma, R., Xia, X., Riess, K., Bauer, R., & Thines, M. (2015). Comparative genomics including the early-diverging smut fungus *Ceraceosorus bombacis* reveals signatures of parallel evolution within plant and animal pathogens of fungi and oomycetes. *Genome Biology and Evolution, 7*, 2781–2798.

Soanes, D., & Richards, T. A. (2014). Horizontal gene transfer in eukaryotic plant pathogens. *Annual Review of Phytopathology, 52*, 583–614.

Sonah, H., Deshmukh, R. K., & Belanger, R. R. (2016). Computational prediction of effector proteins in fungi: opportunities and challenges. *Frontiers in Plant Science, 7*, 126.

Song, T., Ma, Z., Shen, D., Li, Q., Li, W., Su, L., ... Dou, D. (2015). An oomycete CRN effector reprograms expression of plant HSP genes by targeting their promoters. *PLoS Pathogens, 11*, e1005348.

Soyer, J. L., El Ghalid, M., Glaser, N., Ollivier, B., Linglin, J., Grandaubert, J., ... Fudal, I. (2014). Epigenetic control of effector gene expression in the plant pathogenic fungus *Leptosphaeria maculans*. *PLoS Genetics, 10*, e1004227.

Spanu, P. D., Abbott, J. C., Amselem, J., Burgis, T. A., Soanes, D. M., Stuber, K., ... Panstruga, R. (2010). Genome expansion and gene loss in powdery mildew fungi reveal tradeoffs in extreme parasitism. *Science, 330*, 1543–1546.

Sperschneider, J., Dodds, P. N., Gardiner, D. M., Manners, J. M., Singh, K. B., & Taylor, J. M. (2015). Advances and challenges in computational prediction of effectors from plant pathogenic fungi. *PLoS Pathogens, 11*, e1004806.

Sperschneider, J., Gardiner, D. M., Dodds, P. N., Tini, F., Covarelli, L., Singh, K. B., ... Taylor, J. M. (2016). EffectorP: predicting fungal effector proteins from secretomes using machine learning. *New Phytologist, 210*, 743–761.

Sperschneider, J., Williams, A. H., Hane, J. K., Singh, K. B., & Taylor, J. M. (2015). Evaluation of secretion prediction highlights differing approaches needed for oomycete and fungal effectors. *Frontiers in Plant Science, 6*, 1168.

Sperschneider, J., Ying, H., Dodds, P. N., Gardiner, D. M., Upadhyaya, N. M., Singh, K. B., ... Taylor, J. M. (2014). Diversifying selection in the wheat stem rust fungus acts predominantly on pathogen-associated gene families and reveals candidate effectors. *Frontiers in Plant Science, 5*, 372.

Spielmann, T., Montagna, G. N., Hecht, L., & Matuschewski, K. (2012). Molecular make-up of the *Plasmodium* parasitophorous vacuolar membrane. *International Journal of Medical Microbiology, 302*, 179–186.
Stam, R., Jupe, J., Howden, A. J., Morris, J. A., Boevink, P. C., Hedley, P. E., & Huitema, E. (2013). Identification and characterisation CRN effectors in *Phytophthora capsici* shows modularity and functional diversity. *PLoS One, 8*, e59517.
Stam, R., Motion, G., Boevink, P., & Huitema, E. (2013). A conserved oomycete CRN effector targets and modulates tomato TCP14–2 to enhance virulence. *BioRXiv*. http://dx.doi.org/10.1101/001248.
Stergiopoulos, I., Cordovez, V., Okmen, B., Beenen, H. G., Kema, G. H., & de Wit, P. J. (2014). Positive selection and intragenic recombination contribute to high allelic diversity in effector genes of *Mycosphaerella fijiensis*, causal agent of the black leaf streak disease of banana. *Molecular Plant Pathology, 15*, 447–460.
Sun, G., Yang, Z., Kosch, T., Summers, K., & Huang, J. (2011). Evidence for acquisition of virulence effectors in pathogenic chytrids. *BMC Evolutionary Biology, 11*, 195.
Takahara, H., Hacquard, S., Kombrink, A., Hughes, H. B., Halder, V., Robin, G. P., … O'Connell, R. J. (2016). *Colletotrichum higginsianum* extracellular LysM proteins play dual roles in appressorial function and suppression of chitin-triggered plant immunity. *New Phytologist*. http://dx.doi.org/10.1111/nph.13994.
Tanaka, S., Brefort, T., Neidig, N., Djamei, A., Kahnt, J., Vermerris, W., … Kahmann, R. (2014). A secreted *Ustilago maydis* effector promotes virulence by targeting anthocyanin biosynthesis in maize. *eLife, 3*, e01355.
Tanaka, S., Djamei, A., Presti, L. L., Schipper, K., Winterberg, S., Amati, S., … Kahmann, R. (2015). Experimental approaches to investigate effector translocation into host cells in the *Ustilago maydis*/maize pathosystem. *European Journal of Cell Biology, 94*, 349–358.
Torto, T., Li, S., Styer, A., Huitema, E., Testa, A., Gow, N., … Kamoun, S. (2003). EST mining and functional expression assays identify extracellular effector proteins from the plant pathogen *Phytophthora*. *Genome Research, 13*, 1675–1685.
Vallance, J., Le Floch, G., Deniel, F., Barbier, G., Levesque, C. A., & Rey, P. (2009). Influence of *Pythium oligandrum* biocontrol on fungal and oomycete population dynamics in the rhizosphere. *Applied and Environmental Microbiology, 75*, 4790–4800.
Ve, T., Williams, S. J., Catanzariti, A. M., Rafiqi, M., Rahman, M., Ellis, J. G., … Kobe, B. (2013). Structures of the flax-rust effector AvrM reveal insights into the molecular basis of plant-cell entry and effector-triggered immunity. *Proceedings of the National Academy of Sciences of the United States of America, 110*, 17594–17599.
Vlaardingerbroek, I., Beerens, B., Rose, L., Fokkens, L., Cornelissen, B. J., & Rep, M. (2016). Exchange of core chromosomes and horizontal transfer of lineage-specific chromosomes in *Fusarium oxysporum*. *Environmental Microbiology*. http://dx.doi.org/10.1111/1462-2920.13281.
Vlaardingerbroek, I., Beerens, B., Schmidt, S. M., Cornelissen, B. J., & Rep, M. (2016). Dispensable chromosomes in *Fusarium oxysporum f.sp lycopersici*. *Molecular Plant Pathology*. http://dx.doi.org/10.1111/mpp.12440.
Wawra, S., Agacan, M., Boddey, J. A., Davidson, I., Gachon, C. M., Zanda, M., … van West, P. (2012). Avirulence protein 3a (AVR3a) from the potato pathogen *Phytophthora infestans* forms homodimers through its predicted translocation region and does not specifically bind phospholipids. *Journal of Biological Chemistry, 287*, 38101–38109.
Wawra, S., Djamei, A., Albert, I., Nürnberger, T., Kahmann, R., & van West, P. (2013). In vitro translocation experiments with RxLR-reporter fusion proteins of Avr1b from *Phytophthora sojae* and AVR3a from *Phytophthora infestans* fail to demonstrate specific autonomous uptake in plant and animal cells. *Molecular Plant-Microbe Interactions, 26*, 528–536.

Weiberg, A., Wang, M., Lin, F. M., Zhao, H., Zhang, Z., Kaloshian, I., ... Jin, H. (2013). Fungal small RNAs suppress plant immunity by hijacking host RNA interference pathways. *Science, 342*, 118–123.

Whipps, J. M., Hand, P., Pink, D., & Bending, G. D. (2008). Phyllosphere microbiology with special reference to diversity and plant genotype. *Journal of Applied Microbiology, 105*, 1744–1755.

Whisson, S. C., Boevink, P. C., Moleleki, L., Avrova, A. O., Morales, J. G., Gilroy, E. M., ... Birch, P. R. (2007). A translocation signal for delivery of oomycete effector proteins into host plant cells. *Nature, 450*, 115–118.

Win, J., & Kamoun, S. (2008). Adaptive evolution has targeted the C-terminal domain of the RXLR effectors of plant pathogenic oomycetes. *Plant Signaling & Behavior, 3*, 251–253.

Win, J., Krasileva, K. V., Kamoun, S., Shirasu, K., Staskawicz, B. J., & Banfield, M. J. (2012). Sequence divergent RXLR effectors share a structural fold conserved across plant pathogenic oomycete species. *PLoS Pathogens, 8*, e1002400.

Xiong, Q., Ye, W., Choi, D., Wong, J., Qiao, Y., Tao, K., ... Ma, W. (2014). *Phytophthora* suppressor of RNA silencing 2 is a conserved RxLR effector that promotes infection in soybean and *Arabidopsis thaliana*. *Molecular Plant-Microbe Interactions, 27*, 1379–1389.

Yang, L., McLellan, H., Naqvi, S., He, Q., Boevink, P. C., Armstrong, M., ... Birch, P. R. (2016). Potato NPH3/RPT2-like protein StNRL1, targeted by a *Phytophthora infestans* RXLR effector, is a susceptibility factor. *Plant Physiology, 171*, 645–657.

Zhang, D., Burroughs, A. M., Vidal, N. D., Iyer, L. M., & Aravind, L. (2016). Transposons to toxins: the provenance, architecture and diversification of a widespread class of eukaryotic effectors. *Nucleic Acids Research, 44*, 3513–3533.

Zhang, S., & Xu, J. R. (2014). Effectors and effector delivery in *Magnaporthe oryzae*. *PLoS Pathogens, 10*, e1003826.

Zheng, X., McLellan, H., Fraiture, M., Liu, X., Boevink, P. C., Gilroy, E. M., ... Brunner, F. (2014). Functionally redundant RXLR effectors from *Phytophthora infestans* act at different steps to suppress early flg22-triggered immunity. *PLoS Pathogens, 10*, e1004057.

Zuccaro, A., Lahrmann, U., Guldener, U., Langen, G., Pfiffi, S., Biedenkopf, D., ... Kogel, K. H. (2011). Endophytic life strategies decoded by genome and transcriptome analyses of the mutualistic root symbiont *Piriformospora indica*. *PLoS Pathogens, 7*, e1002290.

CHAPTER EIGHT

Commonalities in Symbiotic Plant-Microbe Signalling

R. Holmer[a], L. Rutten[a], W. Kohlen, R. van Velzen, R. Geurts[1]
Wageningen University, Wageningen, The Netherlands
[1]Corresponding author: E-mail: rene.geurts@wur.nl

Contents

Abstract

Plants face the problem that they have to discriminate symbionts from a diverse pool of soil microbes, including pathogens. Studies on different symbiotic systems revealed commonalities in plant-microbe signalling. In this chapter we focus on four intimate symbiotic interactions: two mycorrhizal ones, with arbuscular- and ectomycorrhizal fungi, and two nitrogen-fixing ones, with rhizobium and *Frankia* bacteria. Comparing these systems uncovered commonalities in the way plants attract their symbiotic partners. Especially flavonoids, and in a lesser extent strigolactones, are pivotal plant signals that are perceived by the microsymbiont. In response, signal molecules are exuded by the microbes to trigger symbiotic responses in their host plant. Strikingly, microbes that establish an endosymbiotic relation with their host plant, namely arbuscular mycorrhizal fungi, rhizobium and *Frankia* bacteria, make

[a] Authors contributed equally.

Advances in Botanical Research, Volume 82
ISSN 0065-2296
http://dx.doi.org/10.1016/bs.abr.2016.11.003

use of a symbiotic signalling network that is highly conserved in plants. The use of flavonoids as attractants for symbiotic microbes, in combination with the use of a common plant signalling network to establish endosymbioses, raises questions about how plants manage to discriminate their microbial partners.

1. INTRODUCTION

High throughput sequencing approaches have uncovered an overwhelming diversity of soil microbes. Plants affect this microbial community — directly or indirectly — with their root systems. For example, roots exude substantial amounts of organic and amino acids, polymerized sugars (e.g., mucilage) as well as release border cells and dead root cap cells, which all form a nutrient source for many microbes (Jones, Nguyen, & Finlay, 2009). On top of that more specific secondary metabolites are exuded that manipulate the microbial community by acting as antimicrobial agent or as attractant. Conversely, soil microbes can affect plant growth. For example, microbes can promote plant growth by improving nutrient availability or inducing resistance against biotic and abiotic stresses (Coleman-Derr & Tringe, 2014; Mendes, Garbeva, & Raaijmakers, 2013). On the other hand, pathogenic microbes can induce resource loss and disease. In this complex plant root microbiome network the plant must therefore discriminate between bacteria and fungi that provide an advantage and those that act as commensals or even pathogens. In this chapter we will focus on the molecular communication in a symbiotic context which occurs in plant roots and the rhizosphere. Plants establish several intricate long-term mutualistic relationships with microbes that are hosted intercellularly (ecto) or intracellularly (endo). Here, we will discuss the commonalities of four intimate symbiotic interactions. Thereby we will focus on two key stages of the interaction: attraction of the microbial partner and subsequent microbe-induced signalling to establish a symbiosis.

2. INTIMATE PLANT ROOT–MICROBE SYMBIOSES

Plant root symbioses occur at different levels of engagement, ranging from loosely attached microbes that provide a certain advantage to the plant to bacteria that are intracellularly accommodated as organelle-like structures (Mendes et al., 2013; Van Loon, 2007). The best studied plant root symbioses are those with arbuscular mycorrhizal and ectomycorrhizal fungi and those with rhizobium and *Frankia* nitrogen-fixing bacteria, together encompassing a diverse range of plant and microbial species.

2.1 Mycorrhizal Symbioses

Mycorrhizal symbioses — the symbiotic interactions between some soil fungi and plant roots — can occur in several forms. Of these the ancient arbuscular (endo) mycorrhiza and the much younger forms of ectomycorrhiza are best studied.

Based on fossil records **arbuscular mycorrhizal symbiosis** is estimated to be at least ~400–460 million years old, and evolved in a period that coincides with colonization of terrestrial habitats by plants (Redecker, Kodner, & Graham, 2000; Remy, Taylor, Hass, & Kerp, 1994; Simon, Bousquet, Levesque, & Lalonde, 1993). Still today the vast majority of land plant species establish an arbuscular mycorrhizal symbiosis, underlining the ecological importance of this interaction (Wang & Qiu, 2006). The fungi that establish an arbuscular mycorrhizal symbiosis belong to a distinct taxonomic phylum, the Glomeromycota. This phylum possibly represents more than 1000 species, though only less than 300 have been characterized to a certain level of detail (Redecker et al., 2013). Arbuscular mycorrhizal fungi are obligate biotrophs. Their hyphae penetrate plant roots intercellularly and form intracellular feeding structures — called arbuscules — in root cortical cells (Fig. 1A). Arbuscules are surrounded by a plant-derived membrane, but are largely deprived of plant cell wall material (Balestrini & Bonfante, 2014). At this symbiotic interface nutrients are exchanged. Minerals — especially phosphates and nitrates — taken up by the fungal extraradical mycelium are delivered to the plant in return for carbohydrates. Arbuscules remain functional for several days, after which they collapse and disappear, leading to a reversion of the plant cell to its asymbiotic cortical fate.

Ectomycorrhizal symbiosis can occur between diverse groups of plant and fungal species, as a result of several independent evolutionary events (Martin et al., 2016). Overall this type of symbiosis can occur in about 2% of all land plants, including all dominant tree species in temperate forests, such as pines (*Pinus*), Douglas firs (*Pseudotsuga*), oaks (*Quercus*), willows (*Salix*), beeches (*Fagus*) and birches (*Betula*) (Smith & Read, 2008; Tedersoo, May, & Smith, 2010). Ectomycorrhizal fungi belong to several taxonomic phyla including Basidiomycota, Ascomycota and Zygomycota (Tedersoo & Smith, 2013) which are all closely related to species with a saprotrophic lifestyle. Saprophytic fungi have an extensive repertoire of genes encoding

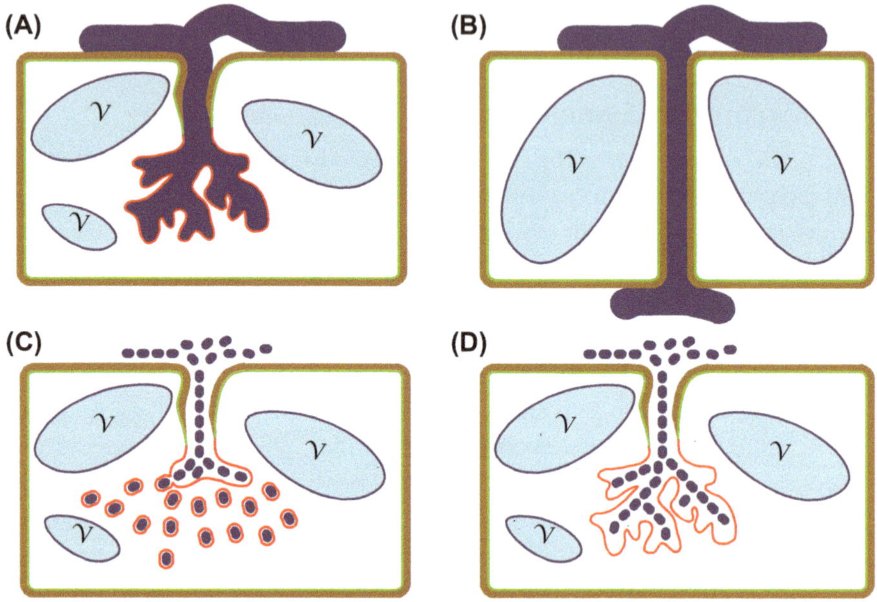

Figure 1 Schematic representation of the cellular mode of infection of the four symbioses discussed in this chapter. *Green lines* indicate plant cell membrane, *red lines* indicate the plant cell membrane–derived symbiotic interface in the form of the periarbuscular membrane for arbuscular mycorrhizal, symbiosomes for rhizobium/legumes and fixation threads for *Frankia*/actinorhizal plants and for rhizobium/*Parasponia*. (A) Hyphae of an endomycorrhizal fungus penetrating the cell and forming a feeding structure named arbuscule. Arbuscules are not surrounded by a plant-derived cell wall. (B) Hyphae of an ectomycorrhizal fungus growing intercellularly. (C) Rhizobium bacteria released within transient organelle-like structures – named symbiosomes – in nodule cells of most legumes. (D) *Frankia* in actinorhizal plants and some rhizobia in *Parasponia* and in some basal legumes infect cells of nodules through fixation threads. Fixation threads are largely deprived of plant cell wall. The bacteria in fixation threads remain in contact with the apoplast. Blue: vacuole (v); purple: microbe; brown: plant cell wall; green: plant plasma membrane; red: plant-derived endosymbiotic membrane.

degrading enzymes that can effectively mobilize resources, in particular nitrogen and phosphorus, from a variety of organic substrates (Plett & Martin, 2011). However, compared to their saprophytic sister clades, ectomycorrhizal fungi only have a limited set of genes encoding plant cell wall–degrading enzymes (e.g., pectin lyases and pectinases) (Kohler et al., 2015). Nevertheless, expression studies indicate that some of the plant cell wall–degrading enzymes that have been maintained may function during symbiosis (Balestrini & Bonfante, 2005).

Despite the diverse and paraphyletic groups of plant and fungal species that establish ectomycorrhizal symbioses, there is a remarkable resemblance in symbiotic phenotypes. The fungi preferentially colonize newly formed lateral roots. Upon hyphal attachment, they modulate root growth which allows them to colonize the root apoplast, forming a so-called Hartig net. The penetration depth of the hyphae is variable, but it typically comprises several layers of cortical cells, excluding the endodermis. In contrast to the endomycorrhizal symbiosis, root cells are not invaded intracellularly (Fig. 1B). Ultimately, many fungal hyphae cover the root surface forming a thick, multilayered 'mantle', insulating the infected lateral root. The molecular mechanisms underlying Hartig net development remain elusive. In 2015, using a combination of genome sequencing and reverse genetic studies, have provided new insights into the early symbiosis signalling. For example, in the fungus *Laccaria bicolor*, it was found that an aquaporin (LbAQP1) is essential for Hartig net development and the expression of effector genes. LbAQP expression is triggered upon direct root contact and functions as a transport facilitator for plant-signalling molecules, most likely H_2O_2, NO or CO_2 (Navarro-Rodenas, Xu, Kemppainen, Pardo, & Zwiazek, 2015). Additionally *L. bicolor* produces auxin (IAA) in its mycelium that triggers auxin-related responses in the plant root (Vayssières et al., 2015). This finding is in line with pioneer work that showed that increased mycorrhizal activity is associated with increased auxin biosynthesis by the fungus (Gea, Normand, Vian, & Gay, 1994). Together, these studies make clear that plant-fungal signalling intertwines with plant auxin homoeostasis and possibly used reactive oxygen species to establish a symbiotic interaction.

2.2 Nitrogen-Fixing Endosymbioses

A selective, though diverse, group of plant species is able to establish an endosymbiosis with nitrogen-fixing (diazotrophic) bacteria. These bacteria belong either to the genus *Frankia* or to the paraphyletic group of bacteria known as rhizobia. *Frankia* and rhizobia strains gained the symbiotic trait by horizontal gene transfer.

The genus *Frankia* is a diverse assemblage of filamentous sporangia-forming actinobacterial species that can be saprophytic, facultative symbiotic or obligatory symbiotic. The *Frankia* genus can be separated in four separate bacterial clusters based on phylogenetic analysis, with only three of them that can establish symbiosis (Gtari, Tisa, & Normand, 2013). Plant species that can form a nitrogen-fixing endosymbiosis with

Frankia bacteria (~230 species known as actinorhizal plants) are dispersed over 25 genera and 8 taxonomic families, suggesting multiple evolutionary origins of this symbiosis (Pawlowski & Demchenko, 2012).

Relative to *Frankia*, rhizobia are even more diverse, representing 15 genera in 8 families of α-, β- and γ-Proteobacteria (Remigi, Zhu, Young, & Masson-Boivin, 2015). Nitrogen-fixing symbiosis with rhizobia is prominent in the legume family (Fabaceae), but can also occur in *Parasponia*, a genus in the Cannabis family (Cannabaceae). Based on the phylogenetic distance between Fabaceae and Cannabaceae it is most probable that — similarly to the actinorhizal symbiosis — there are multiple origins for rhizobial symbiosis (Behm, Geurts, & Kiers, 2014; Geurts, Lillo, & Bisseling, 2012). The formation of specific nodule-like structures (root nodules) by the host plant, in which the bacteria proliferate and fix nitrogen, is common for both types of endosymbiosis with diazotrophic bacteria.

The reason for this may be that rhizobia and *Frankia* bacteria are generally not able to infect differentiated cells of the plant root. Only cells of the future nodule that are mitotically activated by the microsymbiont can be infected, suggesting that these cells are developmentally reprogrammed (Geurts, Xiao, & Reinhold-Hurek, 2016). The nodules are optimized to facilitate growth of the microbial partner, which, once inside nodule cells, differentiates in its symbiotic form and fixes atmospheric nitrogen into ammonia in exchange for carbohydrates.

Variation exists in the way the nitrogen-fixing bacterial partner is hosted. In most legume nodules, rhizobia are hosted in transient organelle-like structures, called symbiosomes. Symbiosomes are released from intracellular infection threads that have guided the rhizobium bacteria from the epidermis towards the newly formed nodule. Hundreds of symbiosomes surrounded by a plant-derived membrane, often containing only one bacterium, can be present in a single nodule cell. This membrane forms a symbiotic interface where nutrient exchanges take place between the bacteria and the cytoplasm of the host cell (Fig. 1C). In *Parasponia* and actinorhizal plants symbiosomes are not formed. Instead, the bacteria remain in thread-like structures, known as fixation threads (Fig. 1D). Fixation threads differ from the penetrating infection thread by a reduction of plant cell wall material. Fixation threads also occur in a few legume species and may represent a more ancestral form of bacterial endosymbiosis than symbiosomes (Behm et al., 2014).

Of the four intimate symbiotic interactions that are central here, three have evolved more than once: the symbioses with rhizobia, *Frankia* and ectomycorrhizal fungi. This suggests an evolutionary advantage of root symbiosis for both partners. Interestingly, several studies indicate that similar mechanisms have been coopted in all four symbiotic interactions. Later we will discuss the commonalities in signalling mechanisms between the four symbioses central in this chapter.

3. RECOGNITION AND ATTRACTION OF SYMBIOTIC PARTNERS

As outlined earlier, not all plants are able to form an intricate root microbe symbiosis; nor are all soil microbes symbiotic. Consequently, symbiotic partners need to recognize each other. Microbes recognize potential host plants by root exudates. Indeed, plants can exude signalling molecules to attract their symbiotic partner. Common signals in symbiotic partner recognition are exuded flavonoids, which play a role in all four symbioses. In addition, it was noted that exuded strigolactones can act as signal molecules, especially in arbuscular mycorrhizal symbiosis. Strikingly, both types of molecules function also as endogenous plant signals.

3.1 Flavonoids Induce Microbial Responses

Flavonoids are a subclass of plant polyphenolic compounds and are a major class of secondary metabolites. As is typical for plant secondary metabolites, flavonoids are diverse; ~9000 chemical structures have so far been reported (Ferrer, Austin, Stewart, & Noel, 2008). Flavonoids are synthesized through the phenylpropanoid pathway. A chalcone synthase produces the chalcone scaffolds from which all other flavonoids are derived and is the first enzyme specific for flavonoid production (Falcone Ferreyra, Rius, & Casati, 2012). A series of enzymatic reactions can alter the chalcone scaffold into a huge diversity of compounds. Flavonoids are typically categorized in subclasses based on these enzymatic reactions. The major subclasses of flavonoids include phlobaphenes, flavones, flavanones, flavonols, aurones, isoflavonoids, anthocyanins and condensed tannins. Many flavonoids are known to have glycosidated forms, i.e., quercitrin is formed by the addition of the deoxy sugar rhamnose to the flavonol quercetin, whereas rutin is formed by the addition of the disaccharide rutinose. Such small modifications can have drastic consequences for the observed effects in symbioses.

The involvement of flavonoids in symbioses has been described for all four types of symbiosis discussed in this review (arbuscular mycorrhizal, ectomycorrhizal, rhizobial and actinorhizal symbioses). For two compounds positive effects in all four symbioses have been described (Fig. 2). Naringenin positively influences arbuscular mycorrhizal colonization (Garg & Singla, 2016) and rhizobium symbiosis (Weston & Mathesius, 2013), enhances spore germination of the ectomycorrhizal fungus *Suillis bovinus* (Kikuchi, Matsushita, Suzuki, & Hogetsu, 2007) and restores *Frankia* nodulation in a chalcone synthase mutant of the actinorhizal plant *Casuarina glauca* (Abdel-Lateif et al., 2013). Quercetin has been reported to stimulate spore germination, hyphal branching and growth of arbuscular mycorrhizal fungi (Bécard, Douds, & Pfeffer, 1992; Tsai & Phillips, 1991), the growth rate of rhizobium bacteria (Hartwig, Joseph, & Phillips, 1991), the actinorhizal nodulation (Sayed & Wheeler, 1999) and it also stimulates the production of the symbiotic effector protein MiSSP7 in the ectomycorrhizal fungus *L. bicolor* (Plett & Martin, 2012).

The molecular mode of action of naringenin and quercetin is not always known. Best studied is the effect of naringenin — and other flavonoids — in rhizobia, where flavonoids target NodD proteins. Rhizobial NodD proteins belong to the class of LysR-type transcriptional regulators that are activated upon the binding of external signals (Honma, Asomaning, & Ausubel, 1990; Mulligan & Long, 1989). Binding of a flavonoid molecule causes a conformational change which results in an increased binding affinity for specific

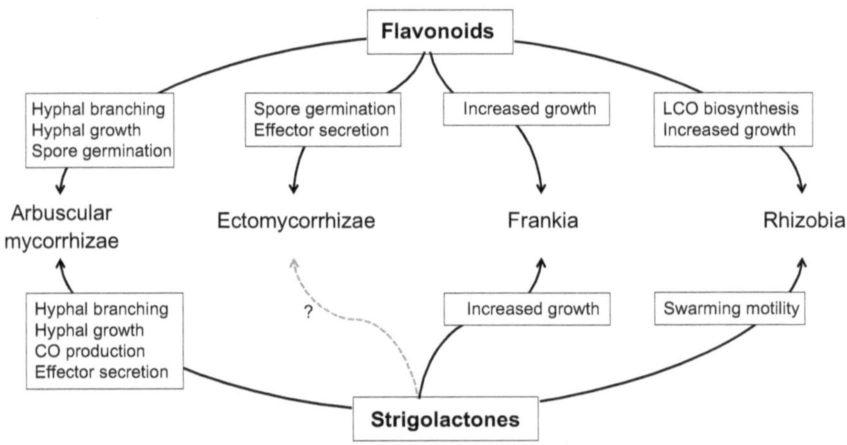

Figure 2 Flavonoids and strigolactones are generic attractants for microsymbionts. For now it remains unclear whether strigolactones have a direct effect on ectomycorrhizal symbiosis. Increased growth of *Frankia* can possibly be attributed to either flavonoids, strigolactones or even other components as total root exudates were used to demonstrate this (Beauchemin et al., 2012). COs, chito-oligosaccharides; LCOs, lipo-chito-oligosaccharides.

cis-regulatory elements. In case of NodD this element is known as the nod box (Chen et al., 2005). Rhizobia generally have several operons that contain a nod box in their promoter region. Most prominent are the genes encoding an ABC transporter (NodI and NodJ) and three genes encoding the enzymes *N*-acetylglucosaminyltransferase (NodC), a chitooligosaccharide deacetylase (NodB) and an *N*-acyltransferase (NodA). These proteins are essential for biosynthesis and secretion of lipo-chito-oligosaccharide molecules (LCOs), which act as potent symbiotic signal molecules (see Section 4) (Limpens, van Zeijl, & Geurts, 2015; Oldroyd, 2013).

In case of arbuscular mycorrhizae and *Frankia* it remains elusive whether flavonoids trigger biosynthesis of similar symbiotic signalling molecules, despite the fact that flavonoids have a positive effect on both symbioses (Auguy et al., 2011; Garg & Singla, 2016). LCOs and short-chain chitin oligomers (tetra and pentameric COs) have been shown to be produced by the mycorrhizal fungus *Rhizophagus irregularis*, but their biosynthetic pathways have not yet been uncovered (Genre et al., 2013; Lin et al., 2014; Maillet et al., 2011; Tisserant et al., 2013). In case of symbiotic *Frankia* species, LCO biosynthesis genes are not common, and only found in a representative of a relatively isolated taxonomic lineage (cluster 2): namely (candidatus) *Frankia datiscae* strain DG1 (Persson et al., 2015). For this strain it was found that *nodA*, *nodB*, *nodC*, *nodI* and *nodJ* are expressed when the bacteria occupy *Datisca glomerata* root nodules (Beauchemin et al., 2012). Therefore, it is most probable that *F. datiscae* LCO signals play a symbiotic role.

Other flavonoids have been described to be involved in one or a few of the discussed symbioses, but were never tested in the other types of symbioses. Nevertheless, these observations can shed an interesting light on the symbiotic role of flavonoids. Especially interesting is the described host specificity in the legume rhizobia interaction (Reddy, María, & Soto, 2007), which in part is determined by recognition of specific flavonoids. Whereas a specific flavonoid can induce expression of the LCO biosynthetic *nodABC* operon in one bacterium, the same compound can have a negative effect in another bacterium. For example, the flavonoid coumestrol positively influences the symbiosis between *Glycine max* and *Sinorhizobium fredii* USDA191 (Kosslak, Bookland, Barkei, Paaren, & Appelbaum, 1987) but negatively influences the symbiosis between *Medicago sativa* and *Sinorhizobium meliloti* 1021 (Zuanazzi et al., 1998). In this context, it is also relevant to note that the composition of root exudates may vary depending on the developmental stage of the root. For example, studies in *M. sativa* indicate that flavonoids with a positive effect on the symbiosis are exuded in the elongation and differentiation zone that is susceptible to rhizobium infection. Whereas repelling flavonoids are exuded in the adjacent regions of the root (i.e., the root tip and the more mature part of the root) (Peters & Long, 1988; Zuanazzi et al., 1998). However, such studies have not been further extended to see whether similar mechanisms exist in other nodulating taxa.

In addition to direct application of flavonoids to microbial cultures, reverse genetic studies in plants have been conducted. In most studies chalcone synthase genes were targeted. Chalcone synthase knockdown in actinorhizal *C. glauca*, or in the legume model *Medicago truncatula* results in impaired nodulation (Abdel-Lateif et al., 2013; Wasson, Pellerone, & Mathesius, 2006). In both plant systems this phenotype can be restored by the application of naringenin. In contrast, no effect was reported on the arbuscular mycorrhizal symbiosis when using a chalcone-synthase double-mutant in maize (Bécard, Taylor, Douds, Pfeffer, & Doner, 1995). The fact that a chalcone synthase maize mutant can be normally mycorrhizal with different fungal species demonstrates that flavonoids are not essential signals for this symbiosis. Nevertheless, flavonoids may act as facultative signals, and may play a role in host selection, by activating certain fungi over others (Ellouze et al., 2012).

The importance of flavonoids in root nodule symbiosis may be the result of the fact that flavonoids also act as endogenous plant signals that control auxin transport (Brown et al., 2001; Mathesius et al., 1998; Wasson et al., 2006). Based on quantitative modelling and experimental studies it is hypothesized that a transient decrease in auxin efflux can lead to formation of a local auxin maximum, which is the onset of nodule development (Deinum, Geurts, Bisseling, & Mulder, 2012; Hirsch, Bhuvaneswari, Torrey, & Bisseling, 1989). Such a function of flavonoids in nodulation is supported by the finding that naringenin can restore nodulation in the *M. truncatula* cytokinin signalling mutant Mt*cre1* (Ng, Perrine-Walker, Wasson, & Mathesius, 2015). This study demonstrates that naringenin not only acts as an attractant of symbiotic microbes, but also functions as an endogenous plant signal, which — in a symbiotic context — acts downstream, or in parallel, to rhizobium-induced cytokinin signalling.

3.2 Dual Role of Strigolactones

Strigolactones are carotenoid-derived terpenoid lactones, often composed of four rings. Three rings form a tricyclic lactone, which is connected to the fourth butenolide ring via an enol ether bridge (Pandey, Sharma, & Pandey, 2016). Strigolactones are known as endogenous plant hormones that control several steps in plant development (Brewer, Koltai, & Beveridge, 2013). Since the discovery of their function in the regulation of shoot branching in 2008, major advances have been made on the identification of the strigolactone biosynthesis and perception pathway. A carotenoid isomerase [named DWARF27 (D27) in most species], two carotenoid cleaving dioxygenases (named CCD7 and CCD8), and a cytochrome P450 (possibly MAX1 in *Arabidopsis thaliana*) are sequentially required to produce the strigolactone backbones: either 4-deoxyorobanchol or 5-deoxystrigol (Alder et al., 2012; Zhang et al., 2014). It is postulated that this backbone can be further decorated to produce the wealth of different strigolactone metabolites (Al-Babili & Bouwmeester, 2015). In plants the strigolactone receptor was identified as an α/β hydrolase [named OsDWARF14 (D14) in rice (*Oryza sativa*) and AtDAD2 in *A. thaliana*]. Together with a specific F-box protein named OsDWARF3/AtMAX2 it forms the SCF E3 ubiquitin ligase complex required for strigolactone signalling (Hamiaux et al., 2012; Zhao et al., 2015).

The discovery that strigolactones stimulate hyphal branching in the arbuscular mycorrhizal fungus *Gigaspora margarita* (Akiyama, Matsuzaki, & Hayashi, 2005) has launched an interest in the involvement of these compounds in symbiotic signalling. The observation that strigolactones induce hyphal branching in *G. margarita* — at very low concentrations — has led to the hypothesis that the induction of hyphal branching must be receptor mediated (Akiyama, Ogasawara, Ito, & Hayashi, 2010). Furthermore, it was found that the synthetic strigolactone analog GR24 triggers mitochondrial activity in the arbuscular mycorrhizal fungi *Rhizophagus intraradices* and *Gigaspora rosea* (Besserer et al., 2006). However, it should be noted that to induce hyphal branching in *G. rosea* besides GR24, also the flavonoid quercetin is needed in the fungal growth medium (Besserer, Bécard, Jauneau, Roux, & Séjalon-Delmas, 2008). As quercetin is known to stimulate arbuscular mycorrhizal growth, hyphal branching and spore germination (Tsai & Phillips, 1991) this suggests that with this specific fungus strigolactones alone might not be sufficient to induce hyphal branching. In an independent experiment increased production of short-chain COs upon application of GR24 was reported for *R. irregularis* (Genre et al., 2013) (Fig. 2). In addition, a putative effector protein (RiSIS1) was identified in a screening of upregulated genes in GR24-treated *R. irregularis* (Tsuzuki, Handa, Takeda, & Kawaguchi, 2016). Using host-induced gene silencing the *RiSIS1* gene was knocked down during infection, which resulted in significant suppression of colonization and stunted arbuscules. This suggests that RiSIS1 is a strigolactone-induced effector protein.

Application of GR24 to four ectomycorrhizal species revealed no effect on hyphal branching (Steinkellner et al., 2007) (Fig. 2). This suggests strigolactones play a less important or different role in this type of symbiosis. In contrast, a negative effect of GR24 was observed on growth and branching of a range of phytopathogenic fungi (Dor, Joel, Kapulnik, Koltai, & Hershenhorn, 2011), including species previously found not to be affected by GR24 (Steinkellner et al., 2007). It should be noted that these effects were only observed when relatively high concentrations of GR24 were used (Dor et al., 2011) and as such it remains unclear whether these concentrations were biologically relevant.

Apart from the beneficial effects in arbuscular mycorrhizal symbiosis, several studies revealed effects of strigolactones in the rhizobium/legume symbiosis (Fig. 2). Exogenous application of GR24 increases *M. sativa* nodule number when inoculated with *S. meliloti* (Soto et al., 2010). Interestingly, the same study reports that the bacterial growth and *nodC* expression are not affected by GR24, leading the authors to hypothesize that the effect of GR24 is on the plant. However, more recently it was suggested that GR24 might affect *S. meliloti* by promoting bacterial swarming motility (Peláez-Vico, Bernabéu-Roda, Kohlen, Soto, & López-Ráez, 2016). In an independent experiment in *M. truncatula* low concentrations (0.1 μM) of GR24 also resulted in increased nodule numbers, but higher concentrations (2−5 μM) resulted in reduced nodule numbers and lateral root density (De Cuyper et al., 2014). Taken together this suggests that strigolactones act mainly as plant hormones involved in developmental programs during rhizobial symbiosis. In line with this, the strigolactone biosynthesis gene *MtD27* was shown to be inducible by rhizobium LCOs 3 h after application and that this induction is regulated by the common symbiotic signalling pathway (Van Zeijl et al., 2015).

Mutants and knockdown experiments of strigolactone biosynthesis genes in several species shed light on the dual role of strigolactones in symbioses. Whereas often symbiotic phenotypes are observed, it is not trivial to decide whether these phenotypes are an effect of a change in direct signalling between host and symbiont, or whether a change in hormonal balance causes a difference in plant developmental program. CCD are among the most studied strigolactone biosynthetic enzymes in a symbiotic context. Mutation or knockdown of *ccd7/8* in several plant species results in reduced mycorrhizal colonization (Gomez-Roldan et al., 2008; Kohlen et al., 2012; Kretzschmar et al., 2012; Liu et al., 2013; Vogel et al., 2010). In addition, nodulation was reported to be impaired in the *Lotus japonicus CCD7* knockdown and both *ccd7* and *ccd8* mutants of pea (Foo, Yoneyama, Hugill, Quittenden, & Reid, 2013; Liu et al., 2013). The importance of strigolactones in mycorrhizal colonization is further supported by the identification of a strigolactone transporter in *Petunia x hybrida*. The knockout of the ABC transporter PhPDR1 resulted in significantly reduced orobanchol levels in root exudates, which had effects on *G. margarita* and *R. irregularis* mycorrhization efficiency. Plants show reduced colonization due to reduced mycorrhizal growth, branching and spore germination (Kretzschmar et al., 2012). The GRAS transcriptional regulators NSP1 and NSP2 were identified as regulators of strigolactone biosynthesis in rice and *M. truncatula* by regulating *D27* expression (Liu et al., 2011). *M. truncatula nsp1* and *nsp2* mutants are not capable of forming nodules (Catoira et al., 2000; Oldroyd & Long, 2003). The *nsp1* mutant and the *nsp1/nsp2* double mutant produce no detectable amounts of strigolactones, whereas the *nsp2* mutant has a reduced and different strigolactone composition. Interestingly, mycorrhizal colonization of the *nsp1/nsp2* double mutant by *R. irregularis* was only mildly reduced (Liu et al., 2011). In addition, the *L. japonicus nsp1* mutant is unable to form nodules; however, infection by the arbuscular mycorrhizal fungus *R. irregularis* was unaffected (Heckmann et al., 2006).

The rice and pea F-box mutants Os*d3*/Ps*rms4* are markedly reduced in mycorrhizal colonization (Foo et al., 2013; Yoshida et al., 2012). This suggests that strigolactone perception *in planta* plays a role in arbuscular mycorrhizal colonization. Strikingly, in the pea Ps*rms4* mutant nodule numbers are increased (Foo et al., 2013), indicating that the effect of strigolactones in nodulation is regulated differently compared to mycorrhization.

Interestingly, a severe mycorrhization phenotype in rice could be complemented by introducing a copy of Os*D14-LIKE* gene (Gutjahr et al., 2015). Os*D14-Like* is paralogous to Os*D14* and has strong similarities with the *A. thaliana* karrikin receptor *AtKAI2*, which is responsible for detecting the smoke compound karrikin. OsD14 and OsD14-LIKE have been reported to have partially overlapping, but also distinct, functions for strigolactone and karrikin responses, as the *Atkai2* mutants are insensitive to karrikins but weakly responsive to strigolactones (Scaffidi et al., 2014). In addition, it was recently demonstrated that in *A. thaliana* AtD14 and AtD14-like have different affinities for specific strigolactone stereoisomers (Scaffidi et al., 2014). This could indicate that the perception of specific strigolactones is regulated by multiple receptor complexes.

Taken together, the involvement of strigolactones in arbuscular mycorrhiza symbiosis is relatively well described, although several details remain unclear. A possible involvement in rhizobial symbiosis is just starting to be discovered. However, given the distinct nature of both symbioses, the mechanisms by which strigolactones functions are likely different between the two. For ectomycorrhizal and actinorhizal symbioses, no clear data on the involvement of strigolactones is available yet. As strigolactones are plant hormones involved in key developmental processes, it is not surprising that ectomycorrhizal hosts were found to possess the strigolactone biosynthetic genes (Garcia, Delaux, Cope, & Ane, 2015).

4. A CONSERVED SIGNALLING PATHWAY FOR ENDOSYMBIOSES

As mentioned earlier, arbuscular mycorrhizal fungi, rhizobia and some basal *Frankia* species produce LCO signals in a symbiotic context, whereas no evidence has been found that LCO signals are playing a role in ectomycorrhizal symbiosis. This suggests that LCO signalling is a feature of microbes that establish an endosymbiosis rather than an ectosymbiosis.

LCOs are prominent signal molecules that are perceived by the host plant and set in motion symbiotic responses. Genetic studies in legumes, rice, *Parasponia* and the actinorhizal plant species *Datisca glutinosa* (nodulated by *Frankia* sp. harbouring LCO biosynthesis genes), but also in *C. glauca*, a species that is nodulated by *Frankia* sp. that lack LCO biosynthesis genes uncovered a common symbiotic signalling network. This conserved symbiotic network stretches from transmembrane receptor kinases to a network of transcription factors that control the readout of symbiotic signalling (Oldroyd, 2013). A hallmark of endosymbiotic signalling is the induction of regular oscillations of the nuclear calcium concentration. To achieve this a complex of nuclear envelope-localized proteins are essential, including a potassium-permeable channel (encoded by *MtDMI1*, *LjCASTOR*, *LjPOLLUX*), a cyclic nucleotide—gated calcium channel and a calcium ATPase (Capoen et al., 2011; Charpentier et al., 2016; Imaizumi-Anraku et al., 2005; Lévy et al., 2004). The induced calcium oscillations are decoded by a calcium-/calmodulin-dependent kinase (CCaMK), which is the onset of a transcriptional network (Soyano & Hayashi, 2014). Besides some common elements, like the CCaMK interacting transcription factor

LjCYCLOPS, the activated network varies between arbuscular mycorrhizal and root nodule symbioses. For example, activation of the *NIN* transcription factor is essential for root nodule formation in legumes and *C. glauca*, whereas it is not for arbuscular mycorrhizal symbiosis (Clavijo et al., 2015; Marsh et al., 2007; Schauser, Roussis, Stiller, & Stougaard, 1999). Conversely, arbuscular mycorrhizal symbiosis requires activation of GRAS transcription regulators such as *MtRAM1* in *Medicago*, which is not essential for root nodule formation (Gobbato et al., 2012). Despite this divergence in transcriptional responses, the common symbiotic signalling genes are conserved in angiosperm and gymnosperm species that form an arbuscular mycorrhizal symbiosis. By contrast, plants that exclusively establish an ectomycorrhizal symbiosis — e.g., *Pinaceae* species — have lost several of these genes (Garcia et al., 2015). This supports the idea that ectomycorrhizal symbioses are founded on different signalling cues than arbuscular mycorrhizal and root nodule endosymbioses.

4.1 LCO Signalling

Most comprehensive studies on symbiotic signalling have been done in the legume model systems *L. japonicus* and *M. truncatula*. Both species have evolved to interact with a specific rhizobial species (*Mesorhizobium loti* for *L. japonicus* and *S. meliloti* for *M. truncatula*). By using these symbiotic models, it was revealed that rhizobium LCOs are specifically recognized by a heteromeric complex of receptor-like kinases (LysM-RLKs) containing Lysine motif (LysM) domains: named LjNFR1 and LjNFR5 in *L. japonicus* and MtLYK3 and MtNFP in *M. truncatula* (Arrighi et al., 2006; Limpens et al., 2003; Madsen et al., 2003; Radutoiu et al., 2003). The LysM domain is a ubiquitous molecular structure of 42—48 amino acids with a symmetrical βααβ folding. LysM domain-containing proteins were first described in bacteria to bind peptidoglycan (Buist, Steen, Kok, & Kuipers, 2008). In legumes LjNFR1/MtLYK3 and LjNFR5/MtNFP harbour three LysM domains in the receptor region which are essential to recognize specific rhizobium LCOs (Broghammer et al., 2012). In addition, it was found in *L. japonicus* that LjNFR5 interacts also with LjSYMRK, an LRR-type receptor that commits an essential function in symbiotic signalling (Antolín-Llovera, Ried, & Parniske, 2014). Interestingly, This symbiotic receptor kinase (SYMRK) is also essential for arbuscular mycorrhizal symbiosis, whereas both LjNFR1/MtLYK3 and LjNFR5/MtNFP only play an additive role in arbuscular mycorrhizal symbiosis (Oldroyd, 2013). Arbuscular mycorrhizal LCOs are known to trigger lateral root formation in *M. truncatula*, a response that is abolished in the *Mtnfp* knockout mutant (Maillet et al., 2011). *Mtlyk3* and *Ljnfr1* mutants display only a reduced level of infection when inoculated with a low dose of arbuscular mycorrhizal spores (Zhang et al., 2015). Nevertheless, transcriptome studies in *M. truncatula* revealed that MtNFP is playing a prominent role in arbuscular mycorrhizal LCO-induced transcriptional changes (Czaja et al., 2012). Two reasons may explain this discrepancy between knockout phenotype and function. Firstly, the weak arbuscular mycorrhizal symbiosis phenotype of the *Ljnr1/Mtlyk3* and *Ljnfr5/Mtnfp* knockout mutants may be the result of gene redundancy in *L. japonicus* and *M. truncatula*. Both rhizobium LCO receptors evolved upon gene duplication events, giving rise to closely related homologues (De Mita, Streng, Bisseling, & Geurts, 2014; Op den Camp, De Mita, et al., 2011; Op den Camp, Streng, et al., 2011; Young et al., 2011). Expression studies of these homologous genes show that they may also function in rhizobium and/or arbuscular mycorrhizal symbiosis (Rasmussen et al., 2016; Young et al., 2011). Secondly, it was found that arbuscular mycorrhizal fungi not only produce LCOs, but also short-chain chitooligosaccharides (tetra and pentameric COs) as symbiotic signals (Genre et al., 2013). Such COs trigger in part similar symbiotic responses as reported for arbuscular mycorrhizal LCOs, though lack the capacity to promote lateral root formation (Maillet et al., 2011). LCO and CO signals may be perceived by different (symbiotic) receptor complexes.

Nonlegume systems provided additional support for a function of *NFR1/LYK3* and *NFR5/NFP* homologous genes in arbuscular mycorrhizal symbiosis. Reverse genetic studies in *Parasponia andersonii* and tomato (*Solanum lycopersicum*) revealed an essential role for putative NFR5/NFP orthologs in arbuscular mycorrhizal symbiosis (Buendia, Wang, Girardin, & Lefebvre, 2016; Op den Camp, De Mita, et al., 2011; Op den Camp, Streng, et al., 2011). In rice (*O.sativa*), it was demonstrated that the putative orthologue of NFR1/LYK3 - CHITIN-ELICITOR RECEPTOR KINASE 1 (OsCERK1)- plays such role (Miyata et al., 2014; Zhang et al., 2015). In *Frankia* no reverse genetic studies on LysM-RKs have been published yet. However, it is tempting to speculate that in actinorhizal plant species that can be nodulated by cluster 2 *Frankia* species, close homologues of *NFR1/LYK3* and/or *NFR5/NFP* play a symbiotic role in LCO perception.

The finding that COs and the chitin innate immune receptor OsCERK1 commit symbiotic functions uncovered a functional overlap between pathogenicity and symbiosis. Subsequent studies in *L. japonicus* and *M. truncatula* revealed four lines of supportive evidence for such dual function of LCO receptors. (1) Rhizobium LCOs transiently trigger defence-related gene expression in an LjNFR1-dependent manner (Nakagawa et al., 2011). (2) MtNFP has a function in defence against fungal and oomycete pathogens (Ben et al., 2013; Rey, Chatterjee, Buttay, Toulotte, & Schornack, 2015; Rey et al., 2013). (3) Ectopic expression of both receptors – LjNFR5-LjNFR1 or MtNFP-MtLYK3 – in *Nicotiana benthamiana* leaves triggers a hypersensitive response (HR) (Broghammer et al., 2012; Pietraszewska-Bogiel et al., 2013). (4) Ectopic expression of *MtNFP* in *M. truncatula* triggers a premature cell death in nodules (Moling et al., 2014). This, and other studies, also made clear that LCO receptors are under tight posttranslational control in legumes, probably to prevent pathogenic responses. For example, in *M. truncatula* nodules MtNFP and MtLYK3 accumulate only in nodule cells where infection takes place, but both receptors are rapidly removed from the membrane surrounding the rhizobium infection thread (Moling et al., 2014). Furthermore, it was found that LCO receptors are located in lipid-raft–like micro-domains in the plasma membrane, which play an important role in complex formation and receptor turnover (Haney & Long, 2010; Lefebvre et al., 2010). Taken together, these data suggest that dual functioning of LCO receptors in defence and symbiosis is a conserved feature in legumes and nonlegume species.

The biological function of the overlap of LCO receptors in symbiotic and innate immune signalling remains unclear. However, a challenging model can be postulated (Limpens et al., 2015). In this model competition between receptors occur to form multimeric complexes that differ in their functioning. Presence of LCOs (and/or short-chain COs) results in preferential formation of symbiotic receptor complexes at the expense of the formation of complexes that act in innate immunity (Fig. 3). In legumes, such innate immune receptor complex has not yet been characterized. However, studies in rice revealed that perception of chitin oligomers requires an additional LysM-domain-containing receptor, which lacks an intracellular kinase domain (Kaku et al., 2006). This chitin elicitor binding protein (OsCEBiP) binds chitin oligomers and forms a heteromeric complex with OsCERK1 to activate chitin-triggered defence responses (Hayafune et al., 2014; Shimizu et al., 2010). Such innate immune receptor complex may also have a function in symbiosis. It is known that several typical innate immune responses, such as calcium influx, production of reactive oxygen species (ROS), and focal exocytosis are associated with rhizobial and arbuscular mycorrhizal infection (Brewin, 2004). Rhizobium triggers formation of infection threads, which

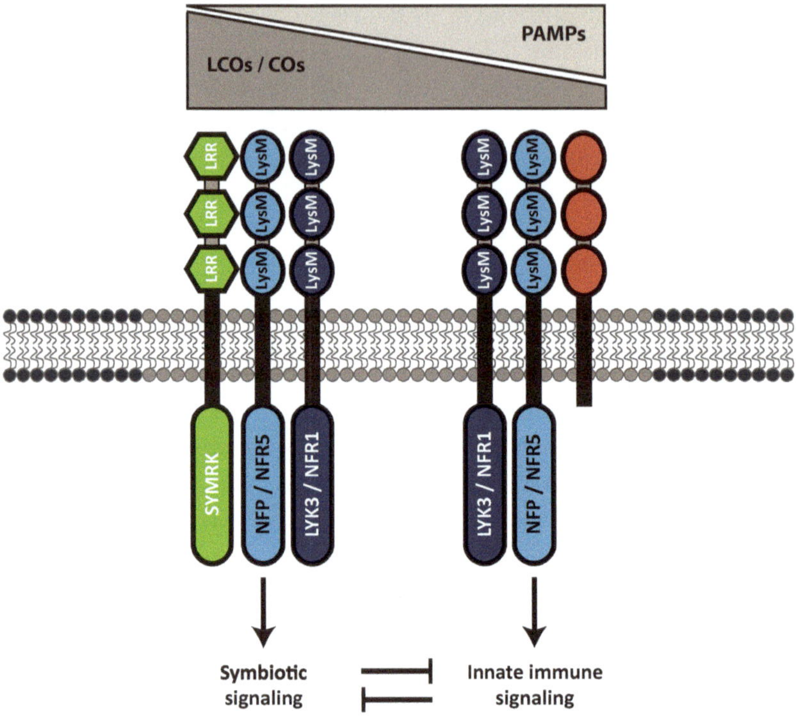

Figure 3 Hypothetical model explaining the dual functioning of LCO receptors in symbiotic and innate immune signalling as uncovered in *Medicago truncatula* and *Lotus japonicus*. Symbiotic signals and pathogen associated molecular patterns (PAMPs) are perceived by NFP/NFR5-NFR1/LYK3 receptor complexes. To commit either symbiotic or innate immune signalling a third receptor is essential. For symbiotic signalling this receptor may be symbiotic receptor kinase (SYMRK), as it interacts with NFP/NFR5. To induce innate immune responses this receptor has not been identified yet, but may have similarities to CeBIP in rice. OsCEBiP binds chitin oligomers and forms a heteromeric complex with the rice homologue of NFR1/LYK3 (OsCERK1) to activate innate immune signalling.

are tip-growing structures. ROS production is thought to facilitate the oxidative cross-linking of the infection thread matrix to allow the formation of a tube-like infection thread (Brewin, 2004). In a scenario that innate immune responses play a symbiotic role, the spatiotemporal regulation of receptor complexes becomes crucial to prevent HR.

4.2 Bypassing LCO Signalling

Besides LCO-mediated signalling, alternative routes occur to mediate symbiotic responses. For example, many *Frankia* species (clusters 1 and 3) do not possess the machinery to produce LCOs (Tisa, Beauchemin, Gtari, Sen, & Wall, 2013). Furthermore, there are some legume lineages — e.g., several Aeschynomene species — that are nodulated by *Bradyrhizobium* strains that lack the highly conserved *nodABC* operon necessary for LCO synthesis (Fabre et al., 2015; Giraud et al., 2007). Nevertheless, studies in actinorhizal plant species *C. glauca* and *Alnus glutinosa* using the non-LCO producing *Frankia* strain Cci3i, revealed that both SYMRK and CCaMK are essential to establish a symbiotic interaction, and activation of symbiotic signalling induces calcium oscillations (Chabaud et al., 2015; Franche et al., 2011; Gherbi et al., 2008; Granqvist et al., 2015). This strongly suggests that the underlying signalling pathway to establish an endosymbiosis is highly conserved, but can be activated by different signalling inputs.

The way non-LCO—producing rhizobia and *Frankia* achieve activation of the common symbiosis signalling pathway may vary. One way is by producing effector-like molecules that are secreted via the type III secretion system (T3SS). This mechanism is used by several rhizobium strains (Okazaki et al., 2016), and studies in soybean revealed that such effectors can bypass NFR1-NFR5-based signalling (Okazaki, Kaneko, Sato, & Saeki, 2013). However, additional mechanisms may also occur. For example, in case of Aeschynomene legumes the common symbiosis signalling pathway can also be activated in a T3SS-independent way (Fabre et al., 2015; Okazaki et al., 2016).

The current hypothesis is that *Frankia* strains of clusters 1 and 3 produce signalling molecules upon host recognition, of which the chemical nature is still poorly understood, but most probably different from LCOs. A first characterization of such signals came from studies on *Frankia* sp. strain CcI3i that nodulates *C. glauca.* The signalling molecules produced by this strain are of low molecular weight, in the range of 500—5000 Da. Moreover, rhizobium and arbuscular mycorrhizal LCOs typically accumulate in the organic fraction upon a butanol extraction, whereas, in the case of *Frankia* CcI3i exudates, only water fractions could induce symbiotic responses (i.e., calcium oscillation). Furthermore, a chitinase treatment on the active water fractions did not affect their signalling capacity (Chabaud et al., 2015). This makes it highly unlikely that this strain produces LCO-type symbiotic signal molecules.

Studies with other *Frankia* strains revealed that, at least within a taxonomic cluster, the symbiotic signals are to a certain level conserved. For example, *A. glutinosa* and *C. glauca* are nodulated by two different *Frankia* strains of the same cluster 1. Despite this strain specificity, *Frankia* sp. strain AC14a that nodulates *A. glutinosa* also induces calcium oscillation responses in *C. glauca*. By contrast, the more distant BCU110501 strain of cluster 3 was unable to induce such responses (Chabaud et al., 2015). This suggests that the symbiotic signals produced by *Frankia* species are partially conserved within a taxonomic cluster, but may differ in a broader phylogenetic context.

5. REPRESSING IMMUNITY

Although innate immune responses may be an integral part of the symbiotic infection process, it is essential that severe immune responses are avoided. Immune responses are controlled by two antagonistic hormones jasmonic acid and salicylic acid. The latter hormone is a major signal in resistance to biotrophic pathogens, whereas defence against necrotrophic mainly relies on jasmonic acid (Pieterse, Van der Does, Zamioudis, Leon-Reyes, & Van Wees, 2012). Both hormones act antagonistically, such that activation of jasmonic acid signalling compromises salicylic acid–dependent innate immune responses, and vice versa.

Studies in legumes suggest that repression of innate immunity is in part controlled by LCO signalling. In alfalfa (*M. sativa*) evidence was found that LCO signalling suppresses salicylic acid–dependent responses. LCO-deficient or incompatible rhizobia induce accumulation of salicylic acid, whereas compatible strains trigger a decrease of this defence hormone (Martinez-Abarca et al., 1998). Similarly, studies in pea (*Pisum sativum*) showed that endomycorrhizal fungi only trigger a transient increase in salicylic acid levels, which is repressed during prolonged colonization. In contrast, in a symbiosis deficient *ccamk* knockout mutant salicylic acid levels remain high upon inoculation with endomycorrhizal fungi, suggesting that this suppression is based on activation of the symbiosis signalling network (Blilou, Ocampo, Garcıa-Garrido, & García-Garrido, 1999). Interestingly, defence responses in nonlegumes (*Zea mays, Setaria viridis*), and even in non–arbuscular mycorrhizal plants (*A. thaliana*) seem to be downregulated upon LCO perception; however, it is currently unclear how this downregulation is linked to JA and SA signalling (Liang et al., 2013; Tanaka et al., 2015).

The Jasmonic acid – salicylic acid balance is in part controlled by DELLA GRAS-type transcriptional regulators (Navarro et al., 2008). DELLAs promote jasmonic acid signalling by binding JAZ (Jasmonate zim-domain) repressor proteins (Hou, Lee, Xia, Yan, & Yu, 2010). JAZ proteins repress jasmonic acid signalling upon binding with the MYC2 transcriptional activator (Boter, Ruı, & Abdeen, 2004; Hou et al., 2010). As MYC2 activity promotes DELLA accumulation, this results in a feedforward loop in jasmonic acid signalling (Wild et al., 2012; Yang et al., 2012). Several experiments indicate that endomycorrhizal fungi and rhizobium exploit this pathway, thereby indirectly reducing salicylic acid responses. *Della* knockout mutants in *M. truncatula* and rice are impaired in nodulation and/or arbuscule formation (Floss, Levy, Lévesque-Tremblay, Pumplin, & Harrison, 2013; Fonouni-Farde et al., 2016; Pimprikar et al., 2016; Yu et al., 2014). These phenotypes can be mimicked by application of gibberellins, whereas ectopic expression of a dominant active *DELLA* allele (*MtDELLA1Δ18*) promotes symbiotic responses (Floss et al., 2013; Jin et al., 2016; Pimprikar et al., 2016). Interestingly, the dominant active allele can also complement the *cyclops* symbiotic signalling mutant (Floss et al., 2013). This is likely due to the fact that in *M. truncatula* the DELLA1 protein was found to be able to form a complex with CYCLOPS and CCaMK, together activating the RAM1 GRAS-type transcriptional regulator (Pimprikar et al., 2016). Taken together this suggests that MtDELLA1, by interacting with JAZ proteins, plays an important role in the LCO-signalling network and the promotion of endomycorrhizal symbiosis through the modulation of jasmonic acid—salicylic acid balance.

Besides LCO triggered repression of immunity, plant immunity can also be manipulated by microbe secreted effector proteins. Studies in arbuscular mycorrhiza and ectomycorrhiza uncovered several small secreted effector proteins that are produced by the arbuscular mycorrhizal fungus *R. irregularis* and the ectomycorrhizal fungus *L. bicolor* (Lin et al., 2014; Martin et al., 2008; Tisserant et al., 2013). The mode of action of two such effector proteins has been characterized.

The *R. irregularis* effector protein RiSP7 is secreted into *M. truncatula* root cells, where it localizes in the nucleus and interacts with a defence controlling ethylene-responsive transcription factor (MtERF19) (Kloppholz, Kuhn, & Requena, 2011). In *M. truncatula* roots this gene is highly expressed upon pathogenic interaction, but only transiently during arbuscular mycorrhizal colonization. Ectopic expression of *RiSP7* in *M. truncatula* roots positively affects mycorrhizal colonization, while reducing defence responses. Intriguingly, RiSP7 has some similarity to the secreted NodO protein of *Rhizobium leguminosarum*, which enhances LCO signalling in the host plant. However, localization studies suggest that NodO localizes in the plant membrane, rather than acting as a nuclear effector (Economou, Hamilton, Johnston, & Downie, 1990; Sutton, Lea, & Downie, 1994).

The ectomycorrhizal fungus *L. bicolor* expresses the *LbMiSSP7* gene encoding a secreted effector protein in response to plant exuded flavonoids (Plett & Martin, 2012). In black cottonwood poplar (*Populus trichocarpa*) it was shown that LbMiSSP7 is secreted in root cells where it localizes in the nucleus. There it stabilizes a JAZ protein (PtJAZ6) by direct interaction (Plett et al., 2014). As outlined above, JAZ proteins are repressors of jasmonic acid–triggered immunity. Generally, JAZ proteins are degraded upon interaction with the F-box protein COI1 (coronatine-insensitive 1). This degradation is triggered by jasmonic acid signalling. LbMiSSP7 interaction to PtJAZ6 affects formation of the JAZ–COI1 complex. This prevents the jasmonic acid–dependent degradation of JAZ, resulting in reduced plant immune responses. Given that jasmonic acid is a negative regulator of ectomycorrhizal symbiosis, counteracting this plant innate immune response promotes the plant–fungus interaction.

6. PERSPECTIVES IN SYMBIOTIC SIGNALLING

Central questions for future research will be on specificity of symbiotic signalling. How can a single symbiotic network that is conserved in most land plants trigger distinct root phenotypes? Since the symbiotic signalling network is basically conserved in most plant species the differences in the readout may be determined by yet unknown factors, such as the hormonal balance and/or the nutrient status of the root. For example, recently it was shown that *M. truncatula* lateral roots have an increased sensitivity to rhizobium LCOs compared with the main root. This indicates that susceptibility of a plant root varies, depending on the developmental and/or nutrient status (Sun et al., 2015).

Additional questions concerning specificity can also be addressed concerning the plant exuded flavonoids that act as attractants for symbiotic microbes. As shown for naringenin, these compounds are perceived by a diverging range of symbionts. Most probably this range extends to other soil borne microbes, most of which will not be symbiotic. Therefore, perhaps exuded flavonoids do not act as specific signals, but rather are more generic signals to which any root microbe can respond. For example, it was reported that exuded flavonoids may play a role also in phosphate and iron acquisition (Cesco, Neumann, Tomasi, Pinton, & Weisskopf, 2010). In addition, the finding that flavonoids — similar to strigolactones — have a dual function, not only act as an attractant, but also function as endogenous plant signal interfering with auxin homoeostasis, provides novel leads in symbiosis research.

Extending the range of model systems that are amenable for molecular genetic studies provided novel insights into symbiotic signalling. Establishment of new protocols for culturing *Frankia* and arbuscular mycorrhizal fungi, host-induced gene silencing to trigger fungal gene expression, transformation of the ectomycorrhizal fungus *L. bicolor*, the actinorhizal plants *Datsica*, *Casuarina*, and the nonlegume rhizobia host *Parasponia*, in combination with microbial genome sequencing, has opened new avenues. Although unravelling symbiotic signalling in these systems is still in its infancy, findings that have been achieved since the turn of the century are already groundbreaking. As mentioned above, it was demonstrated that especially in the endosymbioses (*Frankia*, rhizobium and arbuscular mycorrhiza) commonalities occur in symbiotic signalling (Gherbi et al., 2008; Op den Camp, De Mita, et al., 2011; Op den Camp, Streng, et al., 2011). One such commonality is that symbionts recognize plant-secreted flavonoids and strigolactones. Another common theme is the use of LCO or CO signals of microbial origin of which biosynthesis is activated upon recognition of plant exuded molecules like flavonoids and/or strigolactones. LCOs/COs activate a conserved symbiotic network in plants that controls the diverse signalling output of the different symbiotic interactions (Parniske, 2008). Furthermore, it became apparent that LCO-induced signalling can be bypassed. Especially in *Frankia* this appears to be a common strategy. Nevertheless, first studies indicate that LCO-independent signalling relies on the same symbiotic signalling network as identified in LCO-dependent systems. Uncovering the nature of the non-LCO signal molecules in *Frankia* and rhizobia will add a new building brick in the symbiotic signalling network.

Commonly new insights in the molecular aspects of root symbiosis are mainly generated by studying legume models *M. truncatula* and *L. japonicus*. With new model species in place in combination with next generation sequence technologies, this field will be revolutionized in the years to come.

ACKNOWLEDGEMENTS

R.Geurts is supported by NWO-VICI 865.13.001 and W.Kohlen by NWO-VENI 863.15.010.

REFERENCES

Abdel-Lateif, K., Vaissayre, V., Gherbi, H., Verries, C., Meudec, E., Perrine-Walker, F., … Hocher, V. (2013). Silencing of the chalcone synthase gene in *Casuarina glauca* highlights the important role of flavonoids during nodulation. *The New Phytologist, 199*, 1012–1021. http://dx.doi.org/10.1111/nph.12326.

Akiyama, K., Matsuzaki, K., & Hayashi, H. (June, 2005). Plant sesquiterpenes induce hyphal branching in arbuscular mycorrhizal fungi. *Nature, 435*, 824–827. http://dx.doi.org/10.1038/nature03608.

Akiyama, K., Ogasawara, S., Ito, S., & Hayashi, H. (2010). Structural requirements of strigolactones for hyphal branching in AM fungi. *Plant and Cell Physiology, 51*(7), 1104–1117. http://dx.doi.org/10.1093/pcp/pcq058.

Al-Babili, S., & Bouwmeester, H. J. (2015). Strigolactones, a novel carotenoid-derived plant hormone. *Annual Review of Plant Biology, 66*(1), 161–186. http://dx.doi.org/10.1146/annurev-arplant-043014-114759.

Alder, A., Jamil, M., Marzorati, M., Bruno, M., Vermathen, M., Bigler, P., … Al-Babili, S. (2012). The path from -carotene to carlactone, a strigolactone-like plant hormone. *Science, 335*(6074), 1348–1351. http://dx.doi.org/10.1126/science.1218094.

Antolín-Llovera, M., Ried, M. K., & Parniske, M. (2014). Cleavage of the SYMBIOSIS RECEPTOR- LIKE KINASE ectodomain promotes complex formation with Nod factor receptor 5. *Current Biology, 24*, 422–427.

Arrighi, J. F., Barre, A., Ben Amor, B., Bersoult, A., Soriano, L. C., Mirabella, R., … Gough, C. (2006). The *Medicago truncatula* lysine motif-receptor-like kinase gene family includes NFP and new nodule-expressed genes. *Plant Physiology, 142*(1), 265–279. http://dx.doi.org/10.1104/pp.106.084657.

Auguy, F., Abdel-Lateif, K., Doumas, P., Badin, P., Guerin, V., Bogusz, D., & Hocher, V. (2011). Activation of the isoflavonoid pathway in actinorhizal symbioses. *Functional Plant Biology, 38*(8–9), 690–696. http://dx.doi.org/10.1071/FP11014.

Balestrini, R., & Bonfante, P. (2005). The interface compartment in arbuscular mycorrhizae: a special type of plant cell wall? *Plant Biosystems – An International Journal Dealing with All Aspects of Plant Biology, 139*(1), 8–15. http://dx.doi.org/10.1080/11263500500056799.

Balestrini, R., & Bonfante, P. (June, 2014). Cell wall remodeling in mycorrhizal symbiosis: a way towards biotrophism. *Frontiers in Plant Science, 5*, 237. http://dx.doi.org/10.3389/fpls.2014.00237.

Beauchemin, N. J., Furnholm, T., Lavenus, J., Svistoonoff, S., Doumas, P., Bogusz, D., … Tisa, L. S. (2012). Casuarina root exudates alter the physiology, surface properties, and plant infectivity of *Frankia* sp. strain CcI3. *Applied and Environmental Microbiology, 78*(2), 575–580. http://dx.doi.org/10.1128/AEM.06183-11.

Bécard, G., Douds, D. D., & Pfeffer, P. E. (1992). Extensive in vitro hyphal growth of vesicular-arbuscular mycorrhizal fungi in the presence of CO_2 and flavonols extensive in vitro hyphal growth of vesicular-arbuscular mycorrhizal fungi in the presence of CO_2 and flavonols. *Applied and Environmental Microbiology, 58*(3), 821–825.

Bécard, G., Taylor, L. P., Douds, D. D., Pfeffer, P. E., & Doner, L. W. (1995). Flavonoids are not necessary plant signal compounds in arbuscular mycorrhizal symbioses. *Molecular Plant-Microbe Interactions*. http://dx.doi.org/10.1094/MPMI-8-0252.

Behm, J. E., Geurts, R., & Kiers, E. T. (2014). Parasponia: a novel system for studying mutualism stability. *Trends in Plant Science, 19*(12), 757–763. http://dx.doi.org/10.1016/j.tplants.2014.08.007.

Ben, C., Toueni, M., Montanari, S., Tardin, M.-C., Fervel, M., Negahi, A., … Gentzbittel, L. (2013). Natural diversity in the model legume *Medicago truncatula* allows identifying distinct genetic mechanisms conferring partial resistance to *Verticillium* wilt. *Journal of Experimental Botany, 64*, 317–332. http://dx.doi.org/10.1093/jxb/ers337.

Besserer, A., Bécard, G., Jauneau, A., Roux, C., & Séjalon-Delmas, N. (2008). GR24, a synthetic analog of strigolactones, stimulates the mitosis and growth of the arbuscular mycorrhizal fungus *Gigaspora rosea* by boosting its energy metabolism. *Plant Physiology, 148*(1), 402–413. http://dx.doi.org/10.1104/pp.108.121400.

Besserer, A., Puech-Pagès, V., Kiefer, P., Gomez-Roldan, V., Jauneau, A., Roy, S., … Séjalon-Delmas, N. (2006). Strigolactones stimulate arbuscular mycorrhizal fungi by activating mitochondria. *PLoS Biology, 4*(7), 1239–1247. http://dx.doi.org/10.1371/journal.pbio.0040226.

Blilou, I., Ocampo, J. A., Garcıa-Garrido, M., & García-Garrido, J. M. (1999). Resistance of pea roots to endomycorrhizal fungus or Rhizobium correlates with enhanced levels of endogenous salicylic acid. *Journal of Experimental Botany, 50*(340), 1663–1668. http://dx.doi.org/10.1093/jxb/50.340.1663.

Boter, M., Ruı, O., & Abdeen, A. (2004). Conserved MYC transcription factors play a key role in jasmonate signaling both in tomato and *Arabidopsis*. *Genes and Development, 2*, 1577–1591. http://dx.doi.org/10.1101/gad.297704.tion.

Brewer, P. B., Koltai, H., & Beveridge, C. A. (2013). Diverse roles of strigolactones in plant development. *Molecular Plant, 6*(1), 18–28. http://dx.doi.org/10.1093/mp/sss130.

Brewin, N. J. (2004). Plant cell wall remodelling in the Rhizobium-legume symbiosis. *Critical Reviews in Plant Sciences, 23*(4), 293–316. http://dx.doi.org/10.1080/07352680490480734.

Broghammer, A., Krusell, L., Blaise, M., Sauer, J., Sullivan, J. T., Maolanon, N., … Stougaard, J. (2012). Legume receptors perceive the rhizobial lipochitin oligosaccharide signal molecules by direct binding. *Proceedings of the National Academy of Sciences of the United States of America, 109*(34), 13859–13864. http://dx.doi.org/10.1073/pnas.1205171109.

Brown, D. E., Rashotte, A. M., Murphy, A. S., Normanly, J., Tague, B. W., Peer, W. A., … Muday, G. K. (2001). Flavonoids act as negative regulators of auxin transport in vivo in arabidopsis. *Plant Physiology, 126*(2), 524–535. http://dx.doi.org/10.1104/pp.126.2.524.

Buendia, L., Wang, T., Girardin, A., & Lefebvre, B. (2016). The LysM receptor-like kinase SlLYK10 regulates the arbuscular mycorrhizal symbiosis in tomato. *The New Phytologist, 210*(1), 184–195. http://dx.doi.org/10.1111/nph.13753.

Buist, G., Steen, A., Kok, J., & Kuipers, O. P. (April 2008). LysM, a widely distributed protein motif for binding to (peptido)glycans. *Molecular Microbiology, 68*, 838–847. http://dx.doi.org/10.1111/j.1365-2958.2008.06211.x.

Capoen, W., Sun, J., Wysham, D., Otegui, M. S., Venkateshwaran, M., Hirsch, S., … Oldroyd, G. E. D. (2011). Nuclear membranes control symbiotic calcium signaling of legumes. *Proceedings of the National Academy of Sciences of the United States of America, 108*(34), 14348–14353. http://dx.doi.org/10.1073/pnas.1107912108.

Catoira, R., Galera, C., de Billy, F., Penmetsa, R. V., Journet, E. P., Maillet, F., … Denarie, J. (2000). Four genes of *Medicago truncatula* controlling components of a nod factor transduction pathway. *The Plant Cell, 12*(9), 1647–1666. Retrieved from http://www.ncbi.nlm.nih.gov/pubmed/11006338.

Cesco, S., Neumann, G., Tomasi, N., Pinton, R., & Weisskopf, L. (2010). Release of plant-borne flavonoids into the rhizosphere and their role in plant nutrition. *Plant Soil, 329*, 1–25. http://dx.doi.org/10.1007/s11104-009-0266-9.

Chabaud, M., Gherbi, H., Pirolles, E., Vaissayre, V., Moukouanga, D., Franche, C., … Svistoonoff, S. (2015). Rapid report chitinase-resistant hydrophilic symbiotic factors secreted by Frankia activate both Ca^{2+} spiking and NIN gene expression in the actinorhizal plant *Casuarina glauca*. *The New Phytologist*. http://dx.doi.org/10.1111/nph.13732.

Charpentier, M., Sun, J., Martins, T. V., Radhakrishnan, G. V., Findlay, K., Soumpourou, E., … Oldroyd, G. E. D. (2016). Nuclear-localized cyclic nucleotide–gated channels mediate symbiotic calcium oscillations. *Science, 352*(6289), 1102–1105. http://dx.doi.org/10.1126/science.aae0109.

Chen, X.-C., Feng, J., Hou, B.-H., Li, F.-Q., Li, Q., & Hong, G.-F. (2005). Modulating DNA bending affects NodD-mediated transcriptional control in *Rhizobium leguminosarum. Nucleic Acids Research, 33*(8), 2540–2548. http://dx.doi.org/10.1093/nar/gki537.

Clavijo, F., Diedhiou, I., Vaissayre, V., Brottier, L., Acolatse, J., Moukouanga, D., … Svistoonoff, S. (2015). The *Casuarina* NIN gene is transcriptionally activated throughout *Frankia* root infection as well as in response to bacterial diffusible signals. *The New Phytologist, 208*, 887–903. http://dx.doi.org/10.1111/nph.13506.

Coleman-Derr, D., & Tringe, S. G. (June 2014). Building the crops of tomorrow: advantages of symbiont-based approaches to improving abiotic stress tolerance. *Frontiers in Microbiology, 5*, 1–6. http://dx.doi.org/10.3389/fmicb.2014.00283.

Czaja, L. F., Hogekamp, C., Lamm, P., Maillet, F., Martinez, E.a., Samain, E., … Hohnjec, N. (2012). Transcriptional responses toward diffusible signals from symbiotic microbes reveal MtNFP- and MtDMI3-dependent reprogramming of host gene expression by arbuscular mycorrhizal fungal Lipochitooligosaccharides. *Plant Physiology, 159*(4), 1671–1685. http://dx.doi.org/10.1104/pp.112.195990.

De Cuyper, C., Fromentin, J., Yocgo, R. E., De Keyser, A., Guillotin, B., Kunert, K., … Goormachtig, S. (2014). From lateral root density to nodule number, the strigolactone analogue GR24 shapes the root architecture of *Medicago truncatula. Journal of Experimental Botany, 66*(1), 137–146. http://dx.doi.org/10.1093/jxb/eru404.

De Mita, S., Streng, A., Bisseling, T., & Geurts, R. (2014). Evolution of a symbiotic receptor through gene duplications in the legume-rhizobium mutualism. *The New Phytologist, 201*(3), 961–972. http://dx.doi.org/10.1111/nph.12549.

Deinum, E. E., Geurts, R., Bisseling, T., & Mulder, B. M. (May 2012). Modeling a cortical auxin maximum for nodulation: different signatures of potential strategies. *Frontiers in Plant Science, 3*, 1–19. http://dx.doi.org/10.3389/fpls.2012.00096.

Dor, E., Joel, D. M., Kapulnik, Y., Koltai, H., & Hershenhorn, J. (2011). The synthetic strigolactone GR24 influences the growth pattern of phytopathogenic fungi. *Planta, 234*(2), 419–427. http://dx.doi.org/10.1007/s00425-011-1452-6.

Economou, A., Hamilton, W. D., Johnston, A. W., & Downie, J. A. (1990). The rhizobium nodulation gene nodO encodes a Ca2(+)-binding protein that is exported without N-terminal cleavage and is homologous to haemolysin and related proteins. *The EMBO Journal, 9*(2), 349–354. Retrieved from http://www.ncbi.nlm.nih.gov/pubmed/2303029.

Ellouze, W., Hamel, C., Cruz, A. F., Ishii, T., Gan, Y., Bouzid, S., & St-Arnaud, M. (2012). Phytochemicals and spore germination: at the root of AMF host preference? *Applied Soil Ecology, 60*, 98–104. http://dx.doi.org/10.1016/j.apsoil.2012.02.004.

Fabre, S., Gully, D., Poitout, A., Patrel, D., Arrighi, J.-F., Giraud, E., … Cartieaux, F. (December 2015). The Nod factor-independent nodulation in *Aeschynomene evenia* required the common plant-microbe symbiotic "toolkit". *Plant Physiology, 169*, 2654–2664. http://dx.doi.org/10.1104/pp.15.01134.

Falcone Ferreyra, M. L., Rius, S. P., & Casati, P. (September 2012). Flavonoids: biosynthesis, biological functions, and biotechnological applications. *Frontiers in Plant Science, 3*, 1–15. http://dx.doi.org/10.3389/fpls.2012.00222.

Ferrer, J. L., Austin, M. B., Stewart, C., & Noel, J. P. (2008). Structure and function of enzymes involved in the biosynthesis of phenylpropanoids. *Plant Physiology and Biochemistry, 46*(3), 356–370. http://dx.doi.org/10.1016/j.plaphy.2007.12.009.

Floss, D. S., Levy, J. G., Lévesque-Tremblay, V., Pumplin, N., & Harrison, M. J. (2013). DELLA proteins regulate arbuscule formation in arbuscular mycorrhizal symbiosis. *Proceedings of the National Academy of Sciences of the United States of America, 110*(51), E5025–E5034. http://dx.doi.org/10.1073/pnas.1308973110.

Fonouni-Farde, C., Tan, S., Baudin, M., Brault, M., Wen, J., Mysore, K. S., … Diet, A. (2016). DELLA-mediated gibberellin signalling regulates Nod factor signalling and rhizobial infection. *Nature Communications,* 7, 12636. http://dx.doi.org/10.1038/ncomms12636.

Foo, E., Yoneyama, K., Hugill, C. J., Quittenden, L. J., & Reid, J. B. (2013). Strigolactones and the regulation of pea symbioses in response to nitrate and phosphate deficiency. *Molecular Plant, 6*(1), 76–87. http://dx.doi.org/10.1093/mp/sss115.

Franche, C., Gherbi, H., Benabdoun, M., Svistoonoff, S., Hocher, V., & Bogusz, D. (2011). New insights in the molecular events underlying actinorhizal nodulation in the tropical tree *Casuarina glauca. BMC Proceedings, 5*(Suppl. 7), O33. http://dx.doi.org/10.1186/1753-6561-5-S7-O33.

Garcia, K., Delaux, P.-M., Cope, K. R., & Ane, J.-M. (2015). Molecular signals required for the establishment and maintenance of ectomycorrhizal symbioses. *The New Phytologist, 208,* 79–87. http://dx.doi.org/10.1080/15411790701409895.

Garg, N., & Singla, P. (2016). Stimulation of nitrogen fixation and trehalose biosynthesis by naringenin (Nar) and arbuscular mycorrhiza (AM) in chickpea under salinity stress. *Plant Growth Regulation,* 1–18. http://dx.doi.org/10.1007/s10725-016-0146-2.

Gea, L., Normand, L., Vian, B., & Gay, G. (December 1994). Structural aspects of ectomycorrhiza of *Pinus Pinaster* (Ait) Sol formed by an Iaa overproducer mutant of *Hebeloma cylindrosporum* Romagnesi. *The New Phytologist, 128*(4), 659–670. http://dx.doi.org/10.1111/j.1469-8137.1994.tb04030.x.

Genre, A., Chabaud, M., Balzergue, C., Puech-Pagès, V., Novero, M., Rey, T., … Barker, D. G. (2013). Short-chain chitin oligomers from arbuscular mycorrhizal fungi trigger nuclear Ca^{2+} spiking in *Medicago truncatula* roots and their production is enhanced by strigolactone. *The New Phytologist, 198,* 190–202. http://dx.doi.org/10.1111/nph.12146.

Geurts, R., Lillo, A., & Bisseling, T. (2012). Exploiting an ancient signalling machinery to enjoy a nitrogen fixing symbiosis. *Current Opinion in Plant Biology, 15*(4), 438–443. http://dx.doi.org/10.1016/j.pbi.2012.04.004.

Geurts, R., Xiao, T. T., & Reinhold-Hurek, B. (2016). What does it take to evolve a nitrogen-fixing endosymbiosis? *Trends in Plant Science, 21*(3), 199–208. http://dx.doi.org/10.1016/j.tplants.2016.01.012.

Gherbi, H., Markmann, K., Svistoonoff, S., Estevan, J., Autran, D., Giczey, G., … Bogusz, D. (2008). SymRK defines a common genetic basis for plant root endosymbioses with arbuscular mycorrhiza fungi, rhizobia. *Proceedings of the National Academy of Sciences of the United States of America, 105*(12), 4928–4932. http://dx.doi.org/10.1073/pnas.0710618105.

Giraud, E., Moulin, L., Vallenet, D., Barbe, V., Cytryn, E., Avarre, J.-C., … Sadowsky, M. (2007). Legumes symbioses: absence of Nod genes in photosynthetic bradyrhizobia. *Science, 316*(5829), 1307–1312. http://dx.doi.org/10.1126/science.1139548.

Gobbato, E., Marsh, J. F., Vernié, T., Wang, E., Maillet, F., Kim, J., … Oldroyd, G. E. D. (2012). A GRAS-type transcription factor with a specific function in mycorrhizal signaling. *Current Biology, 22*(23), 2236–2241. http://dx.doi.org/10.1016/j.cub.2012.09.044.

Gomez-Roldan, V., Fermas, S., Brewer, P. B., Puech-Pagès, V., Dun, E. A., Pillot, J.-P., … Rochange, S. F. (2008). Strigolactone inhibition of shoot branching. *Nature, 455*(7210), 189–194. http://dx.doi.org/10.1038/nature07271.

Granqvist, E., Sun, J., Op Den Camp, R., Pujic, P., Hill, L., Normand, P., … Oldroyd, G. E. D. (2015). Bacterial-induced calcium oscillations are common to nitrogen-fixing associations of nodulating legumes and nonlegumes. *The New Phytologist, 207*(3), 551–558. http://dx.doi.org/10.1111/nph.13464.

Gtari, M., Tisa, L. S., & Normand, P. (2013). Diversity of *Frankia* strains, actinobacterial symbionts of actinorhizal plants. In R. Aroca (Ed.), *Symbiotic endophytes* (Vol. 37, pp. 123–148). Berlin, Heidelberg: Springer. http://dx.doi.org/10.1007/978-3-642-39317-4.

Gutjahr, C., Gobbato, E., Choi, J., Riemann, M., Johnston, M. G., Summers, W., … Paszkowski, U. (2015). Rice perception of symbiotic arbuscular mycorrhizal fungi requires the karrikin receptor complex. *Science, 350*(6267), 1521–1524. http://dx.doi.org/10.1126/science.aac9715.

Hamiaux, C., Drummond, R. S. M. S. M., Janssen, B. J. J., Ledger, S. E. E., Cooney, J. M. M., Newcomb, R. D. D., & Snowden, K. C. C. (2012). DAD2 is an α/β hydrolase likely to be involved in the perception of the plant branching hormone, strigolactone. *Current Biology, 22*(21), 2032–2036. http://dx.doi.org/10.1016/j.cub.2012.08.007.

Haney, C. H., & Long, S. R. (2010). Plant flotillins are required for infection by nitrogen-fixing bacteria. *Proceedings of the National Academy of Sciences of the United States of America, 107*(1), 478–483. http://dx.doi.org/10.1073/pnas.0910081107.

Hartwig, U. A., Joseph, C. M., & Phillips, D. A. (1991). Flavonoids released naturally from alfalfa seeds enhance growth rate of *Rhizobium meliloti*. *Plant Physiology, 95*(3), 797–803. http://dx.doi.org/10.1104/pp.95.3.797.

Hayafune, M., Berisio, R., Marchetti, R., Silipo, A., Kayama, M., Desaki, Y., … Shibuya, N. (2014). Chitin-induced activation of immune signaling by the rice receptor CEBiP relies on a unique sandwich-type dimerization. *Proceedings of the National Academy of Sciences of the United States of America, 111*, E404–E413. http://dx.doi.org/10.1073/pnas.1312099111.

Heckmann, A. B., Lombardo, F., Miwa, H., Perry, J. A., Bunnewell, S., Parniske, M., … Downie, J. A. (2006). *Lotus japonicus* nodulation requires two GRAS domain regulators, one of which is functionally conserved in a non-legume. *Plant Physiology, 142*(4), 1739–1750. http://dx.doi.org/10.1104/pp.106.089508.

Hirsch, A. M., Bhuvaneswari, T. V., Torrey, J. G., & Bisseling, T. (1989). Early nodulin genes are induced in alfalfa root outgrowths elicited by auxin transport inhibitors. *Proceedings of the National Academy of Sciences of the United States of America, 86*(4), 1244–1248. http://dx.doi.org/10.1073/pnas.86.4.1244.

Honma, M. A., Asomaning, M., & Ausubel, F. M. (1990). *Rhizobium meliloti* nodD genes mediate host-specific activation of nodABC. *Journal of Bacteriology, 172*(2), 901–911. Retrieved from http://www.ncbi.nlm.nih.gov/pubmed/2298703.

Hou, X., Lee, L. Y. C., Xia, K., Yan, Y., & Yu, H. (2010). DELLAs modulate jasmonate signaling via competitive binding to JAZs. *Developmental Cell, 19*(6), 884–894. http://dx.doi.org/10.1016/j.devcel.2010.10.024.

Imaizumi-Anraku, H., Takeda, N., Charpentier, M., Perry, J., Miwa, H., Umehara, Y., … Hayashi, M. (2005). Plastid proteins crucial for symbiotic fungal and bacterial entry into plant roots. *Nature, 433*(7025), 527–531. http://dx.doi.org/10.1038/nature03237.

Jin, Y., Liu, H., Luo, D., Yu, N., Dong, W., Wang, C., … Wang, E. (2016). DELLA proteins are common components of symbiotic rhizobial and mycorrhizal signalling pathways. *Nature Communications,* 7, 12433. http://dx.doi.org/10.1038/ncomms12433.

Jones, D. L., Nguyen, C., & Finlay, R. D. (2009). Carbon flow in the rhizosphere: carbon trading at the soil-root interface. *Plant and Soil, 321*(1–2), 5–33. http://dx.doi.org/10.1007/s11104-009-9925-0.

Kaku, H., Nishizawa, Y., Ishii-Minami, N., Akimoto-Tomiyama, C., Dohmae, N., Takio, K., ... Shibuya, N. (2006). Plant cells recognize chitin fragments for defence signaling through a plasma membrane receptor. *Proceedings of the National Academy of Sciences of the United States of America, 103*, 11086–11091. http://dx.doi.org/10.1073/pnas.0508882103.

Kikuchi, K., Matsushita, N., Suzuki, K., & Hogetsu, T. (2007). Flavonoids induce germination of basidiospores of the ectomycorrhizal fungus *Suillus bovinus*. *Mycorrhiza, 17*(7), 563–570. http://dx.doi.org/10.1007/s00572-007-0131-8.

Kloppholz, S., Kuhn, H., & Requena, N. (2011). A secreted fungal effector of *Glomus intraradices* promotes symbiotic biotrophy. *Current Biology, 21*(14), 1204–1209. http://dx.doi.org/10.1016/j.cub.2011.06.044.

Kohlen, W., Charnikhova, T., Lammers, M., Pollina, T., Tóth, P., Haider, I., ... López-Ráez, J. A. (2012). The tomato carotenoid cleavage dioxygenase8 (SlCCD8) regulates rhizosphere signaling, plant architecture and affects reproductive development through strigolactone biosynthesis. *The New Phytologist, 196*(2), 535–547. http://dx.doi.org/10.1111/j.1469-8137.2012.04265.x.

Kohler, A., Kuo, A., Nagy, L. G., Morin, E., Barry, K. W., Buscot, F., ... Martin, F. (2015). Convergent losses of decay mechanisms and rapid turnover of symbiosis genes in mycorrhizal mutualists. *Nature Genetics, 47*(4), 410–415. http://dx.doi.org/10.1038/ng.3223.

Kosslak, R. M., Bookland, R., Barkei, J., Paaren, H. E., & Appelbaum, E. R. (1987). Induction of *Bradyrhizobium japonicum* common nod genes by isoflavones isolated from *Glycine max*. *Proceedings of the National Academy of Sciences of the United States of America, 84*(21), 7428–7432. Retrieved from http://www.ncbi.nlm.nih.gov/pubmed/16593884.

Kretzschmar, T., Kohlen, W., Sasse, J., Borghi, L., Schlegel, M., Bachelier, J. B., ... Martinoia, E. (2012). A petunia ABC protein controls strigolactone-dependent symbiotic signalling and branching. *Nature*. http://dx.doi.org/10.1038/nature10873. Nature Publishing Group.

Lefebvre, B., Timmers, T., Mbengue, M., Moreau, S., Hervé, C., Tóth, K., ... Ott, T. (2010). A remorin protein interacts with symbiotic receptors and regulates bacterial infection. *Proceedings of the National Academy of Sciences of the United States of America, 107*(5), 2343–2348. http://dx.doi.org/10.1073/pnas.0913320107.

Lévy, J., Bres, C., Geurts, R., Chalhoub, B., Kulikova, O., Duc, G., ... Debellé, F. (2004). A putative Ca^{2+} and calmodulin-dependent protein kinase required for bacterial and fungal symbioses. *Science, 303*(5662), 1361–1364. http://dx.doi.org/10.1126/science.1093038.

Liang, Y., Cao, Y., Tanaka, K., Thibivilliers, S., Wan, J., Choi, J., ... Stacey, G. (2013). Nonlegumes respond to rhizobial Nod factors by suppressing the innate immune response. *Science, 341*(6152), 1384–1387. http://dx.doi.org/10.1126/science.1242736.

Limpens, E., Franken, C., Smit, P., Willemse, J., Bisseling, T., & Geurts, R. (2003). LysM domain receptor kinases regulating rhizobial Nod factor-induced infection. *Science, 302*(5645), 630–633. http://dx.doi.org/10.1126/science.1090074.

Limpens, E., van Zeijl, A., & Geurts, R. (2015). Lipochitooligosaccharides modulate plant host immunity to enable endosymbioses. *Annual Review of Phytopathology, 53*(1), 311–334. http://dx.doi.org/10.1146/annurev-phyto-080614-120149.

Lin, K., Limpens, E., Zhang, Z., Ivanov, S., Saunders, D. G. O., Mu, D., ... Huang, S. (2014). Single nucleus genome sequencing reveals high similarity among nuclei of an endomycorrhizal fungus. *PLoS Genetics, 10*, e1004078. http://dx.doi.org/10.1371/journal.pgen.1004078.

Liu, J., Novero, M., Charnikhova, T., Ferrandino, A., Schubert, A., Ruyter-Spira, C., ... Cardinale, F. (2013). Carotenoid cleavage dioxygenase 7 modulates plant growth, reproduction, senescence, and determinate nodulation in the model legume *Lotus japonicus*. *Journal of Experimental Botany, 64*(7), 1967–1981. http://dx.doi.org/10.1093/jxb/ert056.

Liu, W., Kohlen, W., Lillo, A., Op den Camp, R., Ivanov, S., Hartog, M., … Geurts, R. (2011). Strigolactone biosynthesis in *Medicago truncatula* and rice requires the symbiotic GRAS-type transcription factors NSP1 and NSP2. *The Plant Cell, 23*(10), 3853–3865. http://dx.doi.org/10.1105/tpc.111.089771.

Madsen, E. B., Madsen, L. H., Radutoiu, S., Sato, S., Kaneko, T., Tabata, S., & Sandal, N. (2003). A receptor kinase gene of the LysM type is involved in legume perception of rhizobial signals. *Nature, 425*, 637–640.

Maillet, F., Poinsot, V., André, O., Puech-Pagès, V., Haouy, A., Gueunier, M., … Dénarié, J. (2011). Fungal lipochitooligosaccharide symbiotic signals in arbuscular mycorrhiza. *Nature, 469*(7328), 58–63. http://dx.doi.org/10.1038/nature09622.

Marsh, J. F., Rakocevic, A., Mitra, R. M., Brocard, L., Sun, J., Eschstruth, A., … Oldroyd, G. E. D. D. (2007). *Medicago truncatula* NIN is essential for rhizobial-independent nodule organogenesis induced by autoactive calcium/calmodulin-dependent protein kinase. *Plant Physiology, 144*(1), 324–335. http://dx.doi.org/10.1104/pp.106.093021.

Martin, F., Aerts, A., Ahrén, D., Brun, A., Danchin, E. G. J., Duchaussoy, F., … Grigoriev, I. V. (2008). The genome of *Laccaria bicolor* provides insights into mycorrhizal symbiosis. *Nature, 452*(7183), 88–92. http://dx.doi.org/10.1038/nature06556.

Martin, F., Kohler, A., Murat, C., Veneault-Fourrey, C., David, S., & Hibbett, D. S. (2016). Unearthing the roots of ectomycorrhizal symbioses. *Nature Reviews Microbiology*. http://dx.doi.org/10.1038/nrmicro.2016.149.

Martinez-Abarca, F., Herrera-Cervera, J. A., Bueno, P., Sanjuan, J., Bisseling, T., & Olivares, J. (1998). Involvement of salicylic acid in the establishment of the *Rhizobium meliloti* -alfalfa symbiosis. *Molecular Plant-Microbe Interactions, 11*, 153–155. http://dx.doi.org/10.1094/MPMI.1998.11.2.153.

Mathesius, U., Schlaman, H. R., Spaink, H. P., Of Sautter, C., Rolfe, B. G., & Djordjevic, M. A. (1998). Auxin transport inhibition precedes root nodule formation in white clover roots and is regulated by flavonoids and derivatives of chitin oligosaccharides. *Plant Journal, 14*(1), 23–34. http://dx.doi.org/10.1046/j.1365-313X.1998.00090.x.

Mendes, R., Garbeva, P., & Raaijmakers, J. M. (2013). The rhizosphere microbiome: significance of plant beneficial, plant pathogenic, and human pathogenic microorganisms. *FEMS Microbiology Reviews, 37*(5), 634–663. http://dx.doi.org/10.1111/1574-6976.12028.

Miyata, K., Kozaki, T., Kouzai, Y., Ozawa, K., Ishii, K., Asamizu, E., … Nakagawa, T. (2014). Bifunctional plant receptor, OsCERK1, regulates both chitin-triggered immunity and arbuscular mycorrhizal symbiosis in rice. *Plant and Cell Physiology, 55*(11), 1864–1872. http://dx.doi.org/10.1093/pcp/pcu129.

Moling, S., Pietraszewska-Bogiel, A., Postma, M., Fedorova, E., Hink, M. A., Limpens, E., … Bisseling, T. (October 2014). Nod factor receptors form heteromeric complexes and are essential for intracellular infection in medicago nodules. *The Plant Cell, 26*, 4188–4199. http://dx.doi.org/10.1105/tpc.114.129502.

Mulligan, J. T., & Long, S. R. (1989). A family of activator genes regulates expression of *Rhizobium meliloti* nodulation genes. *Genetics, 122*(1), 7–18. Retrieved from http://www.ncbi.nlm.nih.gov/pubmed/2731734.

Nakagawa, T., Kaku, H., Shimoda, Y., Sugiyama, A., Shimamura, M., Takanashi, K., … Kouchi, H. (2011). From defence to symbiosis: limited alterations in the kinase domain of LysM receptor-like kinases are crucial for evolution of legume-*Rhizobium* symbiosis. *The Plant Journal, 65*(2), 169–180. http://dx.doi.org/10.1111/j.1365-313X.2010.04411.x.

Navarro, L., Bari, R., Achard, P., Lisón, P., Nemri, A., Harberd, N. P., & Jones, J. D. G. (2008). DELLAs control plant immune responses by modulating the balance of jasmonic acid and salicylic acid signaling. *Current Biology, 18*, 650–655. http://dx.doi.org/10.1016/j.cub.2008.03.060.

Navarro-Rodenas, A., Xu, H., Kemppainen, M., Pardo, A. G., & Zwiazek, J. J. (2015). *Laccaria bicolor* aquaporin LbAQP1 is required for Hartig net development in trembling aspen (*Populus tremuloides*). *Plant, Cell and Environment, 38*(11), 2475–2486. http://dx.doi.org/10.1111/pce.12552.

Ng, J., Perrine-Walker, F., Wasson, A., & Mathesius, U. (2015). The control of auxin transport in parasitic and symbiotic root–microbe interactions. *Plants, 4*(3), 606–643. http://dx.doi.org/10.3390/plants4030606.

Okazaki, S., Kaneko, T., Sato, S., & Saeki, K. (2013). Hijacking of leguminous nodulation signaling by the rhizobial type III secretion system. *Proceedings of the National Academy of Sciences of the United States of America, 110*(42), 17131–17136. http://dx.doi.org/10.1073/pnas.1302360110.

Okazaki, S., Tittabutr, P., Teulet, A., Thouin, J., Fardoux, J., Chaintreuil, C., ... Giraud, E. (2016). *Rhizobium*–legume symbiosis in the absence of Nod factors: two possible scenarios with or without the T3SS. *The ISME Journal, 10*, 64–74. http://dx.doi.org/10.1038/ismej.2015.103.

Oldroyd, G. E. D. (2013). Speak, friend, and enter: signalling systems that promote beneficial symbiotic associations in plants. *Nature Reviews. Microbiology, 11*(4), 252–263. http://dx.doi.org/10.1038/nrmicro2990.

Oldroyd, G. E., & Long, S. R. (2003). Identification and characterization of nodulation-signaling pathway 2, a gene of *Medicago truncatula* involved in Nod factor signaling. *Plant Physiology, 131*(3), 1027–1032. http://dx.doi.org/10.1104/pp.102.010710.

Op den Camp, R. H. M., De Mita, S., Lillo, A., Cao, Q., Limpens, E., Bisseling, T., & Geurts, R. (2011). A phylogenetic strategy based on a legume-specific whole genome duplication yields symbiotic cytokinin type-A response regulators. *Plant Physiology, 157*(December 2011), 2013–2022. http://dx.doi.org/10.1104/pp.111.187526.

Op den Camp, R. H. M., Streng, A., De Mita, S., Cao, Q., Polone, E., Liu, W., ... Geurts, R. (December 2011). LysM-type mycorrhizal receptor recruited for rhizobium symbiosis in nonlegume *Parasponia*. *Science, 331*, 909–912. http://dx.doi.org/10.1126/science.1198181.

Pandey, A., Sharma, M., & Pandey, G. K. (2016). Emerging roles of strigolactones in plant responses to stress and development. *Frontiers in Plant Science,* 7(434). http://dx.doi.org/10.3389/fpls.2016.00434.

Parniske, M. (2008). Arbuscular mycorrhiza: the mother of plant root endosymbioses. *Nature Reviews Microbiology, 6*, 763–775. http://dx.doi.org/10.1038/nrmicro1987.

Pawlowski, K., & Demchenko, K. N. (2012). The diversity of actinorhizal symbiosis. *Protoplasma, 249*(4), 967–979. http://dx.doi.org/10.1007/s00709-012-0388-4.

Peláez-Vico, M. A., Bernabéu-Roda, L., Kohlen, W., Soto, M. J., & López-Ráez, J. A. (2016). Strigolactones in the rhizobium-legume symbiosis: stimulatory effect on bacterial surface motility and down-regulation of their levels in nodulated plants. *Plant Science, 245*, 119–127. http://dx.doi.org/10.1016/j.plantsci.2016.01.012.

Persson, T., Battenberg, K., Demina, I. V., Vigil-Stenman, T., Vanden Heuvel, B., Pujic, P., ... Berry, A. M. (2015). Candidatus Frankia Datiscae Dg1, the actinobacterial microsymbiont of *Datisca glomerata*, expresses the canonical nod genes nodABC in symbiosis with its host plant. *PLoS One, 10*(5), e0127630. http://dx.doi.org/10.1371/journal.pone.0127630.

Peters, N. K., & Long, S. R. (1988). Alfalfa root exudates and compounds which promote or Inhibit induction of *Rhizobium meliloti* nodulation genes. *Plant Physiology, 88*(2), 396–400. Retrieved from http://www.ncbi.nlm.nih.gov/pubmed/16666315.

Pieterse, C. M. J., Van der Does, D., Zamioudis, C., Leon-Reyes, A., & Van Wees, S. C. M. (2012). Hormonal modulation of plant immunity. *Annual Review of Cell and Developmental Biology, 28*, 489–521. http://dx.doi.org/10.1146/annurev-cellbio-092910-154055.

Pietraszewska-Bogiel, A., Lefebvre, B., Koini, M.a., Klaus-Heisen, D., Takken, F. L. W., Geurts, R., … Gadella, T. W. J. (2013). Interaction of *Medicago truncatula* lysin motif receptor-like kinases, NFP and LYK3, produced in *Nicotiana benthamiana* induces defence-like responses. *PLoS One, 8*(6), 16–18. http://dx.doi.org/10.1371/journal.pone.0065055.

Pimprikar, P., Carbonnel, S., Paries, M., Katzer, K., Klingl, V., Bohmer, M. J. J., … Gutjahr, C. (2016). A CCaMK-CYCLOPS-DELLA complex activates transcription of RAM1 to regulate arbuscule branching. *Current Biology, 26*(8), 987–998. http://dx.doi.org/10.1016/j.cub.2016.01.069.

Plett, J. M., Daguerre, Y., Wittulsky, S., Vayssières, A., Deveau, A., Melton, S. J., … Martin, F. (2014). Effector MiSSP7 of the mutualistic fungus *Laccaria bicolor* stabilizes the populus JAZ6 protein and represses jasmonic acid (JA) responsive genes. *Proceedings of the National Academy of Sciences of the United States of America, 111*(22), 8299–8304. http://dx.doi.org/10.1073/pnas.1322671111.

Plett, J. M., & Martin, F. (2011). Blurred boundaries: lifestyle lessons from ectomycorrhizal fungal genomes. *Trends in Genetics, 27*(1), 14–22. http://dx.doi.org/10.1016/j.tig.2010.10.005.

Plett, J. M., & Martin, F. (2012). Poplar root exudates contain compounds that induce the expression of MiSSP7 in *Laccaria bicolor. Plant Signaling and Behavior,* 7(1), 12–15. http://dx.doi.org/10.4161/psb.7.1.18357.

Radutoiu, S., Madsen, L. H., Madsen, E. B., Felle, H. H., Umehara, Y., Gronlund, M., … Stougaard, J. (2003). Plant recognition of symbiotic bacteria requires two LysM receptor-like kinases. *Nature, 425*(6958), 585–592. http://dx.doi.org/10.1038/nature02039.

Rasmussen, S. R., Füchtbauer, W., Novero, M., Volpe, V., Malkov, N., Genre, A., … Radutoiu, S. (May 2016). Intraradical colonization by arbuscular mycorrhizal fungi triggers induction of a lipochitooligosaccharide receptor. *Scientific Reports, 6*, 29733. http://dx.doi.org/10.1038/srep29733.

Reddy, P. M., María, M. R., & Soto, D. L. D. (2007). Flavonoids as signaling molecules and regulators of root nodule development. *Dynamic Soil, Dynamic Plant, Global Sci.*

Redecker, D., Kodner, R., & Graham, L. E. (2000). Glomalean fungi from the Ordovician. *Science, 289*(5486), 1920–1921. http://dx.doi.org/10.1126/science.289.5486.1920.

Redecker, D., Schüßler, A., Stockinger, H., Stürmer, S. L., Morton, J. B., & Walker, C. (2013). An evidence-based consensus for the classification of arbuscular mycorrhizal fungi (Glomeromycota). *Mycorrhiza, 23*, 515–531. http://dx.doi.org/10.1007/s00572-013-0486-y.

Remigi, P., Zhu, J., Young, J. P. W., & Masson-Boivin, C. (2015). Symbiosis within symbiosis: evolving nitrogen-fixing legume symbionts. *Trends in Microbiology*. http://dx.doi.org/10.1016/j.tim.2015.10.007.

Remy, W., Taylor, T. N., Hass, H., & Kerp, H. (December 1994). Four hundred-million-year-old vesicular arbuscular mycorrhizae. *Proceedings of the National Academy of Sciences of the United States of America, 91*, 11841–11843. http://dx.doi.org/10.1073/pnas.91.25.11841.

Rey, T., Chatterjee, A., Buttay, M., Toulotte, J., & Schornack, S. (2015). *Medicago truncatula* symbiosis mutants affected in the interaction with a biotrophic root pathogen. *The New Phytologist, 206*, 497–500. http://dx.doi.org/10.1111/nph.13233.

Rey, T., Nars, A., Bonhomme, M., Bottin, A., Huguet, S., Balzergue, S., ... Jacquet, C. (2013). NFP, a LysM protein controlling Nod factor perception, also intervenes in *Medicago truncatula* resistance to pathogens. *The New Phytologist, 198*, 875–886. http://dx.doi.org/10.1111/nph.12198.

Sayed, W. F., & Wheeler, C. T. (1999). Effect of the flavonoid quercetin on culture and isolation of *Frankia* from *Casuarina* root nodules. *Folia Microbiologica, 44*(1), 59–62. http://dx.doi.org/10.1007/BF02816223.

Scaffidi, A., Waters, M. T., Sun, Y. K., Skelton, B. W., Dixon, K. W., Ghisalberti, E. L., ... Smith, S. M. (2014). Strigolactone hormones and their stereoisomers signal through two related receptor proteins to induce different physiological responses in arabidopsis. *Plant Physiology, 165*(3), 1221–1232. http://dx.doi.org/10.1104/pp.114.240036.

Schauser, L., Roussis, A., Stiller, J., & Stougaard, J. (1999). A plant regulator controlling development of symbiotic root nodules. *Nature, 402*(6758), 191–195. Retrieved from <Go to ISI>://000083716400053.

Shimizu, T., Nakano, T., Takamizawa, D., Desaki, Y., Ishii-Minami, N., Nishizawa, Y., ... Shibuya, N. (2010). Two LysM receptor molecules, CEBiP and OsCERK1, cooperatively regulate chitin elicitor signaling in rice. *Plant Journal, 64*, 204–214. http://dx.doi.org/10.1111/j.1365-313X.2010.04324.x.

Simon, L., Bousquet, J., Levesque, R. C., & Lalonde, M. (1993). Origin and diversification of endomycorrhizal fungi and coincidence with vascular land plants. *Nature, 365*, 67–69.

Smith, S. E., & Read, D. (2008). *Mycorrhizal symbiosis.* (3rd ed.). New York: Academic Press. http://dx.doi.org/10.1016/B978-012370526-6.50015-5.

Soto, M. J., Fernández-Aparicio, M., Castellanos-Morales, V., García-Garrido, J. M., Ocampo, J.a., Delgado, M. J., & Vierheilig, H. (2010). First indications for the involvement of strigolactones on nodule formation in alfalfa (*Medicago sativa*). *Soil Biology and Biochemistry, 42*(2), 383–385. http://dx.doi.org/10.1016/j.soilbio.2009.11.007.

Soyano, T., & Hayashi, M. (2014). Transcriptional networks leading to symbiotic nodule organogenesis. *Current Opinion in Plant Biology, 20*, 146–154. http://dx.doi.org/10.1016/j.pbi.2014.07.010.

Steinkellner, S., Lendzemo, V., Langer, I., Schweiger, P., Khaosaad, T., Toussaint, J. P., & Vierheilig, H. (2007). Flavonoids and strigolactones in root exudates as signals in symbiotic and pathogenic plant-fungus interactions. *Molecules, 12*(7), 1290–1306. http://dx.doi.org/10.3390/12071290.

Sun, J., Miller, J. B., Granqvist, E., Wiley-Kalil, A., Gobbato, E., Maillet, F., ... Oldroyd, G. E. D. (March 2015). Activation of symbiosis signaling by arbuscular mycorrhizal fungi in legumes and rice. *The Plant Cell, 27*, 823–838. http://dx.doi.org/10.1105/tpc.114.131326.

Sutton, J. M., Lea, E. J., & Downie, J. A. (1994). The nodulation-signaling protein NodO from *Rhizobium leguminosarum* biovar viciae forms ion channels in membranes. *Proceedings of the National Academy of Sciences of the United States of America, 91*(21), 9990–9994. Retrieved from http://www.ncbi.nlm.nih.gov/pubmed/7524090.

Tanaka, K., Cho, S. H., Lee, H., Pham, A. Q., Batek, J. M., Cui, S., ... Stacey, G. (2015). Effect of lipo-chitooligosaccharide on early growth of C_4 grass seedlings. *Journal of Experimental Botany, 66*(19), 5727–5738. http://dx.doi.org/10.1093/jxb/erv260.

Tedersoo, L., May, T. W., & Smith, M. E. (2010). Ectomycorrhizal lifestyle in fungi: global diversity, distribution, and evolution of phylogenetic lineages. *Mycorrhiza, 20*(4), 217–263. http://dx.doi.org/10.1007/s00572-009-0274-x.

Tedersoo, L., & Smith, M. E. (2013). Lineages of ectomycorrhizal fungi revisited: foraging strategies and novel lineages revealed by sequences from belowground. *Fungal Biology Reviews, 27*(3–4), 83–99. http://dx.doi.org/10.1016/j.fbr.2013.09.001.

Tisa, L. S., Beauchemin, N., Gtari, M., Sen, A., & Wall, L. G. (2013). What stories can the *Frankia* genomes start to tell us? *Journal of Biosciences, 38*(4), 719–726. http://dx.doi.org/10.1007/s12038-013-9364-1.

Tisserant, E., Malbreil, M., Kuo, A., Kohler, A., Symeonidi, A., Balestrini, R., … Martin, F. (2013). Genome of an arbuscular mycorrhizal fungus provides insight into the oldest plant symbiosis. *Proceedings of the National Academy of Sciences of the United States of America, 110*(50), 20117–20122. http://dx.doi.org/10.1073/pnas.1313452110.

Tsai, S. M., & Phillips, D. A. (1991). Flavonoids released naturally from alfalfa promote development of symbiotic glomus spores in vitro. *Applied and Environmental Microbiology, 57*(5), 1485–1488.

Tsuzuki, S., Handa, Y., Takeda, N., & Kawaguchi, M. (2016). Strigolactone-induced putative secreted protein 1 is required for the establishment of symbiosis by the arbuscular mycorrhizal fungus *Rhizophagus irregularis*. *Molecular Plant-Microbe Interactions, 29*(4), 1–59. http://dx.doi.org/10.1094/MPMI-10-15-0234-R.

Van Loon, L. C. (2007). Plant responses to plant growth-promoting rhizobacteria. *European Journal of Plant Pathology, 119*(3), 243–254. http://dx.doi.org/10.1007/s10658-007-9165-1.

Van Zeijl, A., Liu, W., Xiao, T.ting, Kohlen, W., Yang, W.-C., Bisseling, T., & Geurts, R. R. (2015). The strigolactone biosynthesis gene DWARF27 is co-opted in rhizobium symbiosis. *BMC Plant Biology, 15*(1), 260. http://dx.doi.org/10.1186/s12870-015-0651-x.

Vayssières, A., Pěnčík, A., Felten, J., Kohler, A., Ljung, K., Martin, F. M., & Legue, V. (September 2015). Development of the *Populus-Laccaria bicolor* ectomycorrhiza modifies root auxin metabolism, signalling and response. *Plant Physiology, 169*, 114.255620. http://dx.doi.org/10.1104/pp.114.255620.

Vogel, J. T., Walter, M. H., Giavalisco, P., Lytovchenko, A., Kohlen, W., Charnikhova, T., … Klee, H. J. (2010). SlCCD7 controls strigolactone biosynthesis, shoot branching and mycorrhiza-induced apocarotenoid formation in tomato. *The Plant Journal : For Cell and Molecular Biology, 61*(2), 300–311. http://dx.doi.org/10.1111/j.1365-313X.2009.04056.x.

Wang, B., & Qiu, Y. L. (2006). Phylogenetic distribution and evolution of mycorrhizas in land plants. *Mycorrhiza, 16*(5), 299–363. http://dx.doi.org/10.1007/s00572-005-0033-6.

Wasson, A. P., Pellerone, F. I., & Mathesius, U. (2006). Silencing the flavonoid pathway in *Medicago truncatula* inhibits root nodule formation and prevents auxin transport regulation by rhizobia. *The Plant Cell, 18*(7), 1617–1629. http://dx.doi.org/10.1105/tpc.105.038232.

Weston, L. A., & Mathesius, U. (2013). Flavonoids: their structure, biosynthesis and role in the rhizosphere, including allelopathy. *Journal of Chemical Ecology, 39*(2), 283–297. http://dx.doi.org/10.1007/s10886-013-0248-5.

Wild, M., Davière, J.-M., Cheminant, S., Regnault, T., Baumberger, N., Heintz, D., … Achard, P. (2012). The arabidopsis DELLA RGA-LIKE3 is a direct target of MYC2 and modulates jasmonate signaling responses. *The Plant Cell, 24*, 3307–3319. http://dx.doi.org/10.1105/tpc.112.101428.

Yang, D.-L., Yao, J., Mei, C.-S., Tong, X.-H., Zeng, L.-J., Li, Q., … He, S. Y. (2012). Plant hormone jasmonate prioritizes defence over growth by interfering with gibberellin signaling cascade. *Proceedings of the National Academy of Sciences of the United States of America, 109*(19), E1192–E1200. http://dx.doi.org/10.1073/pnas.1201616109.

Yoshida, S., Kameoka, H., Tempo, M., Akiyama, K., Umehara, M., Yamaguchi, S., … Shirasu, K. (2012). The D3 F-box protein is a key component in host strigolactone responses essential for arbuscular mycorrhizal symbiosis. *The New Phytologist, 196*(4), 1208–1216. http://dx.doi.org/10.1111/j.1469-8137.2012.04339.x.

Young, N. D., Debellé, F., Oldroyd, G. E. D., Geurts, R., Cannon, S. B., Udvardi, M. K., ... Roe, B. A. (2011). The *Medicago* genome provides insight into the evolution of rhizobial symbioses. *Nature, 480*(7378), 520–524. http://dx.doi.org/10.1038/nature10625.

Yu, N., Luo, D., Zhang, X., Liu, J. J., Wang, W., Jin, Y., ... Wang, E. (2014). A DELLA protein complex controls the arbuscular mycorrhizal symbiosis in plants. *Cell Research, 24*(1), 130–133. http://dx.doi.org/10.1038/cr.2013.167.

Zhang, X., Dong, W., Sun, J., Feng, F., Deng, Y., He, Z., ... Wang, E. (2015). The receptor kinase CERK1 has dual functions in symbiosis and immunity signalling. *The Plant Journal, 81*(2), 258–267. http://dx.doi.org/10.1111/tpj.12723.

Zhang, Y., van Dijk, A. D. J., Scaffidi, A., Flematti, G. R., Hofmann, M., Charnikhova, T., ... Bouwmeester, H. J. (2014). Rice cytochrome P450 MAX1 homologs catalyze distinct steps in strigolactone biosynthesis. *Nature Chemical Biology, 10*(12), 1028–1033. http://dx.doi.org/10.1038/nchembio.1660.

Zhao, L.-H., Zhou, X. E., Yi, W., Wu, Z., Liu, Y., Kang, Y., ... Xu, H. E. (2015). Destabilization of strigolactone receptor DWARF14 by binding of ligand and E3-ligase signaling effector DWARF3. *Cell Research, 25*(11), 1219–1236. http://dx.doi.org/10.1038/cr.2015.122.

Zuanazzi, J. A. S., Clergeot, P. H., Quirion, J. C., Husson, H. P., Kondorosi, A., & Ratet, P. (1998). Production of *Sinorhizobium meliloti* nod gene activator and repressor flavonoids from *Medicago sativa* roots. *Molecular Plant-Microbe Interactions, 11*(8), 784–794. http://dx.doi.org/10.1094/MPMI.1998.11.8.784.

PART THREE

Plant Communication With Animals

CHAPTER NINE

Plant–Pollinator Communication

P. Wester[1], K. Lunau
Heinrich-Heine-University, Düsseldorf, Germany
[1]Corresponding author: E-mail: Westerpetra3@gmail.com

Contents

Abstract

The communication between flowers and pollinators is the essential feature of sexual reproduction in zoophilous flowering plants and helps to ensure pollen transfer between flowers of conspecific plants. The plants' side of communication between flowers and pollinators includes the display or concealment of primary attractants, such as nectar, pollen or other kinds of floral rewards, and of secondary attractants of flowers and inflorescences to enable detection and discrimination by pollinators. These secondary attractants constitute the signalling apparatus of flowers with their visual, olfactory, gustatory and tactile signals addressed to potential pollinators. The flower visitors' side includes the various sensory capabilities of different flower visitors and their abilities to handle flowers. Because the attributes of flower visitors needed to handle flowers differ largely among bees, flies, beetles, birds, bats and others, many flowering plants have evolved flowers adapted to one particular group of pollinators and consequently emit signals to attract their specific pollinators. Selective attraction of pollinators and deterrence of flower antagonists by means of specific signals, innate and learnt preferences of flower visitors as well as sensory exploitation make communication between flowers and pollinators a highly complex and diversified interaction.

Advances in Botanical Research, Volume 82
ISSN 0065-2296
http://dx.doi.org/10.1016/bs.abr.2016.10.004

Figure 1 Floral attractants. (A) Black nectar of the bird-pollinated *Melianthus comosus* flower; (B) *Trigona spinipes* stealing pollen from a *Schizocentron elegans* pollen sac; (C) bird-pollinated *Axinaea costaricensis* flower with staminal appendages as food bodies (photograph by Juan Francisco Morales: https://melas-centroamerica.com/axinaea-costaricensis); (D) orchid bee *Eulaema* cf. *cingulata* collecting perfume at *Spathiphyllum cannifolium*; (E) *Gorteria diffusa* with raised black spots on the ray florets mimicking

1. INTRODUCTION

Flowers are sensory billboards displaying visual, olfactory, gustatory and tactile signals to attract pollinators (Chittka, 1997; Raguso, 2004a, Fig. 1) and at the same time to repel herbivores and exclude nectar robbers and pollen thieves (Junker & Blüthgen, 2010, Fig. 2). The detectability of floral signals from a distance and at proximity (Hempel de Ibarra, Giurfa, & Vorobyev, 2001), the distinctness of floral signals compared to that of coflowering species (van der Kooi, Pen, Staal, Stavenga, & Elzenga, 2016), the guidance towards landing platforms on the flowers or floral rewards by means of floral guides (Lunau, Wacht, & Chittka, 1996) are tasks of floral signals which might require complex properties.

Flowers' communication signals are very diverse and include honest signalling of expectable rewards (Knauer & Schiestl, 2015) such as emptying of replenishable nectar reward and pollen (Ohashi & Thomson, 2005) and advertising fertilization by floral colour change (Weiss, 1995). The flower visitors' marking of emptied flowers with scent marks (Gawleta, Zimmermann, & Eltz, 2005; Wilms & Eltz, 2008) adds flower visitor-derived signals to flowers. In addition, the competition of flowering plants for pollinators (Harder & Aizen, 2010; Knight et al., 2005) and the competition of pollinators for limited floral resources (Mitchell, Flanagan, Brown, Waser, & Karron, 2009) result in a highly complex communication system between flowers and flower visitors (Ruxton & Schaefer, 2013).

The interactions between flowering plants and their animal pollinators are based on communication. Plant–pollinator communication systems need a tuning of communicative features. Flowering plants have to adapt to the sensory capabilities of pollinators, and pollinators have to adjust their sensory systems to the floral signalling. The sensory capabilities of pollinating animals including not only beetles, bees, flies, butterflies and other insects, but also vertebrates such as birds, reptiles, bats and nonflying mammals are

bombyliid flies *Megapalpus nitidus* of which both sexes feed on nectar, (F) but only males try to copulate with the fly marks; (G) stingless bee *Tetragonisca angustula* collecting the red resin from a *Clusia fluminensis* flower, note the resin collected at the bee's hindleg; (H) *Macropis europaea* collecting oil from the flowers of *Lysimachia vulgaris*, note the mixture of pollen and oil at their hindlegs; (I) *Stapelia gigantea* flower imitating rotten meat with brownish colour, purple markings, hairiness and foul smell, attracting calliphorid flies, depositing eggs on the flower; (J) flower of *Serapias* orchid offering shelter for a bee.

Figure 2 Floral filters. (A) Selective attraction of brood-site seeking flies in *Aristolochia cretica*; (B) visual claw marks on *Symphytum* flowers as indicators of previous visits; (C) the red and UV-absorbing colour and black-and-white beetle-sized markings of *Papaver rhoeas* flowers are selectively attractive to mate-seeking East-Mediterranean scarabaeid beetles preferring red colours; (D) mechanical filter inhibits access to floral resources of *Trollius europaeus* except for pollinating anthomyiid flies; (E) *Centaurea* flower head is mechanically protected against egg-laying gall-producing flies; (F) although densely covered with pollen, the bumblebee worker is unable to harvest the spiny pollen grains of *Alcea rosea*; (G) the sticky nectar and green colour of *Whiteheadia bifolia* flowers excludes many flower visitors, but not pollinating Cape rock elephant-shrews *Elephantulus edwardii*; (H) only specialized bees are able to open *Polygala myrtifolia* flowers by pressing the bearded crest.

extremely diverse and mostly not congruent with the human sensory system (Glover, 2014; Schaefer & Ruxton, 2011).

The floral resources sought for by flower visitors are diverse as are the sensorial capabilities to detect and identify these resources; besides, pollen and nectar flowers offer resin, fatty oils, perfume, starch, egg-laying substrate, heat and shelter (Willmer, 2011). Thereby the composition and concentration of the ingredients of floral rewards vary and are adapted to distinct groups of pollinators.

Knowledge about plant–pollinator communication is a useful tool for the study of pollination of distinct plant species, pollination networks (Olesen, Bascompte, Dupont, & Jordano, 2007), pollinator decline (Potts et al., 2010), pollination management and effects of climate change on crop pollination (Klein et al., 2007), and effects of neophytes and neozoons on pollination effectivity of native plants (Chittka & Schürkens, 2001; Junker et al., 2011). Neophytes and neozoons are plants and animals, respectively, that have colonized new areas mostly due to human activities beginning with the discovery of the Americas by Columbus.

To cover the enormous number of publications about communication between plants and pollinators, we focus on research of the last decade and three main topics which are the signalling of flowers as sensory billboards (Raguso, 2004a), the sensorial capabilities of pollinators determining flower detection and discrimination as well as sensory bias and preferences in flower visitors (Chittka & Raine, 2006). These topics cover a lot of new and exciting themes of plant–pollinators communication such as evolution of floral signals (Schiestl & Johnson, 2013), multimodal signalling (Leonard, Dornhaus, & Papaj, 2011), floral filters of pollinators and antagonists (Johnson & Steiner, 2000), manipulation of flower visitors to increase pollen transfer (Pohl, Watolla, & Lunau, 2008) and search strategies of pollinators (Morawetz, Chittka, & Späthe, 2014).

2. SENSORIA, SENSORY BIAS AND PREFERENCES OF POLLINATORS

2.1 Sensory Modalities

Flowers are sensory billboards and emit multimodal signals. All pollinators can see, smell, taste and feel flowers and its attractants; however, they do it in many different ways as compared to humans (Chittka & Thomson, 2001; Lunau & Maier, 1995; Raguso, 2008). Moreover, some animals that pollinate flowers possess sensory receptors in other organs than humans; for example, bees and hoverflies can taste with the tarsi of their legs (De Brito Sanchez et al., 2014; Wacht, Lunau, & Hansen, 2000), the antennae of bees have sensilla sensitive to olfactory, gustatory and mechanical stimuli (De Brito Sanchez, 2011). Hawkmoths can smell with the tip of their proboscis (Haverkamp et al., 2016). In contrast to insects, vertebrates are able to move their eyes and fixate target objects. Some sensory capabilities of pollinating animals are still not known to play a role in flower visitation such as the perception of electric fields (Clarke, Whitney, Sutton, & Robert, 2013; Sutton, Clarke, Morley, & Robert, 2016), polarization pattern in the dorsal visual field (Foster et al., 2014), and floral iridescence (Lunau, 2016; Whitney, Reed, Rands, Chittka, & Glover, 2016). The principal differences between the sensory modalities in the communication between flowers and pollinators are intriguing. Visual signals of flowers travel with the speed of light and provide complex real-time information about flowers and their visual environment, while olfactory signals travel with the speed of the wind, only downwind, and are uncertain and delayed due to turbulences and travel duration. Although each species of flowering plants has a distinct scent and colour, the spectral reflectance curves of flower colours fall into few distinct categories of colours (Chittka, 1997), e.g., named by colour hues such as blue, green, red, blue-green, yellow and purple, whereas floral scents are often composed of many volatiles emitted in distinct concentrations and combinations and constitute a scent bouquet (Raguso, 2004a, 2008). Flower colours are thus optimal cues to easily detect the next similarly coloured target, whereas floral scents are more likely used to identify a particular target (Raguso, 2008).

Gustatory and tactile signals of flowers are perceived only on contact with the sense organs of flower visitors. The specific taste of sugar, i.e., mono- and disaccharides, is the most reliable cue to identify nectar and to check nectar concentration; this holds for all nectar-seeking flower visitors including mammals (Scott, 2004), hummingbirds (Baldwin et al., 2014), honeybees (De Brito Sanchez, 2011), butterflies (Inoue, Asaoka, Seta, Imaeda, & Ozaki, 2009), bats (Ayala-Berdon, Rodríguez-Peña, García Leal, Stoner, & Schondube, 2013) and hoverflies (Wacht et al., 2000), whereas similarly reliable chemical cues of pollen, by which pollen as such is identified, are not known (but see Lunau, Piorek, Krohn, & Pacini, 2015; Wacht, Lunau, & Hansen, 1996). However, nectar-feeding is complex, due to differences in volume and concentration and additional nutritive ingredients (Baker & Baker, 1990), e.g., amino acids (Carter, Shafir, Yehonatan, Palmer, & Thornburg, 2006), or toxins (Adler, 2001). Thus nectar might be regarded not as a pure attractant, but rather as a trait for the manipulation of flower visitors (Johnson, Hargreaves, & Brown, 2006; Pyke, 2016).

Surface textures of flower petals serve as multifunctional traits. Bees and hawkmoths can use flower petal microtexture as a tactile cue (Glover & Martin, 1998; Goyret & Raguso, 2006; Kevan & Lane, 1985). Additionally, flower microtexture serves various tasks, including grip for flower visitors, wettability, heating, altering spectral reflectance properties and gloss (Papiorek, Junker, & Lunau, 2014; Whitney et al., 2011). The nanotextures of some flowers, e.g., *Hibiscus trionum*, display unique visual floral signals with a steadily changing colour hue that might increase the flower's detectability (Whitney, Chittka, Bruce, & Glover, 2016). In buzz-pollinated flowers, pollen is indetectably hidden in poricidal anthers, and tactile cues of stamens are likely to trigger buzzing and to aid sensing of pollen release (Burkart, Schlindwein, & Lunau, 2014; De Luca & Vallejo-Marín, 2013).

The role of multimodal flower stimuli is less explored (Giurfa, Núnez, & Backhaus, 1994; Goyret, 2010; Kunze & Gumbert, 2001; Leonard & Masek, 2014; Raguso & Willis, 2005; Roy & Raguso, 1997) and might be very complex due to deviant responses of different flower visitors (Junker & Blüthgen, 2010), synergistic effects (Balkenius, Rosén, & Kelber, 2006; Dötterl, Glück, Jürgens, Woodring, & Aas, 2014; Riffell & Alarcón, 2013), effects of spatial and temporal continuity (Goyret, Markwell, & Raguso, 2007) and impact on learning behaviour (Goyret, Pfaff, Raguso, & Kelber, 2008). The relative attractive and deterrent effects of floral signals act as a floral filter to selectively attract pollinators and to deter antagonistic flower visitors at the same time (Junker & Blüthgen, 2010; Kessler, Gase, & Baldwin, 2008).

2.2 Number and Sensitivity Range of Receptors

The sensorium of pollinators, i.e., the number of receptor types in a given sense organ and the threshold of sensitivity of the sensory cells, explain only little of how pollinators might perceive flowers. The localization of flowers by chemical and visual signals is largely different. Eyes provide precise information about the spatial position of target flowers with each photoreceptor unit viewing at a different sector in the visual field, whereas noses and antennae can be very sensitive to detect specific volatiles, but are unable to detect the position of the scent source except for its upwind direction.

The number of photoreceptor types in the eyes of diurnal and nocturnal flower visitors is largely different. A minimum of two types of photoreceptors and their associated neural processing are needed for colour vision (von Frisch, 1914; Peitsch et al., 1992). Vertebrates possess very sensitive rods for colour-blind nocturnal vision and mostly two types of cones for diurnal colour vision. But even strictly nocturnal bats have two cone types in the rods-dominated retina indicating the ability of colour vision during twilight (Müller et al., 2009). Also rodents possess two types of cones sensitive in the ultraviolet/blue and green range of wavelengths indicating possible limited colour discrimination in nonflying mammals (Jacobs, Fenwick, & Williams, 2001). Most visual systems of flower-visiting animals are based on three different types of photoreceptors allowing for trichromatic colour vision (Lunau & Maier, 1995). Bees possess 3 types of photoreceptors for colour vision: some butterflies and flies possess 4, 5 and even up to 15 different types of photoreceptors (Chen, Awata, Matsushita, Yang, & Arikawa, 2016; Lunau, 2014; Marshall & Arikawa, 2014). Birds are tetrachromatic and sensitive in red, green and blue and additionally in the violet or ultraviolet range of wavelengths (Odeen & Hastad, 2013). In birds, each photoreceptor is associated with a coloured oil droplet acting as a long-pass filter; this combination has been shown to improve colour discrimination (Vorobyev, 2003).

Honeybees and bumblebees evaluate the input of only one type of photoreceptors if viewing target objects under small visual angles (Dyer, Späthe, & Prack, 2008; Giurfa, Vorobyev, Brandt, Posner, & Menzel, 1997) and thus are colour-blind under these conditions. Because this photoreceptor type is maximally sensitive in the green range of wavelengths, distant or small target objects are seen by their green contrast.

The number of visual sampling units corresponds to the number of photoreceptors in vertebrates, i.e., the number of cones under photopic and the number of rods under scotopic light conditions, and to the number of ommatidia in insects which is responsible for the spatial resolution (Land & Nilsson, 2001). Both, vertebrates and invertebrates, have eye regions of superior spatial resolution, termed fovea centralis in vertebrates and acute zones in insects (Späthe & Chittka, 2003). Due to the limited number of visual sampling units, e.g., ommatidia (honeybee: 3000), insects have much poorer visual acuity as compared to humans (total number of the three types of cones: 4,500,000) with dramatic consequences for sensing the details of floral colour patterns (Hempel de Ibarra, Langridge, & Vorobyev, 2015).

Olfaction in pollinators is based on the filtering of the air downwind from a flower. A large number of sensory cells sensitive to floral volatiles are generally facilitating the perception of scent compounds for two reasons. A large number of different types of odorant receptors enable a flower visitor to sense the complexity of floral scents (DeMaria & Ngai, 2010), and a large number of one particular type of odorant receptor enable the effective perception of one distinct volatile. Noteworthy, none of the main nutrients, i.e., proteins, carbohydrates and lipids, can be identified by a characteristic odour (Jürgens & Shuttleworth, 2015) making them difficult to detect via olfaction unless volatile decomposition products are emitted. Generally, the size of the olfactory epithelia in vertebrates' noses and the size of insects' antennae bearing odorant sensilla in insects correlate with their capabilities to recognize scents (Hansson & Stensmyr, 2011).

The number of odorant receptor types in animals surpasses that of photoreceptors by far (Robertson, Warr, & Carlson, 2003). Honeybees have more than 150 odorant receptor types (Robertson & Wanner, 2006). Although birds have been considered to have a poor olfaction or retain scent information very poorly (Byers, Bradshaw, & Riffell, 2014; Cronk & Ojeda, 2008; Faegri & van der Pijl, 1979; Goldsmith & Goldsmith, 1982; Ioalé & Papi, 1989; Roper, 1999), hummingbirds possess more than 50 odorant receptors (Steiger, Fidler, Valcu, & Kempenaers, 2008) and are capable to use scent marks for flower visitation in an experimental setting (Goldsmith & Goldsmith, 1982). Especially for nocturnal flower visitors, scent is important; for example, bats (Jones, Teeling, & Rossiter, 2013) and mice (Skinner & Chimimba, 2005; Stoddart, 1980) have a well-developed sense of smell.

The similarities and differences between mammalian and insect olfactory systems are striking (Bargmann, 2006). For the perception of floral scents, researchers regard the number and concentration of perceived volatile substances per time unit as important cues, in addition to the dynamic of scent emission (Ayasse, Paxton, & Tengö, 2001; Raguso, 2008). A specific problem of floral scent analysis is that not all volatile compounds of floral scents are active for all pollinators. Gas chromatography–mass spectrometry in combination with electroantennography (GC–MS/EAG) are applied to test whether distinct substances are smelled by insect flower visitors including bees (Milet-Pinheiro, Herz, Dötterl, & Ayasse, 2016; Sandoz, 2011), flies (Chen et al., 2015; Jhumur, Dötterl, & Jürgens, 2007), butterflies (Andersson, 2003), hawkmoths (Hoballah et al., 2005), moths (Dötterl et al., 2006), butterflies (Ômura, Honda, & Hayashi, 2000), beetles (Bartlet et al., 2004; Steenhuisen, Jürgens, & Johnson, 2013) and cockroaches (Vlasáková, Kalinová, Gustafsson, & Teichert, 2008). Behavioural choice tests in olfactometers allow to find evidence for odour preferences for floral volatiles, e.g., in bees (Junker & Blüthgen, 2010), wasps (Shuttleworth & Johnson, 2009), nonflying mammals (Johnson, Burgoyne, Harder, & Dötterl, 2011) and bats (Von Helversen, Winkler, & Bestman, 2000). For example, 3-hexanone, is found in plants pollinated by flying and nonflying mammals (Johnson et al., 2011), many bird-pollinated flowers are only poorly scented, and bees prefer the so-called sweet and pleasant scent dominated by terpenoids.

Several studies found that floral scents possess a double function in attracting pollinators and in repelling floral antagonists and thus act as floral filters (Junker & Blüthgen, 2010; Junker, Höcherl, & Blüthgen, 2010). Synergistic effects of floral volatiles have been demonstrated to play a role in the deterrence of flower antagonists, but not in the attraction of pollinators (Byers et al., 2014).

Some flower visitors leave scent marks or visual marks indicating their visits probably to avoid revisits before the nectar reward of the flower is replenished (Eltz, 2006; Goulson, Chapman, & Hughes, 2001; Saleh, Ohashi, Thomson, & Chittka, 2006). For scent marks, it has been demonstrated that flower visitors have to learn their meaning. In laboratory settings, scent marks may indicate reward, because artificial flower permanently offer sugar water, while in natural flowers scent marks indicate a visited and thus emptied flower until nectar replenishment (Witjes & Eltz, 2007). The scent marks are left passively on the flower by bees rather than actively deposited (Wilms & Eltz, 2008) and are understood from nonconspecific insects (Gawleta et al., 2005). One part of the scent marks emits from the flower and thus represents a cue for flower visitors, while another part of the scent marks is dissolved in the wax layer of the flower and might be used by researchers as an indicator of the visitation history of individual flowers (Witjes & Eltz, 2009; Witjes, Witsch, & Eltz, 2011).

Taste perception plays a role in the detection of nectar but also admixed nutritious substances and bitter-tasting substances (Gardener & Gillman, 2002; Johnson et al., 2006; Kessler & Baldwin, 2007; Nepi, Guarnieri, & Pacini, 2003; Nicolson, Lerch-Henning, Welsford, & Johnson, 2015). Taste perception of pollen is less well studied with the exception of that of hoverflies (Wacht et al., 2000) and bumblebees (Rüdenauer, Späthe, & Leonhardt, 2015).

Orientation of nocturnal as compared to diurnal pollinators relies on particular sensory modalities. Flower-visiting bats use echolocation to detect flowers (von Helversen, Holderied, & von Helverson, 2003; Simon, Holderied, Koch, & von Helversen, 2011) which is weird for humans. Nocturnal insects have improved vision in dim light conditions, e.g., nocturnal *Megalopta* bees (Greiner, Ribi, & Warrant, 2004), or even maintain colour vision under scotopic light conditions, e.g., hawkmoths (Kelber, Balkenius, & Warrant, 2002).

2.3 Sensory Bias

The detectability of flowers for pollinators is strictly dependent of its sensory capabilities. However, it has only rarely been studied which flower properties are more salient, because pollinators trade accuracy off against speed (Chittka & Späthe, 2007; Ings & Chittka, 2008) and because quantitative variation in the pollinator sense organs impacts target detection (Späthe & Chittka, 2003). Moreover, quantitative variation of floral signals has been rarely studied; an exception is *Centaurea cyanea*, for which it was demonstrated that pollinators can select quantitative variation in floral colouration and could exert a selection pressure for gradual evolution of flower colouration (Renoult, Thomann, Schaefer, & Cheptou, 2013).

It is often assumed that the senses of pollinators and floral signalling by plants reciprocally select for each other leading to coevolution: this hypothesis remains little investigated (Chittka & Menzel, 1992; Chittka, Späthe, Schmidt, & Hickelsberger, 2001; Ramírez et al., 2011). However, most sensory capabilities of flower-visiting animals have evolved in other contexts than flower visitation such as communication with conspecifics during courtship and mating, marking territories and finding oviposition substrates (Jürgens, Wee, Shuttleworth, & Johnson, 2013; Schiestl & Johnson, 2013) and thus have evolved prior to flower-visiting behaviour (Chittka, 1996; Schaefer & Ruxton, 2009). Many findings support a scenario of preexisting chemical communication in insects leading to their selecting plants that produce specific floral scent compounds (Ayasse & Dötterl, 2014; Ramírez et al., 2011; Schiestl & Dötterl, 2012). This phenomenon may explain the convergent evolution of floral traits in response to similar pollinator groups, leading to pollination syndromes in floral signalling (Fenster, Armbruster, Wilson, Dudash, & Thomson, 2004; Schiestl & Johnson, 2013; Vogel, 2012). Schiestl and Dötterl (2012) explicitly tested whether plants of the Araceae family may have evolved floral traits that fitted preexisting preferences in the pollinators. They found multiple evidences for a sensory bias scenario rather than a coevolution scenario. The evolution of volatile organic compounds in the pollinating scarab beetles predated the evolution of similar compounds in the flowers.

Sensory bias can also emerge because some floral signals are faster/better learnt or memorized than others (Menzel, 1985; Raine & Chittka, 2007a). Generally, complex signals and multimodal signals are learnt faster and memorized better (Katzenberger, Lunau, & Junker, 2013; Kulahci, Dornhaus, & Papaj, 2008). From laboratory settings, it is known that the learning performance of bumblebees is altered by absolute conditioning as compared to differential conditioning (Dyer & Chittka, 2004; Giurfa, 2004). Absolute conditioning refers to training with only rewarding stimuli, whereas differential conditioning refers to training with rewarding stimuli and nonrewarding or punishing distractor stimuli, e.g., bitter-tasting or holding flowers simulating ambushing crab spiders, and results in finer discrimination (Dyer & Chittka, 2004). The learning behaviour of flower visitors is dependent of the visitation of rewarding and nonrewarding flowers and of visits associated with a risk caused by ambushing predators (Dukas, 2001; Reader, HigginsonBarnard, & Gilbert, 2006), unfavourable treatment by flower with explosive pollination mechanism (Palmer-Jones & Forster, 1972) or bitter-tasting nectar (Johnson et al., 2006). In laboratory settings, the punishment can be simulated by robot crab spiders (Ings & Chittka, 2008, 2009) or nectar admixed with bitter-tasting quinine (Wright et al., 2010).

Peak shift is a behavioural response bias arising from discrimination learning in which animals display a directional, but limited preference for novel stimuli that are more different from previously nonrewarding or punishing stimuli (Dyer & Murphy, 2009; Lynn, Cnaani, & Papaj, 2005). The term refers to the shifted preference of the animal. If animals are trained with two similar stimuli in such a way that one is rewarding (S+) and one punishing (S−), then following training animals show a greatest preference not for the S+, but for a novel stimulus that is slightly more different from the S− than the S+ (Andrew et al., 2014). The peak shift phenomenon has been demonstrated for colour hues (Martínez-Harms, Márquez, Menzel, & Vorobyev, 2014) and olfactory stimuli (Andrew et al., 2014). Multimodal stimuli, i.e., scent aiding a colour discrimination task, have been shown to reduce the amount of peak shift due to a reduced uncertainty in colour discrimination (Leonard et al., 2011).

2.4 Preference

True preferences for floral traits have evolved independently from sensory bias and are known for olfactory as well as for visual flower signals. Innate preferences are shown by naïve and unexperienced animals, but even experienced flower visitors are known to rely on innate preferences (Lunau & Maier, 1995; Rohde, Papiorek, & Lunau, 2013). Learnt preferences are only shown by experienced animals. The finding that distinct pollinators are frequently observed at flowers of various species displaying the same trait does not necessarily indicate that these flower visitors possess an innate preference for this specific trait (Lunau & Maier, 1995).

Specialist flower visitors mostly use olfactory key signals to determine their specific food plants. For example, oil-collecting bees use diacetin (glycerol diacetate) as a reliable private-channel cue to detect oil-producing flowers (Schäffler et al., 2015). Oligolectic bees that are specialized on *Campanula* bellflowers identify their pollen food plants by means of two indicative cues, the blue colour hue and spiroacetals as volatiles that are only rarely found in other flowers than bellflowers (Milet-Pinheiro et al., 2013; Milet-Pinheiro, Ayasse, & Dötterl, 2015). Even generalist flower visitors rely on floral signals to initially detect and approach flowers. Honeybees and bumblebees prefer blue flowers (Gumbert, 2000) as an indicator of a superior nectar standing crop (Raine & Chittka, 2007b). This finding is supported by studies of spontaneous preferences, learning speed and learning capacity (Giurfa, Núnez, Chittka, & Menzel, 1995). Contrarily, Lunau et al. (1996) found that naïve and nontrained bumblebees prefer saturated colours and colours contrasting strongly against the background. Finding solid evidence for the most preferred colour parameter in bees seems difficult, because even in laboratory settings the colour parameters, e.g., green contrast, colour contrast, colour intensity, dominant wavelength, spectral purity and others cannot be varied independently.

Flower visitors develop short-period preferences for distinct flowers and bypass closer, but deviant flowers (Raine, Ings, Dornhaus, Saleh, & Chittka, 2006) to optimize foraging for floral resources; this phenomenon is known as flower constancy (Chittka, Thomson, & Waser, 1999). Various hypotheses have been expressed to explain this behaviour. Temeles et al. (2016) found resource partitioning in a hummingbird based on sex-specific floral colour preferences. Some flower visitors possess outstanding learning capabilities. For example, honeybees can use images of human faces to discriminate flowers (Dyer, Neumeyer, & Chittka, 2005); however, other studies indicate limitations of learning capabilities: Bumblebees easily learn subtle differences in the dominant wavelength of colour stimuli, but spontaneously prefer more spectrally pure colours of trained ones (Rohde et al., 2013). *Eristalis* hoverflies possess an innate proboscis reflex to pure yellow colours, but the flies cannot be conditioned to extend the proboscis towards other colour hues than yellow (Lunau & Wacht, 1994).

Signalling of flowers is complex; many flowers display floral colour patterns comprising a large-sized peripheral and a small-sized central colour. Visual colour patterns of flowers are often paralleled by olfactory patterns (Dobson & Bergström, 2000; Dötterl & Jürgens, 2005; Raguso & Pichersky, 1999). The smallest component of floral colour patterns is often interpreted as a floral guide directing the flower visitors towards the floral reward or directing them towards the nectar reward while distracting them from pollen (Koski & Ashman, 2014; Leonard & Papaj, 2011; Lunau, 2000, 2007; Orbán & Plowright, 2014). UV-bull's eyes, i.e., ultraviolet reflectance patterns mostly displayed by yellow flowers that reflect UV in the periphery but absorb UV in the centre part, are almost invisible for humans, but they offer strong bee-visible colour contrast and have been analysed frequently (Koski & Ashman, 2014; Lunau, 2007; Silberglied, 1979). Flower guides have also unexpected tasks. Floral guides draw the flower visitors' attention towards the access of floral reward and by this way help to prevent nectar robbing (Leonard, Brent, Papaj, & Dornhaus, 2013) and play a role in plant defence due to toxic and UV-absorbing flavonoids (Gronquist et al., 2001).

Hoverflies, e.g., *Eristalis tenax*, possess an innate proboscis reflex that is triggered either by chemical stimuli of pollen and nectar or by visual colour stimuli of pollen (Wacht et al., 1996, 2000). The spectral reflectance in a range of wavelength between 510 nm and 600 nm (yellow) elicits the proboscis extension (Lunau & Wacht, 1994), whereas admixed ultraviolet and blue light inhibits the proboscis extension (Lunau, 2014). Floral guides can have different impact on the handling time, the time needed by a flower visitor to exploit the flower: While black lines decrease the handling time, yellow dots increase the handling time and intensify the movements on the flower (Dinkel & Lunau, 2001).

The relationships between pollinators and flowering plants are not very tight over evolutionary times (Smith, Ané, & Baum, 2008). Even shifts between pollinator guilds of flowering plants are not rare. In the nectar-rewarding terrestrial orchid *Gymnadenia odoratissima*, the phenotypic selection of floral traits, e.g., colour intensity of flowers, flower size and scent, varied with altitude and pollinator fauna (Gross, Sun, & Schiestl, 2016), suggesting geographically structured selection that is leading to regional divergence in floral traits. Pollinator-mediated selection of floral traits in *Gymnadenia conopsea* is very complex due to spatial pattern (Chapurlat, Ågren, & Sletvold, 2015).

Flower colours can also discourage flower visitors. Red flowers are frequently linked to pollination by birds (Porsch, 1931), although naïve flower-visiting birds do not exhibit a preference for red colours (Lunau & Maier, 1995; Lunau, Papiorek, Eltz, & Sazima, 2011), but bird-pollinated flowering plants might benefit from a colour that excludes nectar robbing bees, functioning as a floral filter (Bergamo, Rech, Brito, & Sazima, 2015; Lunau et al., 2011). Thereby, the avoided colour needs not to be inconspicuous for bees as long as flowers of other colours are easier to find (Rodríguez-Gironés, & Santamaría, 2004). Yellow bird-pollinated flowers do not possess the ultraviolet bull's eye pattern that is typical for bee-pollinated flowers and by this way exclude bees in their quest to find the floral reward (Papiorek et al., 2016). Furthermore, floral guides, guiding insects to the centre of the flowers, are mostly absent in bird-pollinated plants (Papiorek et al., 2016).

3. FLOWERS AS SENSORY BILLBOARDS

Flowers often use animals as pollen vectors such as different insects, birds, bats, and small nonflying mammals (Faegri & van der Pijl, 1979; Fleming & Muchhala, 2007; Wester, 2010; Wester, Stanway, & Pauw, 2009). Animals visit flowers to satisfy their needs such as subsistence (food, shelter, temperature regulation) and reproduction (mating and parental care) (Faegri & van der Pijl, 1979). To lure animals, flowers use different attractants. Primary attractants function as lure such as nectar, pollen, perfume, oil, sexual partners or brood substrate, addressing the animal's needs. Primary attractants can also be purely deceiving; most plant species offering brood substrate are cheaters and potential sexual partners are always faked. Secondary attractants such as colour (hue, saturation, brightness, contrast, pattern), gloss, shape or texture, scent, taste, or movement are stimuli offering information about what to find, addressing drives. Such secondary attractants work mostly in combination to advertise the specific primary attractants (Raguso, 2004a).

3.1 Food as Attractant

Primary attractants for foraging animals are mainly nectar (Fig. 1A), but also pollen (Fig. 1B), nutritive pseudopollen or food bodies (Fig. 1C). In open flowers, nectar is freely accessible and can attract attention by visual cues such as glitter (Sandvik & Totland, 2003) or colouration, the latter also inside the flower (Hansen, Beer, & Müller, 2006; Hansen, Olesen, Mione, Johnson, & Müller, 2007; Johnson et al., 2006; Zhang, Cai, et al., 2012; Zhang, Larson-Rabin, Li, & Wang, 2012, Fig. 1A). For nectar and pollen, irrespectively of being freely accessible or hidden from view within the flower, scent can play an important role as an attractant (Dobson & Bergström, 2000; Howell & Alarcón, 2007; Raguso, 2004b). Also visual cues such as colour or colour guides (Hansen, Van der Niet, & Johnson, 2012), shape (Herrera, 1993; Smith et al., 1996) or tactile cues in the form of flower texture (Glover & Martin, 1998; Goyret & Raguso, 2006; Whitney, Chittka, Bruce, & Glover, 2009) can guide the visitor to the food source. Even acoustic cues such as the specific concave shape of flowers can advertise food sources in flowers to echolocating bats (von Helversen & von Helversen, 1999; von Helversen, Holderied, & von Helversen, 2003). Recently, it has been shown that humidity might be an indicator of nectar at close range (von Arx, Goyret, Davidowith, & Raguso, 2012) and that the taste of nectar or pollen provides the animals with gustatory information of food quality (Gardener & Gillman, 2002; Kessler & Baldwin, 2007; Wacht et al., 2000). Food bodies of different floral parts (bracts, petals, stamens, secretions) are rare and mostly eaten by fruit-eating vertebrates such as birds or bats. Furthermore, Sazima et al. (2001) discovered glistening, sugar-containing jelly pellets of *Combretum lanceolatum* (Combretaceae) that attracted birds. The jelly has been interpreted as nectar secretion resulting in gelatinization of cell wall material. Recently, in *Axinaea* (Melastomataceae) it was discovered that also staminal appendages attract pollinating birds (Fig. 1C). The sugar-rich bulbous appendages serve as food and cause explosive pollen release as bellow organs after being grabbed by the birds' beaks (Dellinger et al., 2014). These food bodies probably mimic fruits by colour (or gloss), shape and taste. Pseudopollen may function as food as it contains protein or starch (Davies & Turner, 2004) and in *Lagerstroemia indica* (Lythraceae) it contains more pores compared to real pollen which might facilitate digestion (Nepi et al., 2003).

3.2 Shelter and Temperature as Attractants

Insects often search for shelter in cavities and these are sometimes offered by flowers (Fig. 1J), e.g., bees in *Oncocyclus* irises (Iridaceae; Monty, Saad, & Mahy, 2006; Sapir, Shmida, & Ne'eman, 2006) or beetles in female flowers of *Leucadendron xanthoconus* (Proteaceae), that are more cup-shaped and enclosing compared to male flowers, for protection against rain (Hemborg & Bond, 2005). Temperature increase in flowers is stated to be induced by floral metabolic processes (Seymour, White, & Gibernau, 2003), highly absorptive surfaces (e.g., dark colour; Sapir et al., 2006, but see Vereecken et al., 2013), parabolic reflection (Kevan, 1975) or heliotropic movements (Orueta, 2002) as well as nectar yeast catabolism of sugar (Herrera & Pozo, 2010). Mostly, the flowers themselves benefit directly from temperature regulation (e.g., pollen tube growth; Galen & Stanton, 2003). Only in some cases temperature increase has been demonstrated to be a direct energy reward to insect visitors (Seymour et al., 2003). Although it has been shown that bees are able to choose flowers based on their heat and to associate floral colour with heat (Dyer, Whitney, Arnold, Glover, & Chittka, 2006), direct attraction of flower visitors by floral heat seems to be rare and only be known in connection with brood-substrate mimicry (see below, Angioy et al., 2004; but see Kite et al., 1998). Heat as attractant can also work indirectly via enhancing the evaporation of floral scent (Gottsberger, 2012; but see Seymour & Schultze-Motel, 1999).

3.3 Reproduction Demands as Attractants

The animals' reproduction demands encompass mating behaviour and parental care. Volatile chemicals (scent, **perfume**) are collected by insects, for example, male orchid bees (euglossines) from different fragrant objects (Fig. 1D). Flowers of mostly orchids exploit this behaviour by attracting euglossines as their exclusive pollinators. The male bees collect the floral scents, mix a species-specific perfume from scents of different sources and store it in their hindleg pockets to spray it during display behaviour (Eltz, Roubik, & Lunau, 2005; Eltz, Roubik, & Whitten, 2003; Hetherington-Rauth & Ramírez, 2016; Mitko et al., 2016; Zimmermann, Roubik, & Eltz, 2006). Flowers are used as **mating rendezvous sites** attracting mainly beetles with colour and scent including aggregation pheromones (Gottsberger, 1989; Karremans et al., 2015; Keasar et al., 2010; Steenhuisen & Johnson, 2012). Flowers mimic insect **mating partners** with scent, colour, gloss and structure (see chapter: Mimicry and Deception in Pollination by

Lunau & Wester, 2017; Figs 1E,F and 2E). Parental care behaviour exploited by flowers includes nest-building, oviposition and feeding the offspring. **Nest-building** bees pollinate flowers when they collect nectar, pollen and antibacterial or antifungal resins (Fig. 1G), or oil (Fig. 1H), and maybe wax to build their brood cells (Armbruster, Gong, & Huang, 2011; Lokvam & Braddock, 1999; Schäffler & Dötterl, 2011; Singer, 2002). Insects searching for **oviposition sites** are deceived by flowers mimicking brood substrate with scent, heat as well as visual and tactile cues (see chapter: Mimicry and Deception in Pollination by Lunau & Wester, 2017; Figs 1I, 2A and 3D). Besides nectar and pollen, bees collect nonvolatile fatty oils to **feed their offspring** (Dumri et al., 2008; Pauw, 2006; Sérsic, 2004; Tavares Carneiro, Camillo Aguiar, Feitosa Martins, Machado, & Alves-dos-Santos, 2015, Fig. 1H). Fatty oil production has been discovered in the South African *Diascia longicornis* (Scrophulariaceae) by Vogel (1969) who found oil in the two long spurs of the flowers and predicted the existence of specialized bees collecting the oil with their long forelegs. Shortly later, the bee with modified front tarsi was indeed discovered and described as *Rediviva emdeorum* (Vogel, 1984). Later, more interactions between different *Diascia* species with varying spur length and *Rediviva* species with different front leg lengths as well as other oil plants of different families were found (Steiner & Whitehead, 1990). Today, oil plants and oil-collecting bees of several genera and families are known in different continents (Renner & Schaefer, 2010). Floral scent plays a major role in attracting the bees to oil flowers (Dötterl & Schäffler, 2007; Steiner, Kaiser, & Dötterl, 2011). Recently, diacetin, a volatile, was identified as a floral signal compound shared by unrelated oil plants from around the world and as a key volatile used by oil-collecting bees to locate their host flowers. As only oil-collecting bees are able to detect diacetin, it is the first demonstrated private communication channel in a pollination system (Schäffler et al., 2015). Rare are oviposition sites that are also used as larval food (e.g., yuccas, figs and a few other plants): some insects (moths, wasps, flies) that are attracted by scent or/and visual cues, actively or passively pollinate flowers, lay eggs in a part of the ovaries and the insects' larvae develop in the plants' ovary or fruits (Fleming & Holland, 1998; Herre, Jandér, & Machado, 2008; Kato, Takimura, & Kawakita, 2003; Pellmyr, 2003; Song et al., 2014).

3.4 Deterrence or Change in Attractiveness

Besides attraction, also deterrence or change in attractiveness is important in the communication between plants and flower-visiting animals. For example, secondary compounds in nectar can cause gustatory deterrence

(Adler, 2001; Heil, 2011; Kessler & Baldwin, 2007). Phenolics in the bitter-tasting nectar of *Aloe vryheidensis* (Asphodelaceae) and *Erythrina caffra* (Fabaceae) filter out ineffective pollinating honeybees and sunbirds, but are accepted by effective pollinating frugivorous and insectivorous birds (Johnson et al., 2006; Nicolson et al., 2015).

Anthesis at times when only pollinators are active can reduce nectar or pollen theft by other flower visitors. Flowers of the orchid *Cirrhaea dependens* (Orchidaceae) open early in the morning and offer floral scent only for a short time when their pollinating oil-collecting Euglossine bees are active (Pansarin, Bittrich, & Amaral, 2006). Many plants, pollinated by bats or moths, open their flowers at night and present nectar and scent in the dark in concert with their pollinators' activity (Faegri & van der Pijl, 1979; Fleming, Geiselman, & Kress, 2009; Hahn & Brühl, 2016).

Changes in floral signals after pollination can guide the animals to visit only sexually viable flowers. This can be achieved by wilting, abscising or curling petals, but also by folding one petal (banner petal in *Caesalpinia eriostachys*; Fabaceae) down over the stamens, reducing attractiveness (van Doorn, 1997; Jones & Buchmann, 1974). The same is accomplished by floral colour change combined with stopping nectar production, thereby retaining long-distance attractiveness of the whole inflorescence (Ida & Kudo, 2003; Weiss, 1991). Another possibility is the change or decrease of floral scent (Raguso, 2004c; Tollsten, 2008; see chapter: Mimicry and Deception in Pollination by Lunau & Wester, 2017), even in concert with colour change (Raguso & Weiss, 2015).

This signalling complexity is even increased by flowers that undergo individual changes in signalling caused by wilting, ripening of stamens and flowering phases due to dichogamy (Pohl et al., 2008). By contrast, some flowering plants possess distinct floral morphs and require pollen flow between the morphs. For example, flower morphs, that are associated with dicliny and heterostyly, are characterized by differences in pollen presentation and reward, but require intermorphic pollen transfer. In these plants flowers display conspicuous signals that are overriding the differences among conspecific flowers of different morphs (Pohl et al., 2008). In general, the flower visitors' discrimination between flowers of coflowering species is important to reduce wastage of pollen through visits to flowers of other species.

ACKNOWLEDGEMENTS

We thank Volker Bittrich, Vinicius Brito, Leandro Freitas, Marcelo Monge, Marcus Nadruz, Andre Rech, Fabrício Schmitz Meyer, Gu Shimizu and Martina Wolowski for their help with the identification of neotropical bees and plants.

REFERENCES

Adler, L. S. (2001). The ecological significance of toxic nectar. *Oikos, 91*(3), 409–420.

Andersson, S. (2003). Antennal responses to floral scents in the butterflies *Inachis io, Aglais urticae,* (Nymphalidae), and *Gonepteryx rhamni* (Pieridae). *Chemoecology, 13*, 13–20.

Andrew, S. C., Perry, C. J., Barron, A. B., Berthon, K., Peralta, V., & Cheng, K. (2014). Peak shift in honey bee olfactory learning. *Animal Cognition, 17*(5), 1177–1186. http://dx.doi.org/10.1007/s10071-014-0750-3.

Angioy, A.-M., Stensmyr, M. C., Urru, I., Puliafito, M., Collu, I., & Hansson, B. S. (2004). Function of the heater: the dead horse arum revisited. *Proceedings of the Royal Society London B, 271*(Suppl. 3), S13–S15.

Armbruster, W. S., Gong, Y.-B., & Huang, S.-Q. (2011). Are pollination "syndromes" predictive? Asian *Dalechampia* fit Neotropical models. *American Naturalist, 178*(1), 135–143.

von Arx, M., Goyret, J., Davidowith, G., & Raguso, R. A. (2012). Floral humidity as a reliable sensory cue for profitability assessment by nectar-foraging hawkmoths. *Proceedings of the National Academy of Sciences of the United States of America, 109*(24), 9471–9476.

Ayala-Berdon, J., Rodríguez-Peña, N., García Leal, C., Stoner, K. E., & Schondube, J. E. (2013). Sugar gustatory thresholds and sugar selection in two species of Neotropical nectar-eating bats. *Comparative Biochemistry and Physiology Part A: Molecular Integrative Physiology, 164*(2), 307–313. http://dx.doi.org/10.1016/j.cbpa.2012.10.019.

Ayasse, M., & Dötterl, S. (2014). The role of preadaptations or evolutionary novelties for the evolution of sexually deceptive orchids. *New Phytologist, 203*(3), 710–712.

Ayasse, M., Paxton, R. J., & Tengö, J. (2001). Mating behavior and chemical communication in the order Hymenoptera. *Annual Review of Entomology, 46*, 31–78.

Baker, H. G., & Baker, I. (1990). The predictive value of nectar chemistry to the recognition of pollinator types. *Israel Journal of Botany, 39*, 157–166.

Baldwin, M. W., Toda, Y., Nakagita, T., O'Connell, M. J., Klasing, K. C., Misaka, T., … Liberles, S. D. (2014). Evolution of sweet taste perception in hummingbirds by transformation of the ancestral umami receptor. *Science, 345*(6199), 929–933. http://dx.doi.org/10.1126/science.1255097.

Balkenius, A., Rosén, W., & Kelber, A. (2006). The relative importance of olfaction and vision in a diurnal and a nocturnal hawkmoth. *Journal of Comparative Physiology A, 192*(4), 431–437. http://dx.doi.org/10.1007/s00359-005-0081-6.

Bargmann, C. I. (2006). Comparative chemosensation from receptors to ecology. *Nature, 444*(7117), 295–301.

Bartlet, E., Blight, M. M., Pickett, J. A., Smart, L. E., Turner, G., & Woodcock, C. M. (2004). Orientation and feeding responses of the pollen beetle, *Meligethes aeneus*, to candytuft, *Iberis amara*. *Journal of Chemical Ecology, 30*(5), 913–925.

Bergamo, P. J., Rech, A. R., Brito, V. L. G., & Sazima, M. (2015). Flower colour and visitation rates of *Costus arabicus* support the 'bee avoidance' hypothesis for red-reflecting hummingbird-pollinated flowers. *Functional Ecology, 30*(5), 710–720.

Burkart, A., Schlindwein, C., & Lunau, K. (2014). Assessment of pollen reward and pollen availability in *Solanum stramoniifolium* and *Solanum paniculatum* for buzz-pollinating carpenter bees. *Plant Biology, 16*, 503–507.

Byers, K. J. R. P., Bradshaw, H. D., & Riffell, J. A. (2014). Three floral volatiles contribute to differential pollinator attraction in monkeyflowers (*Mimulus*). *Journal of Experimental Biology, 217*(4), 614–623. http://dx.doi.org/10.1242/jeb.092213.

Carter, C., Shafir, S., Yehonatan, L., Palmer, R. G., & Thornburg, R. (2006). A novel role for proline in plant floral nectars. *Die Naturwissenschaften, 93*(2), 72–79.

Chapurlat, E., Ågren, J., & Sletvold, N. (2015). Spatial variation in pollinator-mediated selection on phenology, floral display and spur length in the orchid *Gymnadenia conopsea*. *New Phytologist, 208*(4), 1264–1275.

Chen, P., Awata, H., Matsushita, A., Yang, E., & Arikawa, K. (2016). Extreme spectral richness in the eye of the common bluebottle butterfly, *Graphium sarpedon*. *Frontiers in Ecology and Evolution, 4*(18). http://dx.doi.org/10.3389/fevo.2016.00018.
Chen, G., Ma, X.-K., Jürgens, A., Lu, J., Liu, E.-X., Sun, W.-B., & Cai, X. H. (2015). Mimicking livor mortis: a well-known but unsubstantiated color profile in sapromyiophily. *Journal of Chemical Ecology, 41*(9), 808–815.
Chittka, L. (1996). Does bee colour vision predate the evolution of flower colour? *Die Naturwissenschaften, 83*(3), 136–138.
Chittka, L. (1997). Bee color vision is optimal for coding flower colors, but flower colors are not optimal for being coded – why? *Israel Journal of Plant Sciences, 45*, 115–127.
Chittka, L., & Menzel, R. (1992). The evolutionary adaptation of flower colors and the insect pollinators' color vision systems. *Journal of Comparative Physiology A, 171*(2), 171–181.
Chittka, L., & Raine, N. E. (2006). Recognition of flowers by pollinators. *Current Opinion in Plant Biology, 9*, 428–435.
Chittka, L., & Schürkens, S. (2001). Successful invasion of a floral market. *Nature, 411*, 653.
Chittka, L., & Späthe, J. (2007). Visual search and the importance of time in complex decision making by bees. *Arthropod-Plant Interactions, 1*, 37–44.
Chittka, L., Späthe, J., Schmidt, A., & Hickelsberger, A. (2001). Adaptation, constraint, and chance in the evolution of flower color and pollinator color vision. In L. Chittka, & J. D. Thomson (Eds.), *Cognitive ecology of pollination* (pp. 106–126). Cambridge: Cambridge University Press.
Chittka, L., & Thomson, J. D. (2001). *Cognitive ecology of pollination*. Cambridge: Cambridge University Press.
Chittka, L., Thomson, J. D., & Waser, N. M. (1999). Flower constancy, insect psychology, and plant evolution. *Die Naturwissenschaften, 86*(8), 361–377.
Clarke, D., Whitney, H., Sutton, G., & Robert, D. (2013). Detection and learning of floral electric fields by bumblebees. *Science, 340*(6128), 66–69.
Cronk, Q. C. B., & Ojeda, I. (2008). Bird-pollinated flowers in an evolutionary and molecular context. *Journal of Experimental Botany, 59*(4), 15–27.
Davies, K. L., & Turner, M. P. (2004). Pseudopollen in *Dendrobium unicum* Seidenf. (Orchidaceae): reward or deception? *Annals of Botany, 94*(1), 129–132.
De Brito Sanchez, M. G. (2011). Taste perception in honey bees. *Chemical Senses, 36*(8), 675–692.
De Brito Sanchez, M. G., Lorenzo, E., Su, S., Liu, F., Zhan, Y., & Giurfa, M. (2014). The tarsal taste of honey bees: behavioral and electrophysiological analyses. *Frontiers in Behavioral Neuroscience, 8*, 25. http://dx.doi.org/10.3389/fnbeh.2014.00025.
De Luca, P. A., & Vallejo-Marín, M. (2013). What's the 'buzz' about? The ecology and evolutionary significance of buzz-pollination. *Current Opinion in Plant Biology, 16*(4), 429–435.
Dellinger, A. S., Penneys, D. S., Staedler, Y. M., Fragner, L., Weckwerth, W., & Schönenberger, J. (2014). A specialized bird pollination system with a bellows mechanism for pollen transfer and staminal food body rewards. *Current Biology, 24*(14), 1615–1619.
DeMaria, S., & Ngai, J. (2010). The cell biology of smell. *The Journal of Cell Biology, 191*(3), 443–452.
Dinkel, T., & Lunau, K. (2001). How drone flies (*Eristalis tenax* L, Syrphidae, Diptera) use floral guides to locate food sources. *Journal of Insect Physiology, 47*(10), 1111–1118.
Dobson, H. E. M., & Bergström, G. (2000). The ecology and evolution of pollen odors. *Plant Systematics and Evolution, 222*(1), 63–87.
van Doorn, W. G. (1997). Effects of pollination on floral attraction and longevity. *Journal of Experimental Biology, 48*(314), 1615–1622.

Dötterl, S., Glück, U., Jürgens, A., Woodring, J., & Aas, G. (2014). Floral reward, advertisement and attractiveness to honey bees in dioecious *Salix caprea*. *PLoS One, 9*(3), e93421. http://dx.doi.org/10.1371/journal.pone.0093421.

Dötterl, S., & Jürgens, A. (2005). Spatial fragrance patterns in flowers of *Silene latifolia*: lilac compounds as olfactory nectar guides? *Plant Systematics and Evolution, 255*(1–2), 99–109.

Dötterl, S., Jürgens, A., Seifert, K., Laube, T., Weißbecker, B., & Schütz, S. (2006). Nursery pollination by a moth in *Silene latifolia*: the role of odours in eliciting antennal and behavioural responses. *New Phytologist, 169*(4), 707–718.

Dötterl, S., & Schäffler, I. (2007). Flower scent of floral oil producing *Lysimachia punctata* as attractant for the oil-bee *Macropis fulvipes*. *Journal of Chemical Ecology, 33*(2), 441–445.

Dukas, R. (2001). Effects of perceived danger on flower choice by bees. *Ecology Letters, 4*, 327–333.

Dumri, K., Seipold, L., Schmidt, J., Gerlach, G., Dötterl, S., Ellis, A. G., & Wessjohanna, L. A. (2008). Non-volatile floral oils of *Diascia* spp. (Scrophulariaceae). *Phytochemistry, 69*(6), 1372–1383.

Dyer, A. G., & Chittka, L. (2004). Fine colour discrimination requires differential conditioning in bumblebees. *Die Naturwissenschaften, 91*(5), 224–227.

Dyer, A. G., & Murphy, A. H. (2009). Honeybees choose "incorrect" colors that are similar to target flowers in preference to novel colors. *Israel Journal of Plant Sciences, 57*(3), 203–210.

Dyer, A. G., Neumeyer, C., & Chittka, L. (2005). Honeybee (*Apis mellifera*) vision can discriminate between and recognise images of human faces. *Journal of Experimental Biology, 208*, 4709–4714.

Dyer, A. G., Späthe, J., & Prack, S. (2008). Comparative psychophysics of bumblebee and honeybee colour discrimination and object detection. *Journal of Comparative Physiology A, 194*(7), 617–627.

Dyer, A. G., Whitney, H. M., Arnold, S. E. J., Glover, B. J., & Chittka, L. (2006). Bees associate warmth with floral colour. *Nature, 442*(7102), 525.

Eltz, T. (2006). Tracing pollinator footprints on natural flowers. *Journal of Chemical Ecology, 32*(5), 907–915.

Eltz, T., Roubik, D. W., & Lunau, K. (2005). Experience-dependent choices ensure species-specific fragrance accumulation in male orchid bees. *Behavioral Ecology and Sociobiology, 59*(1), 149–156.

Eltz, T., Roubik, D. W., & Whitten, W. M. (2003). Fragrances, male display and mating behaviour of *Euglossa hemichlora* – a flight cage experiment. *Physiological Entomology, 28*(4), 251–260.

Faegri, K., & van der Pijl, L. (1979). *The principles of pollination ecology* (3rd ed.). Oxford: Pergamon Press.

Fenster, C. B., Armbruster, W. S., Wilson, P., Dudash, M. R., & Thomson, J. D. (2004). Pollination syndromes and floral specialization. *Annual Review of Ecology, Evolution, and Systematics, 35*(1), 375–403.

Fleming, T. H., Geiselman, C., & Kress, W. J. (2009). The evolution of bat pollination: a phylogenetic perspective. *Annals of Botany, 104*(6), 1017–1043.

Fleming, T. H., & Holland, J. N. (1998). The evolution of obligate pollination mutualisms: senita cactus and senita moth. *Oecologia, 114*(3), 368–375.

Fleming, T. H., & Muchhala, N. (2007). Nectar-feeding bird and bat niches in two worlds: pantropical comparisons of vertebrate pollination systems. *Journal of Biogeography, 35*(5), 764–780.

Foster, J. J., Sharkey, C. R., Gaworska, A. V. A., Roberts, N. W., Whitney, H. M., & Partridge, J. C. (2014). Bumblebees learn polarization patterns. *Current Biology, 24*(12), 1415–1420.

von Frisch, K. (1914). Der Farbensinn und Formensinn der Biene. *Zoologische Jahrbücher Abteilung für allgemeine Zoologie und Physiologie, 37*(1), 1–187.
Galen, C., & Stanton, M. L. (2003). Sunny-side up: flower heliotropism as a source of parental environmental effects on pollen quality and performance in the snow buttercup, *Ranunculus adoneus* (Ranunculaceae). *American Journal of Botany, 90*(5), 724–729.
Gardener, M. C., & Gillman, M. P. (2002). The taste of nectar – a neglected area of pollination ecology. *Oikos, 98*(3), 552–557.
Gawleta, N., Zimmermann, Y., & Eltz, T. (2005). Repellent foraging scent recognition across bee families. *Apidologie, 36*(3), 325–330.
Giurfa, M. (2004). Conditioning procedure and color discrimination in the honeybee *Apis mellifera. Die Naturwissenschaften, 91*(5), 228–231.
Giurfa, M., Núnez, J., & Backhaus, W. (1994). Odour and colour information in the foraging choice behaviour of the honeybee. *Journal of Comparative Physiology A, 175*(6), 773–779. http://dx.doi.org/10.1007/bf00191849.
Giurfa, M., Núnez, J., Chittka, L., & Menzel, R. (1995). Colour preferences of flower-naive honeybees. *Journal of Comparative Physiology A, 177*(3), 247–259.
Giurfa, M., Vorobyev, M., Brandt, R., Posner, B., & Menzel, R. (1997). Discrimination of coloured stimuli by honeybees: alternative use of achromatic and chromatic signals. *Journal of Comparative Physiology A, 180*(3), 235–243.
Glover, B. (2014). *Understanding flowers & flowering – An Integrated approach* (2nd ed.). Oxford: Oxford University Press.
Glover, B. J., & Martin, C. (1998). The role of petal cell shape and pigmentation in pollination success in *Antirrhinum majus. Heredity, 80*(6), 778–784.
Goldsmith, K. M., & Goldsmith, T. H. (1982). Sense of smell in the black-chinned hummingbird. *Condor, 84*, 237–238.
Gottsberger, G. (1989). Beetle pollination and flowering rhythm of *Annona* spp. (Annonaceae) in Brazil. *Plant Systematics and Evolution, 167*(3/4), 165–187.
Gottsberger, G. (2012). How diverse are Annonaceae with regard to pollination? *Botanical Journal of the Linnean Society, 169*(1), 245–261.
Goulson, D., Chapman, J. W., & Hughes, W. (2001). Discrimination of unrewarding flowers by bees; direct detection of rewards and use of repellent scent marks. *Journal of Insect Physiology, 14*(5), 669–678.
Goyret, J. (2010). Look and touch: multimodal sensory control of flower inspection movements in the nocturnal hawkmoth *Manduca sexta. Journal of Experimental Biology, 213*, 3676–3682. http://dx.doi.org/10.1242/jeb.045831.
Goyret, J., Markwell, P. M., & Raguso, R. A. (2007). The effect of decoupling olfactory and visual stimuli on the foraging behavior of *Manduca sexta. Journal of Experimental Biology, 210*(8), 1398–1405. http://dx.doi.org/10.1242/jeb.02752.
Goyret, J., Pfaff, M., Raguso, R. A., & Kelber, A. (2008). Why do *Manduca sexta* feed from white flowers? Innate and learnt colour preferences in a hawkmoth. *Die Naturwissenschaften, 95*, 569–576. http://dx.doi.org/10.1007/s00114-008-0350-7.
Goyret, J., & Raguso, R. A. (2006). The role of mechanosensory input in flower handling efficiency and learning by *Manduca sexta. Journal of Experimental Biology, 209*(11), 1585–1593.
Greiner, B., Ribi, W. A., & Warrant, E. J. (2004). Retinal and optical adaptations for nocturnal vision in the halictid bee *Megalopta genalis. Cell Tissue Reearch, 316*(3), 377–390.
Gronquist, M., Bezzerides, A., Attygalle, A. B., Meinwald, J., Eisner, M., & Eisner, T. (2001). Attractive and defensive functions of the ultraviolet pigments of a flower (*Hypericum calycinum*). *Proceedings of the National Academy of Sciences of the United States of America, 98*(24), 13745–13750.

Gross, K., Sun, M., & Schiestl, F. P. (2016). Why do floral perfumes become different? Region-specific selection on floral scent in a terrestrial orchid. *PLoS One, 11*(2), e0147975. http://dx.doi.org/10.1371/journal.pone.0147975.

Gumbert, A. (2000). Color choices by bumble bees (*Bombus terrestris*): innate preferences and generalization after learning. *Behavioral Ecology and Sociobiology, 48*(1), 36–43.

Hahn, M., & Brühl, C. A. (2016). The secret pollinators: an overview of moth pollination with a focus on Europe and North America. *Arthropod-Plant Interactions, 10*(1), 21–28.

Hansen, D. M., Beer, K., & Müller, C. B. (2006). Mauritian coloured nectar no longer a mystery: a visual signal for lizard pollinators. *Biology Letters, 2*(2), 165–168.

Hansen, D. M., Olesen, J. M., Mione, T., Johnson, S. D., & Müller, C. B. (2007). Coloured nectar: distribution, ecology, and evolution of an enigmatic floral trait. *Biological Reviews of the Cambridge Philosophical Society, 82*(1), 83–111.

Hansen, D. M., Van der Niet, T., & Johnson, S. D. (2012). Floral signposts: testing the significance of visual 'nectar guides' for pollinator behaviour and plant fitness. *Proceedings of the Royal Society B, 279*(1729), 634–639.

Hansson, B. S., & Stensmyr, M. C. (2011). Evolution of insect olfaction. *Neuron, 72*(5), 698–711. http://dx.doi.org/10.1016/j.neuron.2011.11.003.

Harder, L. D., & Aizen, M. A. (2010). Floral adaptation and diversification under pollen limitation. *Philosophical Transactions of the Royal Society B, 365*(1539), 529–543. http://dx.doi.org/10.1098/rstb.2009.0226.

Haverkamp, A., Yon, F., Keesey, I. W., Mißbach, C., König, C., Hansson, B. S., ... Kessler, D. (2016). Hawkmoths evaluate scenting flowers with the tip of their proboscis. *eLife, 5*, e15039. http://dx.doi.org/10.7554/eLife.15039.

Heil, M. (2011). Nectar: generation, regulation and ecological functions. *Trends in Plant Science, 16*(4), 191–200.

von Helversen, D., & von Helversen, O. (1999). Acoustic guide in bat-pollinated flower. *Nature, 398*(6730), 759–760.

von Helversen, D., Holderied, M. W., & von Helversen, O. (2003). Echoes of bat pollinated bell-shaped flowers: conspicuous for nectar-feeding bats? *Journal of Experimental Biology, 206*(6), 1025–1034.

Hemborg, A. M., & Bond, W. J. (2005). Different rewards in female and male flowers can explain the evolution of sexual dimorphism in plants. *Biological Journal of the Linnean Society, 85*(1), 97–109.

Hempel de Ibarra, N., Giurfa, M., & Vorobyev, M. (2001). Detection of coloured patterns by honeybees through chromatic and achromatic cues. *Journal of Comparative Physiology A, 187*(3), 215–224.

Hempel de Ibarra, N., Langridge, K. V., & Vorobyev, M. (2015). More than colour attraction: behavioural functions of flower patterns. *Current Opinion in Insect Science, 12*, 64–70. http://dx.doi.org/10.1016/j.cois.2015.09.005.

Herre, E. A., Jandér, K. C., & Machado, C. A. (2008). Evolutionary ecology of figs and their associates: recent progress and outstanding puzzles. *Annual Review of Ecology, Evolution, and Systematics, 39*, 439–458.

Herrera, C. M. (1993). Selection on floral morphology and environmental determinants of fecundity in a hawk moth-pollinated violet. *Ecological Monographs, 63*(3), 251–275.

Herrera, C. M., & Pozo, M. I. (2010). Nectar yeasts warm the flowers of a winter-blooming plant. *Proceedings of the Royal Society B, 277*(1689), 1827–1834.

Hetherington-Rauth, M. C., & Ramírez, S. R. (2016). Evolution and diversity of floral scent chemistry in the euglossine bee-pollinated orchid genus *Gongora*. *Annals of Botany*. http://dx.doi.org/10.1093/aob/mcw072.

Hoballah, M. E., Stuurman, J., Turlings, T. C., Guerin, P. M., Connétable, S., & Kuhlemeier, C. (2005). The composition and timing of flower odour emission by

wild *Petunia axillaris* coincide with the antennal perception and nocturnal activity of the pollinator *Manduca sexta*. *Planta, 222*(1), 141–150.
Howell, A. D., & Alarcón, R. (2007). *Osmia* bees (Hymenoptera: Megachilidae) can detect nectar rewarding flowers using olfactory cues. *Animal Behaviour, 74*(2), 199–205.
Ida, T. Y., & Kudo, G. (2003). Floral color change in *Weigela middendorffiana* (Caprifoliaceae): reduction of geitonogamous pollination by bumble bees. *American Journal of Botany, 90*(12), 1751–1757.
Ings, T. C., & Chittka, L. (2008). Speed-accuracy tradeoffs and false alarms in bee responses to cryptic predators. *Current Biology, 18*(19), 1520–1524.
Ings, T. C., & Chittka, L. (2009). Predator crypsis enhances behaviourally mediated indirect effects on plants by altering bumblebee foraging preferences. *Proceedings of the Royal Society B: Biological Sciences, 276*(1664), 2031–2036.
Inoue, T. A., Asaoka, K., Seta, K., Imaeda, D., & Ozaki, M. (2009). Sugar receptor response of the food-canal taste sensilla in a nectar-feeding swallowtail butterfly, *Papilio xuthus*. *Die Naturwissenschaften, 96*(3), 355–363. http://dx.doi.org/10.1007/s00114-008-0483-8.
Ioalé, P., & Papi, F. (1989). Olfactory bulb size, odor discrimination and magnetic insensitivity in hummingbirds. *Physiology and Behavior, 45*(5), 995–999.
Jacobs, G. H., Fenwick, J. A., & Williams, G. A. (2001). Cone-based vision of rats for ultraviolet and visible lights. *Journal of Experimental Biology, 204*, 2439–2446.
Jhumur, U. S., Dötterl, S., & Jürgens, A. (2007). Electrophysiological and behavioural responses of mosquitoes to the volatiles of *Silene otites* (Caryophyllaceae). *Arthropod-Plant Interactions, 1*(4), 245–254.
Johnson, S. D., Burgoyne, P. M., Harder, L. D., & Dötterl, S. (2011). Mammal pollinators lured by the scent of a parasitic plant. *Proceedings of the Royal Society B: Biological Sciences, 278*(1716), 2303–2310. http://dx.doi.org/10.1098/rspb.2010.2175.
Johnson, S. D., Hargreaves, A. L., & Brown, M. (2006). Dark, bitter-tasting nectar functions as a filter of flower visitors in a bird-pollinated plant. *Ecology, 87*(11), 2709–2716.
Johnson, S. D., & Steiner, K. E. (2000). Generalization versus specialization in plant pollination systems. *Trends in Ecology and Evolution, 15*, 190–193.
Jones, C. E., & Buchmann, S. L. (1974). Ultraviolet floral patterns as functional orientation cues in hymenopterous pollination systems. *Animal Behaviour, 22*(2), 481–485.
Jones, G., Teeling, E. C., & Rossiter, S. J. (2013). From the ultrasonic to the infrared: molecular evolution and the sensory biology of bats. *Frontiers in Physiology, 4*. http://dx.doi.org/10.3389/fphys.2013.00117.
Junker, R. R., & Blüthgen, N. (2010). Floral scents repel facultative flower visitors, but attract obligate ones. *Annals of Botany, 105*(5), 777–782.
Junker, R. R., Dähler, C. C., Curtis, C., Dötterl, S., Keller, A., & Blüthgen, N. (2011). Hawaiian ant-flower networks: nectar-thieving ants prefer undefended native over introduced plants with floral defences. *Ecological Monographs, 81*(2), 295–311. http://dx.doi.org/10.1890/10-1367.1.
Junker, R. R., Höcherl, N., & Blüthgen, N. (2010). Responses to olfactory signals reflect network structure of flower-visitor interactions. *Journal of Animal Ecology, 79*(4), 818–823.
Jürgens, A., & Shuttleworth, A. (2015). Carrion and dung mimicry in plants. In M. E. Benbow, J. K. Tomberlin, & A. M. Tarone (Eds.), *Carrion ecology, evolution, and their applications* (pp. 361–386). Boca Raton, FL: CRC Press.
Jürgens, A., Wee, S. L., Shuttleworth, A., & Johnson, S. D. (2013). Chemical mimicry of insect oviposition sites: a global analysis of convergence in angiosperms. *Ecology Letters, 16*(9), 1157–1167. http://dx.doi.org/10.1111/ele.12152.
Karremans, A. P., Pupulin, F., Grimaldi, D., Beentjes, K. K., Butôt, R., Fazzi, G. E., … Gravendeel, B. (2015). Pollination of *Specklinia* by nectar-feeding

Drosophila: the first reported case of a deceptive syndrome employing aggregation pheromones in Orchidaceae. *Annals of Botany, 116*(3), 437–455.
Kato, M., Takimura, A., & Kawakita, A. (2003). An obligate pollination mutualism and reciprocal diversification in the tree genus *Glochidion* (Euphorbiaceae). *Proceedings of the National Academy of Sciences of the United States of America, 100*(9), 5264–5267.
Katzenberger, T. D., Lunau, K., & Junker, R. R. (2013). Salience of multimodal flower cues manipulates initial responses and facilitates learning performance of bumblebees. *Behavioural Ecology and Sociobiology, 67*, 1587–1599.
Keasar, T., Harari, A. R., Sabatinelli, G., Keith, D., Dafni, A., Shavit, O., ... Shmida, A. (2010). Red anemone guild flowers as focal places for mating and feeding by Levant glaphyrid beetles. *Biological Journal of the Linnean Society, 99*(4), 808–817.
Kelber, A., Balkenius, A., & Warrant, E. J. (2002). Scotopic colour vision in nocturnal hawkmoths. *Nature, 419*, 922–925.
Kessler, D., & Baldwin, I. T. (2007). Making sense of nectar scents: the effects of nectar secondary metabolites on floral visitors of *Nicotiana attenuata*. *The Plant Journal, 49*(5), 840–854.
Kessler, D., Gase, K., & Baldwin, T. (2008). Field experiments with transformed plants reveal the Sense of floral scents. *Science, 321*, 1200–1202.
Kevan, P. (1975). Sun-tracking solar furnaces in high arctic flowers: significance for pollination and insects. *Science, 189*(4204), 723–726.
Kevan, P. G., & Lane, M. A. (1985). Flower petal microtexture is a tactile cue for bees. *Proceedings of the National Academy of Sciences of the United States of America, 82*(14), 4750–4752.
Kite, G. C., Hetterscheid, W. L. A., Lewis, M. J., Boyce, P. C., Ollerton, J., Cocklin, E., ... Simmonds, M. S. J. (1998). Inflorescence odours and pollinators of *Arum* and *Amorphophallus* (Araceae). In S. J. Owens, & P. J. Rudall (Eds.), *Reproductive biology* (pp. 295–315). Kew: Royal Botanic Gardens.
Klein, A. M., Vaissière, B. E., Cane, J., Steffan-Dewenter, I., Cunningham, S. A., Kremen, C., & Tscharntke, T. (2007). Importance of pollinators in changing landscapes for world crops. *Proceedings of the Royal Society B, 274*, 303–313.
Knauer, A. C., & Schiestl, F. P. (2015). Bees use honest floral signals as indicators of reward when visiting flowers. *Ecology Letters, 18*(2), 135–143. http://dx.doi.org/10.1111/ele.12386.
Knight, T. M., Steets, J. A., Vamosi, J. C., Mazer, S. J., Burd, M., Campbell, D. R., & Ashman, T.-L. (2005). Pollen limitation of plant reproduction: pattern and process. *Annual Review of Ecology, Evolution and Systematics, 36*, 467–497.
van der Kooi, C. J., Pen, I., Staal, M., Stavenga, D. G., & Elzenga, J. T. M. (2016). Competition for pollinators and intra-communal spectral dissimilarity of flowers. *Plant Biology, 18*, 56–62. http://dx.doi.org/10.1111/plb.12328.
Koski, M. H., & Ashman, T.-L. (2014). Dissecting pollinator responses to a ubiquitous ultraviolet floral pattern in the wild. *Functional Ecology, 28*, 868–877.
Kulahci, I. G., Dornhaus, A., & Papaj, D. R. (2008). Multimodal signals enhance decision making in foraging bumble-bees. *Proceedings of the Royal Society B: Biological Sciences, 275*, 797–802.
Kunze, J., & Gumbert, A. (2001). The combined effect of color and odor on flower choice behavior of bumble bees in flower mimicry systems. *Behavioral Ecology, 12*(4), 447–456.
Land, M. F., & Nilsson, D.-E. (2001). *Animal eyes*. Oxford: Oxford University Press.
Leonard, A. S., Brent, J., Papaj, D. R., & Dornhaus, A. (2013). Floral nectar guide patterns discourage nectar robbing by bumble bees. *PLoS One, 8*(2), e55914. http://dx.doi.org/10.1371/journal.pone.0055914.

Leonard, A. S., Dornhaus, A., & Papaj, D. R. (2011). Flowers help bees cope with uncertainty: signal detection and the function of floral complexity. *The Journal of Experimental Biology, 214*(1), 113–121. http://dx.doi.org/10.1242/jeb.047407.

Leonard, A. S., & Masek, P. (2014). Multisensory integration of colors and scents: insights from bees and flowers. *Journal of Comparative Physiology A, 200*, 463–474.

Leonard, A. S., & Papaj, D. R. (2011). 'X'marks the spot: the possible benefits of nectar guides to bees and plants. *Functional Ecology, 25*(6), 1293–1301. http://dx.doi.org/10.1111/j.1365-2435.2011.01885.x.

Lokvam, J., & Braddock, J. F. (1999). Anti-bacterial function in the sexually dimorphic pollinator rewards of *Clusia grandiflora* (Clusiaceae). *Oecologia, 119*(4), 534–540.

Lunau, K. (2000). The ecology and evolution of visual pollen signals. *Plant Systematics and Evolution, 222*(1), 89–111.

Lunau, K. (2007). Stamens and mimic stamens as components of floral colour patterns. *Botanische Jahrbücher für Systematik, Pflanzengeschichte und Pflanzengeographie, 127*, 13–41.

Lunau, K. (2014). Visual ecology of flies with particular reference to colour vision and colour preferences. *Journal of Comparative Physiology A, 200*(6), 497–512.

Lunau, K. (2016). Flower colour: how bumblebees handle colours with perceptually changing hues. *Current Biology, 26*, R229–R246.

Lunau, K., & Maier, E. J. (1995). Innate colour preference of flower visitors. *Journal of Comparative Physiology A, 177*(1), 1–19.

Lunau, K., Papiorek, S., Eltz, T., & Sazima, M. (2011). Avoidance of achromatic colours by bees provides a private niche for hummingbirds. *Journal of Experimental Biology, 214*(9), 1607–1612.

Lunau, K., Piorek, V., Krohn, O., & Pacini, E. (2015). Just spines – mechanical defence of malvaceous pollen against collection by corbiculate bees. *Apidologie, 46*(2), 144–149.

Lunau, K., & Wacht, S. (1994). Optical releasers of the innate proboscis extension in the hoverfly *Eristalis tenax* L. (Syrphidae, Diptera). *Journal of Comparative Physiology A, 174*(5), 575–579.

Lunau, K., Wacht, S., & Chittka, L. (1996). Colour choices of naive bumble bees and their implications for colour perception. *Journal of Comparative Physiology A, 178*, 477–489.

Lunau, K., & Wester, P. (2017). Mimicry and deception in pollination. In G. Becard (Ed.), *How plants communicate with their biotic environment* (Vol. 82, pp. 259–279).

Lynn, S. K., Cnaani, J., & Papaj, D. (2005). Peak shift discrimination learning as a mechanism of signal evolution. *Evolution; International Journal of Organic Evolution, 59*, 1300–1305.

Marshall, J., & Arikawa, K. (2014). Unconventional colour vision. *Current Biology, 24*(24), 1150–1154. http://dx.doi.org/10.1016/j.cub.2014.10.025.

Martínez-Harms, J., Márquez, N., Menzel, R., & Vorobyev, M. (2014). Visual generalization in honeybees: evidence of peak shift in color discrimination. *Journal of Comparative Physiology A, 200*(4), 317–325.

Menzel, R. (1985). Learning in honey bees in an ecological and behavioral context. In B. Hölldobler, & M. Lindauer (Eds.), *Experimental behavioral ecology* (pp. 55–74). Stuttgart: Gustav Fischer.

Milet-Pinheiro, P., Ayasse, M., Dobson, H. E. M., Schlindwein, C., Francke, W., & Dötterl, S. (2013). The chemical basis of host-plant recognition in a specialized bee pollinator. *Journal of Chemical Ecology, 39*(11–12), 1347–1360. http://dx.doi.org/10.1007/s10886-013-0363-3.

Milet-Pinheiro, P., Ayasse, M., & Dötterl, S. (2015). Visual and olfactory floral cues of *Campanula* (Campanulaceae) and their significance for host recognition by an oligolectic bee pollinator. *PLoS One, 10*(6). http://dx.doi.org/10.1371/journal.pone.0128577.

Milet-Pinheiro, P., Herz, K., Dötterl, S., & Ayasse, M. (2016). Host choice in a bivoltine bee: how sensory constraints shape innate foraging behaviors. *BMC Ecology, 16*(20). http://dx.doi.org/10.1186/s12898-016-0074-z.

Mitchell, R. J., Flanagan, R. J., Brown, B. J., Waser, N. M., & Karron, J. D. (2009). New frontiers in competition for pollination. *Annals of Botany, 103*(9), 1403–1413.

Mitko, L., Weber, M. G., Ramírez, S. R., Hedenström, E., Wcislo, W. T., & Eltz, T. (2016). Olfactory specialization for perfume collection in male orchid bees. *Journal of Experimental Biology, 219*(10), 1467–1475.

Monty, A., Saad, L., & Mahy, G. (2006). Bimodal pollination system in rare endemic *Oncocyclus* irises (Iridaceae) of Lebanon. *Canadian Journal of Botany, 84*(8), 1327–1338.

Morawetz, L., Chittka, L., & Späthe, J. (2014). Strategies of the honeybee *Apis mellifera* during visual search for vertical targets presented at various heights: a role for spatial attention? *F1000Research, 3*, 174. http://dx.doi.org/10.12688/f1000research.4799.1.

Müller, B., Glösmann, M., Peichl, L., Knop, G. C., Hagemann, C., & Ammermüller, J. (2009). Bat eyes have ultraviolet-sensitive cone photoreceptors. *PLoS One, 4*(7), e6390. http://dx.doi.org/10.1371/journal.pone.0006390.

Nepi, M., Guarnieri, M., & Pacini, E. (2003). "Real" and feed pollen of *Lagerstroemia indica*: ecophysiological differences. *Plant Biology, 5*(3), 311–314.

Nicolson, S. W., Lerch-Henning, S., Welsford, M., & Johnson, S. D. (2015). Nectar palatability can selectively filter bird and insect visitors to coral tree flowers. *Evolutionary Ecology, 29*(3), 405–417.

Odeen, A., & Hastad, O. (2013). The phylogenetic distribution of ultraviolet sensitivity in birds. *BMC Evolutionary Biology, 13*, 36. http://dx.doi.org/10.1186/1471-2148-13-36.

Ohashi, K., & Thomson, J. D. (2005). Efficient harvesting of renewing resources. *Behavioral Ecology, 16*, 592–605.

Olesen, J. M., Bascompte, J., Dupont, Y. L., & Jordano, P. (2007). The modularity of pollination networks. *Proceedings of the National Academy of Sciences of the United States of America, 104*(50), 19891–19896.

Ômura, H., Honda, K., & Hayashi, N. (2000). Floral scent of *Osmanthus fragrans* discourages foraging behavior of cabbage butterfly, *Pieris rapae*. *Journal of Chemical Ecology, 26*(3), 655–666. http://dx.doi.org/10.1023/A:1005424121044.

Orbán, L. L., & Plowright, C. M. S. (2014). Getting to the start line: how bumblebees and honeybees are visually guided towards their first floral contact. *Insectes Sociaux, 61*(4), 325–336. http://dx.doi.org/10.1007/s00040-014-0366-2.

Orueta, D. (2002). Thermal relationships between *Calendula arvensis* inflorescences and *Usia aurata* bombyliid flies. *Ecology, 83*(11), 3073–3085.

Palmer-Jones, T., & Forster, I. W. (1972). Measures to increase the pollination of lucerne (*Medicago sativa* Linn.). *New Zealand Journal of Agricultural Research, 15*(1), 186–193. http://dx.doi.org/10.1080/00288233.1972.10421294.

Pansarin, E. R., Bittrich, V., & Amaral, M. C. E. (2006). At daybreak – reproductive biology and isolating mechanisms of *Cirrhaea dependens* (Orchidaceae). *Plant Biology, 8*(4), 494–502.

Papiorek, S., Junker, R. R., Alves-dos-Santos, I., Melo, G. A. R., Amaral-Neto, L. P., Sazima, M., … Lunau, K. (2016). Bees, birds and yellow flowers: pollinator-dependent convergent evolution of UV-patterns. *Plant Biology, 18*(1), 46–55.

Papiorek, S., Junker, R. R., & Lunau, K. (2014). Gloss, colour and grip: multifunctional epidermal cell shapes in bee- and bird-pollinated flowers. *PLoS One, 9*(11), e112013.

Pauw, A. (2006). Floral syndromes accurately predict pollination by a specialized oil-collecting bee (*Rediviva peringueyi*, Melittidae) in a guild of South African orchids (Coryciinae). *American Journal of Botany, 93*(6), 917–926.

Peitsch, D., Fietz, A., Hertel, H., de Souza, J., Ventura, D. F., & Menzel, R. (1992). The spectral input systems of hymenopteran insects and their receptor-based colour vision. *Journal of Comparative Physiology A, 170*(1), 23–40.

Pellmyr, O. (2003). Yuccas, yucca moths, and coevolution: a review. *Annals of the Missouri Botanical Garden, 90*(1), 35–55.

Pohl, M., Watolla, T., & Lunau, K. (2008). Anther-mimicking floral guides exploit a conflict between innate preference and learning in bumblebees (*Bombus terrestris*). *Behavioral Ecology and Sociobiology, 63*, 295–302.

Porsch, O. (1931). Grellrot als Vogelblumenfarbe. *Biologia Generalis, 2*, 647–674.

Potts, S. G., Biesmeijer, J. C., Kremen, C., Neumann, P., Schweiger, O., & Kunin, W. E. (2010). Global pollinator declines: trends, impacts and drivers. *Trends in Ecology and Evolution, 25*, 345–353.

Pyke, G. H. (2016). Floral nectar: pollinator attraction or manipulation? *Trends in Ecology and Evolution, 31*(5), 339–341.

Raguso, R. A. (2004a). Flowers as sensory billboards: progress towards an integrated understanding of floral advertisement. *Current Opinion in Plant Biology, 7*(4), 434–440.

Raguso, R. A. (2004b). Why are some floral nectars scented? *Ecology, 85*(6), 1486–1494.

Raguso, R. A. (2004c). Why do flowers smell? The chemical ecology of fragrance-driven pollination. In R. T. Cardé, & J. G. Miller (Eds.), *Advances in insect chemical ecology* (pp. 151–178). Cambridge: Cambridge University Press.

Raguso, R. A. (2008). Wake up and smell the roses: the ecology and evolution of floral scent. *Annual Review of Ecology Evolution, and Systematics, 39*(1), 549–569.

Raguso, R. A., & Pichersky, E. (1999). New perspectives in pollination biology: floral fragrances. A day in the life of a linalool molecule: chemical communication in a plant-pollinator system. Part 1: linalool biosynthesis in flowering plants. *Plant Species Biology, 14*(2), 95e120.

Raguso, R. A., & Weiss, M. R. (2015). Concerted changes in floral colour and scent, and the importance of spatio-temporal variation in floral volatiles. *Journal of the Indian Institute of Science, 95*(1), 69–92.

Raguso, R. A., & Willis, M. A. (2005). Synergy between visual and olfactory cues in nectar feeding by wild hawkmoths, *Manduca sexta*. *Animal Behaviour, 69*(2), 407–418. http://dx.doi.org/10.1016/j.anbehav.2004.04.015.

Raine, N. E., & Chittka, L. (2007a). The adaptive significance of sensory bias in a foraging context: floral colour preferences in the bumblebee *Bombus terrestris*. *PLoS One, 2*(6), e556. http://dx.doi.org/10.1371/journal.pone.0000556.

Raine, N. E., & Chittka, L. (2007b). Flower constancy and memory dynamics in bumblebees (Hymenoptera: Apidae: *Bombus*). *Entomologia Generalis, 29*(2–4), 179–199.

Raine, N. E., Ings, T. C., Dornhaus, A., Saleh, N., & Chittka, L. (2006). Adaptation, genetic drift, pleiotropy, and history in the evolution of bee foraging behavior. *Advances in the Study of Behavior, 36*, 305–354.

Ramírez, S. R., Eltz, T., Fujiwara, M. K., Gerlach, G., Goldman-Huertas, B., Tsutsui, N. D., & Pierce, N. E. (2011). Asynchronous diversification in a specialized plant-pollinator mutualism. *Science, 333*, 1742–1746.

Reader, T., Higginson, A. D., Barnard, C. J., & Gilbert, F. S. (2006). The effects of predation risk from crab spiders on bee foraging behaviour. *Behavioral Ecology, 17*(6), 933–939.

Renner, S. S., & Schaefer, H. (2010). The evolution and loss of oil-offering flowers: new insights from dated phylogenies for angiosperms and bees. *Philosophical Transactions of the Royal Society B, 365*(1539), 423–435.

Renoult, J. P., Thomann, M., Schaefer, H. M., & Cheptou, P.-O. (2013). Selection on quantitative colour variation in *Centaurea cyanus*: the role of the pollinator's visual system. *Journal of Evolutionary Biology, 26*(11), 2415–2427.

Riffell, J. A., & Alarcón, R. (2013). Multimodal floral signals and moth foraging decisions. *PLoS One, 8*(8), e72809. http://dx.doi.org/10.1371/journal.pone.0072809.
Robertson, H. M., & Wanner, K. W. (2006). The chemoreceptor superfamily in the honey bee, *Apis mellifera*: expansion of the odorant, but not gustatory, receptor family. *Genome Research, 16*(11), 1395–1403. http://dx.doi.org/10.1101/gr.5057506.
Robertson, H. M., Warr, C. G., & Carlson, J. R. (2003). Molecular evolution of the insect chemoreceptor gene superfamily in *Drosophila melanogaster*. *Proceedings of the National Academy of Sciences of the United States of America, 100*(2), 14537–14542.
Rodríguez-Gironés, M. A., & Santamaría, L. (2004). Why are so many bird flowers red? *PLoS Biology, 2*(10), 1515–1519.
Rohde, K., Papiorek, S., & Lunau, K. (2013). Bumblebees (*Bombus terrestris*) and honeybees (*Apis mellifera*) prefer similar colours of higher spectral purity over trained colours. *Journal of Comparative Physiology A, 199*, 197–210.
Roper, T. J. (1999). Olfaction in birds. *Advances in the Study of Behaviour, 28*, 247–332.
Roy, B. A., & Raguso, R. A. (1997). Olfactory versus visual cues in a floral mimicry system. *Oecologia, 109*(3), 414–426. http://dx.doi.org/10.1007/s004420050101.
Rüdenauer, F. A., Spaëthe, J., & Leonhardt, S. D. (2015). How to know which food is good for you: bumblebees use taste to discriminate between different concentrations of food differing in nutrient content. *Journal of Experimental Biology, 218*(14), 2233–2240. http://dx.doi.org/10.1242/jeb.118554.
Ruxton, G., & Schaefer, H. M. (2013). Game theory, multi-modal signalling and the evolution of communication. *Behavioral Ecology and Sociobiology, 67*, 1417–1423.
Saleh, N., Ohashi, K., Thomson, J. D., & Chittka, L. (2006). Facultative use of the repellent scent mark in foraging bumblebees: complex versus simple flowers. *Animal Behaviour, 71*(4), 847–854.
Sandoz, J. C. (2011). Behavioral and neurophysiological study of olfactory perception and learning in honeybees. *Frontiers in Systems Neuroscience, 5*, 98. http://dx.doi.org/10.3389/fnsys.2011.00098.
Sandvik, S. M., & Totland, Ø. (2003). Quantitative importance of staminodes for female reproductive success in *Parnassia palustris* under contrasting environmental conditions. *Canadian Journal of Botany, 81*(1), 49–56.
Sapir, Y., Shmida, A., & Ne'eman, G. (2006). Morning floral heat as a reward to pollinators of the *Oncocyclus* irises. *Oecologia, 147*(1), 53–59.
Sazima, M., Vogel, S., do Prado, A. L., de Oliveira, D. M., Franz, G., & Sazima, I. (2001). The sweet jelly of *Combretum lanceolatum* flowers (Combretaceae): a cornucopia resource for bird pollinators in the Pantanal, western Brazil. *Plant Systematics and Evolution, 227*(3), 195–208.
Schaefer, H. M., & Ruxton, G. D. (2009). Deception in plants: mimicry or perceptual exploitation? *Trends in Ecology and Evolution, 24*(12), 676–685.
Schaefer, H. M., & Ruxton, G. D. (2011). *Plant-animal communication*. Oxford: Oxford University Press.
Schäffler, I., & Dötterl, S. (2011). A day in the life of an oil bee: phenology, nesting, and foraging behavior. *Apidologie, 42*(3), 409–424.
Schäffler, I., Steiner, K. E., Haid, M., van Berkel, S. S., Gerlach, G., Johnson, S. D., ... Dötterl, S. (2015). Diacetin, a reliable cue and private communication channel in a specialized pollination system. *Scientific Reports, 5*, 12779.
Schiestl, F. P., & Dötterl, S. (2012). The evolution of floral scent and olfactory preferences in pollinators, coevolution or pre-existing bias? *Evolution; International Journal of Organic Evolution, 66*(7), 2042–2055.
Schiestl, F. P., & Johnson, S. D. (2013). Pollinator-mediated evolution of floral signals. *Trends in Ecology and Evolution, 28*(5), 307–315. http://dx.doi.org/10.1016/j.tree.2013.01.019.

Scott, K. (2004). The sweet and the bitter of mammalian taste. *Current Opinion in Neurobiology, 14*(4), 423–427.

Sérsic, A. N. (2004). Pollination biology in the genus *Calceolaria* L. (Calceolariaceae). *Stapfia, 82*, 1–121.

Seymour, R. S., & Schultze-Motel. (1999). Respiration, temperature regulation and energetics of thermogenic inflorescences of the dragon lily *Dracunculus vulgaris* (Araceae). *Proceedings of the Royal Society of London Series B, 266*(1432), 1975–1983.

Seymour, R. S., White, C. R., & Gibernau, M. (2003). Environmental biology: heat reward for insect pollinators. *Nature, 426*(6964), 243–244.

Shuttleworth, A., & Johnson, S. D. (2009). A key role for floral scent in a wasp-pollination system in *Eucomis* (Hyacinthaceae). *Annals of Botany, 103*(5), 715–725.

Silberglied, R. E. (1979). Communication in the ultraviolet. *Annual Review of Ecology and Systematics, 10*(1), 373–398.

Simon, R., Holderied, M. W., Koch, C. U., & von Helversen, O. (2011). Floral acoustics: conspicuous echoes of a dish-shaped leaf attract bat pollinators. *Science, 333*(6042), 631. http://dx.doi.org/10.1126/science.1204210.

Singer, R. B. (2002). The pollination mechanism in *Trigonidium obtusum* Lindl (Orchidaceae: Maxillariinae): sexual mimicry and trap-flowers. *Annals of Botany, 89*(2), 157–163.

Skinner, J. D., & Chimimba, C. T. (2005). *The mammals of the Southern African subregion* (3rd ed.). Cambridge: Cambridge University Press.

Smith, C. E., Stevens, J. T., Temeles, E. J., Ewald, P. W., Hebert, R. J., & Bonkovsky, R. L. (1996). Effect of floral orifice width and shape on hummingbird-flower interactions. *Oecologia, 106*(4), 482–492.

Smith, S. D., Ané, C., & Baum, D. A. (2008). The role of pollinator shifts in the floral diversification of *Iochroma* (Solanaceae). *Evolution; International Journal of Organic Evolution, 62*(4), 793–806. http://dx.doi.org/10.1111/j.1558–5646.2008.00327.x.

Song, B., Chen, G., Stöcklin, J., Pent, D.-L., Niu, Y., Li, Z.-M., & Sun, H. (2014). A new pollinating seed-consuming mutualism between *Rheum nobile* and a fly fungus gnat, *Bradysia* sp., involving pollinator attraction by a specific floral compound. *New Phytologist, 203*(4), 1109–1118.

Späthe, J., & Chittka, L. (2003). Interindividual variation of eye optics and single object resolution in bumblebees. *Journal of Experimental Biology, 206*, 3447–3453. http://dx.doi.org/10.1242/jeb.00570.

Steenhuisen, S.-L., & Johnson, S. D. (2012). Evidence for beetle pollination in the African grassland sugarbushes (*Protea*: Proteaceae). *Plant Systematics and Evolution, 298*(5), 857–869.

Steenhuisen, S. L., Jürgens, A., & Johnson, S. D. (2013). Effects of volatile compounds emitted by *Protea* species (Proteaceae) on antennal electrophysiological responses and attraction of 10 cetoniine beetles. *Journal of Chemical Ecology, 39*, 438–446.

Steiger, S. S., Fidler, A. E., Valcu, M., & Kempenaers, B. (2008). Avian olfactory receptor gene repertoires: evidence for a well-developed sense of smell in birds? *Proceedings of the Royal Society B: Biological Sciences, 275*(1649), 2309–2317. http://dx.doi.org/10.1098/rspb.2008.060.

Steiner, K. E., Kaiser, R., & Dötterl, S. (2011). Strong phylogenetic effects on floral scent variation of oil-secreting orchids in South Africa. *American Journal of Botany, 98*(10), 1663–1679.

Steiner, K. E., & Whitehead, V. B. (1990). Pollinator adaption to oil-secreting flowers – *Rediviva* and *Diascia*. *Evolution; International Journal of Organic Evolution, 44*(6), 1701–1707.

Stoddart, D. M. (1980). *The ecology of vertebrate olfaction*. London: Chapman & Hall.

Sutton, G. P., Clarke, D., Morley, E. L., & Robert, D. (2016). Mechanosensory hairs in bumble bees (*Bombus terrestris*) detect weak electric fields. *Proceedings of the National*

Academy of Sciences of the United States of America. http://dx.doi.org/10.1073/pnas.1601624113 (preview).
Tavares Carneiro, L., Camillo Aguiar, A. J., Feitosa Martins, C., Machado, I. C., & Alves-dos-Santos, I. (2015). *Krameria tomentosa* oil flowers and their pollinators: bees specialized on trichome elaiophores exploit its epithelial oil glands. *Flora, 215*, 1–8.
Temeles, E. J., Newman, J. T., Newman, J. H., Cho, S. Y., Mazzotta, A. R., & Kress, W. J. (2016). Pollinator competition as a driver of floral divergence: an experimental test. *PLoS One, 11*(1), e0146431. http://dx.doi.org/10.1371/journal.pone.0146431.
Tollsten, L. (2008). A multivariate approach to post-pollination changes in the floral scent of *Platanthera bifolia* (Orchidaceae). *Nordic Journal of Botany, 13*(5), 495–499.
Vereecken, N. J., Dorchin, A., Dafni, A., Hötling, S., Schulz, S., & Watts, S. (2013). A pollinators' eye view of a shelter mimicry system. *Annals of Botany, 111*(6), 1155–1165.
Vlasáková, B., Kalinová, B., Gustafsson, M. H., & Teichert, H. (2008). Cockroaches as pollinators of *Clusia* aff. *sellowiana* (Clusiaceae) on inselbergs in French Guiana. *Annals of Botany, 102*(3), 295–304. http://dx.doi.org/10.1093/aob/mcn092.
Vogel, S. (1969). Flowers offering fatty oil instead of nectar. In *Proceedings of the XI International Botanical Congress, 24 August–2 September 1969 and International Wood Chemistry Symposium, 2–4. September 1969, Seattle, WA, USA*. Abstract No. 229.
Vogel, S. (1984). The *Diascia* flower and its bee-and-oil-based symbiosis in southern Africa. *Acta Botanica Neerlandica, 33*(4), 509–518.
Vogel, S. (2012). *Floral-biological syndromes as elements of diversity within tribes in the flora of South Africa*. Aachen: Shaker.
Von Helversen, O., Winkler, L., & Bestman, H. J. (2000). Sulphur containing 'perfumes' attract flower-visiting bats. *Journal of Comparative Physiology A, 186*(2), 143–153.
Vorobyev, M. (2003). Coloured oil droplets enhance colour discrimination. *Proceedings of the Royal Society B: Biological Sciences, 270*(1521), 1255–1261. http://dx.doi.org/10.1098/rspb.2003.2381.
Wacht, S., Lunau, K., & Hansen, K. (1996). Optical and chemical stimuli control pollen feeding in the hoverfly *Eristalis tenax* L. (Syrphidae; Diptera). *Entomologia Experimentalis et Applicata, 80*(1), 50–53.
Wacht, S., Lunau, K., & Hansen, K. (2000). Chemosensory control of pollen ingestion in the hoverfly *Eristalis tenax* by labellar taste hairs. *Journal of Comparative Physiology A, 186*(2), 193–203.
Weiss, M. R. (1991). Floral colour change as cues for pollinators. *Nature, 354*(6350), 227–229.
Weiss, M. R. (1995). Floral color change: a widespread functional convergence. *American Journal of Botany, 82*(2), 167–185.
Wester, P. (2010). Sticky snack for sengis: the Cape rock elephant-shrew, *Elephantulus edwardii* (Macroscelidea) as a pollinator of the Pagoda lily, *Whiteheadia bifolia* (Hyacinthaceae). *Die Naturwissenschaften, 97*(12), 1107–1112.
Wester, P., Stanway, R., & Pauw, A. (2009). Mice pollinate the Pagoda Lily, *Whiteheadia bifolia* (Hyacinthaceae) – first field observations with photographic documentation of rodent pollination in South Africa. *South African Journal of Botany, 75*(4), 713–719.
Whitney, H. M., Bennett, K., Dorling, M., Sandbach, L., Prince, D., Chittka, L., & Glover, B. J. (2011). Why do so many petals have conical epidermal cells? *Annals of Botany, 108*(4), 609–616. http://dx.doi.org/10.1093/aob/mcr065.
Whitney, H. M., Chittka, L., Bruce, T. J. A., & Glover, B. J. (2009). Conical epidermal cells allow bees to grip flowers and increase foraging efficiency. *Current Biology, 19*(11), 948–953.

Whitney, H. M., Reed, A., Rands, S. A., Chittka, L., & Glover, B. J. (2016). Flower iridescence increases object detection in the insect visual system without compromising object identity. *Current Biology, 26*, 802–808.
Willmer, P. (2011). *Pollination and floral ecology*. Princeton: Princeton University Press.
Wilms, J., & Eltz, T. (2008). Foraging scent marks of bumblebees: footprint cues rather than pheromone signals. *Die Naturwissenschaften, 95*(2), 149–153.
Witjes, S., & Eltz, T. (2007). Influence of scent deposits on flower choice: experiments in an artificial flower array with bumblebees. *Apidologie, 38*(1), 12–18.
Witjes, S., & Eltz, T. (2009). Hydrocarbon footprints as a record of bumblebee flower visitation. *Journal of Chemical Ecology, 35*(11), 1320–1325.
Witjes, S., Witsch, K., & Eltz, T. (2011). Reconstructing the pollinator community and predicting seed set from hydrocarbon footprints on flowers. *Oecologia, 165*(4), 1017–1029.
Wright, G. A., Mustard, J. A., Simcock, N. K., Ross-Taylor, A. A. R., McNicholas, L. D., Popescu, A., & Marion-Poll, F. (2010). Parallel reinforcement pathways for conditioned food aversions in the honeybee. *Current Biology, 20*(24), 2234–2240. http://dx.doi.org/10.1016/j.cub.2010.11.040.
Zhang, F.-P., Cai, X.-H., Wang, H., Ren, Z.-X., Larson-Rabin, Z., & Li, D.-Z. (2012). Dark purple nectar as a foraging signal in a bird-pollinated Himalayan plant. *New Phytologist, 193*(1), 188–195.
Zhang, F.-P., Larson-Rabin, Z., Li, D.-Z., & Wang, H. (2012). Colored nectar as an honest signal in plant-animal interactions. *Plant Signaling and Behavior,* 7(7), 811–812.
Zimmermann, Y., Roubik, D. W., & Eltz, T. (2006). Species-specific attraction to pheromonal analogues in orchid bees. *Behavioral Ecology and Sociobiology, 60*(6), 833–843.

CHAPTER TEN

Mimicry and Deception in Pollination

K. Lunau[1], P. Wester
Heinrich-Heine-University, Düsseldorf, Germany
[1]Corresponding author: E-mail: lunau@uni-duesseldorf.de

Contents

Abstract

Mimicry and deception in pollination represent extended plant–pollinator communication systems including two senders of mistakable signals. Flower mimicry is a highly diverse phenomenon including Batesian mimicry, in which a nonrewarding flower mimics a rewarding one, and signal standardization, also known as Mullerian mimicry, in which two unrelated rewarding flowers display similar signals. In these floral mimicry systems, even experienced flower visitors are deceived. Some flowering plants exploit the pollinators' innate preferences of finding rewarding flowers or floral rewards: In some cases the flowers fit the innate search image of flower visitors so close that no real model exists. In other cases, only the floral reward is mimicked. Particularly, pollen and stamen mimicry systems are very common, in which flowers mimic visual and tactile signals of pollen and stamens, and by means of this fake stamens and pollen deceive or misdirect pollen-eating and pollen-collecting flower visitors that innately respond to these signals. In another type of floral mimicry, animal-pollinated flowers attract flower visitors by mimicking features related to aspects other than flower visitation of their pollinators, e.g., sexually deceptive flowers deceiving mate-seeking male insects, and brood-substrate mimicking flowers deceiving females searching for egg-laying substrates. These deceptive flowers attract specific pollinators by means of sensory exploitation of a well-developed innate behavioural response for distinct volatiles and other cues that have evolved in the context of mating or egg-laying behaviour. Floral mimicry is always beneficial for the mimic, but may impose costs for the deceived pollinators. Then the deceived pollinators likely evolve mechanisms not being deceived and the flowering plants to continue deception, and deception becomes trickier over evolutionary times.

Advances in Botanical Research, Volume 82
ISSN 0065-2296
http://dx.doi.org/10.1016/bs.abr.2016.10.005

1. INTRODUCTION

Flowers of animal-pollinated plants are advertising structures displaying visual, olfactory, gustatory and tactile signals to attract pollinators by promising reward. The discrepancy between promise and reward ranges, however, from being neglectable to extreme. Some flowers openly display their floral reward or at least honestly signal expectable reward (Knauer & Schiestl, 2015). Semihonest are flowers that honestly signal their potential reward, but have been emptied recently such that the flower visitor gets less than expected (Ohashi & Thomson, 2005). Deceptive flowers offer no reward at all (Schäfer & Ruxton, 2009).

The mimicry concept is based on three protagonists, i.e., model, mimic and signal receiver, and on deception (Wickler, 1968). True mimicry, as in a narrow sense, involves the deception of a signal receiver which is unable to discriminate between the model and the mimic of a signal (Lunau, 2011, p. 160; Wickler, 1968). The deceived signal receiver responds in a similar way to the model and to the mimic, but will not get the expected reward. The decisive imitative trait is a signal copy, the mimic's signal. Also signal standardization refers to two senders of a similar signal that a signal receiver does not discriminate. Contrary to true mimicry, there is no need for the signal receiver to discriminate between the two senders, because both not only display similar signals, but also offer similar rewards (Benitez-Vieyra, Hempel de Ibarra, Wertlen, & Cocucci, 2007). This latter phenomenon is often termed Mullerian mimicry in contrast to the former Batesian mimicry (Dafni, 1984; Roy & Widmer, 1999). The key question in these flower mimicry systems is whether both species, model and mimic, are equally likely to induce flower-constant foraging behaviour. Flower constancy, i.e., the temporary foraging specialization for one plant species, is only beneficial for the flower constant flower visitor, if the density of the target plant species is high. Thus, in the case of two species with similar flowers the rare species benefits more from signal uniformity than the frequent one. Johnson, Alexandersson, and Peter Linder (2003) have proposed the term quasi-Batesian mimicry for those flower mimicry systems where a rewarding species is not common enough to induce foraging constancy by pollinators and thus benefits from mimicking the flower signals of more common food plants.

The origin of the flower visitors' preference for distinct floral traits is a key feature of mimicry and related phenomena. The sensory bias concept is grounded in the contribution of any sensory bias to establish preferences and has been originally developed for mate preferences (Endler, 1992;

Endler & Basolo, 1998; see also chapters: Plant–Pollinator Communication by Wester & Lunau, 2017). Transferring the concept of sensory bias to the field of sensory interactions between flowers and pollinators means that pollinators respond more strongly to certain, more pertinent, floral signals (Raine & Chittka, 2007) due to distinctive features in the sensory systems that determine the salience of signals; however, no assumptions are made how the preference for distinct signals has evolved. Such preferences might have evolved due to preferences in a context other than flower visitation or might be caused by distinct properties of sensory receptors. Raguso (2008) explains, based on Endler and Basolo (1998), that sensory exploitation is a model of signal evolution in which the properties of a receiver's (pollinator's) sensory system lead to innate preference or preexisting bias, often before the producer (flower) evolves such a signal. It is obvious that the delimitation between mimicry and sensory bias in the communication between flower visitors and flowers is not always strict (see Ayasse & Dötterl, 2014; Ellis, Johnson, & Benkman, 2010; Gaskett, 2011), because mimicry denotes the limited discrimination ability of deceived signal receivers, whereas sensory bias, and the related concepts of sensory drive, sensory trap and sensory exploitation denote the origin of preferences in signal receivers.

2. HONESTY AND DECEPTION IN POLLINATION

The mutualistic relationship between flowering plants and pollinators is mostly based on trading of floral resources and pollination service. But the coevolution between flowering plants and pollinating bees might have been less shaped by mutual benefits than by reciprocal exploitation (Bailes, Ollerton, Pattrick, & Glover, 2015). Flowers signal and promise reward to flower visitors; however, the standing crop of nectar or pollen in flowers varies in dependence to previously emptied floral resources. Remote-sensing flower visitors preferably use direct indicators of the amount of reward, e.g., visible or sniffable properties of nectar or pollen (Dobson, Danielson, & van Wesep, 1999; Nicholls & Hempel de Ibarra, 2014; Zhang, Larson-Rabin, Li, & Wang, 2012) or indirect indicators of reward, e.g., flower size (Armbruster, Antonsen, & Pélabon, 2005), floral scents (Knauer & Schiestl, 2015; Wright & Schiestl, 2009), olfactory foodprints (Wilms & Eltz, 2008) or visual claw marks left by previous visitors (Matsumura & Washitani, 2000) to avoid time- and energy-consuming landings on empty flowers (Harder, 1990). Other reliable indicators such as flower humidity are discussed (von Arx, 2013; Schäfer, Schäfer, & Levey, 2004).

Self-incompatible flowering plants with polymorphic flowers, e.g., diclinous or heterostylous flowers, require visits of both morphs, because only intermorphic pollen will result in fertilization (Barrett, 1992; Willson, 1982). Likewise, dichogamous flowers require pollen flow from flowers of the staminate to those of the pistillate flowering phase (Dai & Galloway, 2011). Flower visitors benefit from developing preferences for one distinct morph or flowering phase due to differences in pollen reward or handling time (Cawoy et al., 2006; De Jong et al., 2011; Le Corff, Agren, & Schemske, 1998; Wolfe & Barrett, 1987). Pollen presentation provides the most apparent discrimination signals for flowering phases and floral morphs. The flowering plants thus benefit from concealing any indications of morph or flowering phase by means of displaying inconspicuous real pollen and conspicuous pollen-mimicking structures, respectively (Pohl, Watolla, & Lunau, 2008). The nectarless *Begonia* flowers offer no nectar, and thus pollen-collecting bees are deceived by stamen mimicry of the pistillate flowers (Schemske & Agren, 1995; Wyatt & Sazima, 2011). In squirting cucumber, *Ecballium elaterium*, the staminate flowers offer both, nectar and pollen, whereas the female flowers are fully deceptive with the measurable consequence of reduced flower visits (Fahn & Shimony, 2001; Rust, Vaissiere, & Westrich, 2003). In diclinous flowering plants guidance of pollinators from staminate flowers towards pistillate flowers is possible if staminate flowers open earlier in the daytime (or flowering season) as compared to pistillate flowers. The larger size of staminate flowers as compared to pistillate flowers probably refers to the number of flower visits needed for reproductive success; staminate flowers benefit from using as many pollen vectors as possible, whereas pistillate flowers need a number of deposited pollen grains comparable to the number of ovules that could be deposited on the stigma by a single pollinator (Barrett & Hough, 2013; Costich & Meagher, 2001; Delph, Galloway, & Stanton, 1996; Forero-Montaña & Zimmerman, 2010; Ish-Am, 2008). Likewise, the pollen colour dimorphism in the heterostylous trimorphic purple loosestrife, *Lythrum salicaria,* contributes to signalling uniformity. Purple loosestrife has three morphs displaying two kinds of three types of stamens each, the long stamens display inconspicuous green pollen and the middle and short stamens display conspicuous yellow pollen. The presentation of the stamens has been interpreted as to conceal the identity of each of the three morphs and the associated differences in pollen rewards, and to guide the pollinators to the centre of the flower where all three morphs equally reward with nectar, since the long stamens are inconspicuous and the short and middle stamens are very conspicuous and display their yellow pollen close to the nectary (Lunau, 1995).

Deception is an indispensable aspect of mimicry and means that the flower visitor does not get the expected reward (Dafni, 1984; Lunau, 2011, p. 13; Wickler, 1968). Flowers benefit from deception in terms of no costs for producing the reward, but also in terms of more selective attraction of suitable pollinators. Moreover, a deceived pollinator will fly a longer distance until the next visit to a similar flower and thus promotes pollen flow distance or outcrossing distance, respectively (Schiestl, 2005; Whitehead, Linde, & Peakall, 2015).

Flowers support flower visitors by means of emitting variable signals as semihonest indicators of reward such as floral colour change (Weiss, 1991), odour change (Theis & Raguso, 2005) or both (Raguso, 2004). Floral colour changes are regarded as a semihonest signal, since they reliably indicate changes in nectar production, but not the nectar standing crop (Ohashi, Makino, & Arikawa, 2015). Floral colour changes may affect the entire petals or might be restricted to floral guides (Weiss, 1991). Floral colour changes are regarded as a plant signalling strategy to maintain the attractiveness of inflorescences despite wilting and falling off of flowers following fertilization and to indicate rewarding individual flowers at close range (Gori, 1989). For the tropical tree *Tibouchina pulchra*, Brito, Weynans, Sazima, and Lunau (2015) proposed a higher spatial level of floral colour change. They hypothesized that flowering trees act as huge flowers and that single flowers act as floral guides since these trees are widely scattered in the rainforest and need to attract pollinating bees over large distances; thus prechange and postchange flowers contribute to attraction from large distance and prechange flowers direct approaching pollinators to these rewarding flowers. Another aspect of honest signalling in colour changing flowers was highlighted by Lunau (1996). He reported that bees must not learn the meaning of the prechange and postchange colour of bee-pollinated flowers, since the colour that is associated with more reward is that colour that inexperienced flower visitors innately prefer over that colour that is associated with less reward.

3. NO-, ONE- AND MULTIPLE-MODEL MIMICRY

Mimicry involves more than the imitation of signals and is based on the deception of a signal receiver that cannot fully or not at all discriminate between a model and a mimic signal. Deception and dishonest floral signals represent thus an obligatory aspect of mimicry (Lunau, 2011, p. 13; Wickler, 1968). In some cases, the three protagonists of flower mimicry systems, the honestly signalling (and rewarding) model, the signal copying (nonrewarding) mimic or the deceived signal receiver and potential pollinator are represented by single and distinct species. This kind of one-model flower mimicry is characterized by its specificity. Many known examples of one-model flower mimicry include nectarless orchids (Jersáková, Johnson, & Kindlmann, 2006), e.g., the deceptive orchid (*Disa nivea*) successfully exploits a reciprocally specialized mutualism between a nectar-producing plant (*Zaluzianskya microsiphon*) and its long-proboscid fly pollinator (*Prosoeca ganglbaueri*) (Anderson, Johnson, & Carbutt, 2005). Also in most sexually deceptive orchids, the models are the females of a single bee species that are mimicked by the orchid flowers (Schiestl et al., 1999).

In some cases, rewardless flowers do not resemble the flowers of any other species, but still are regarded as protagonists in no-model mimicry system (Nilsson, 1992). Some spring flowering orchids, e.g., *Calypso bulbosa*, have no coflowering model species (Tuomi, Lamsa, Wannas, Abeli, & Jakalaniem, 2015). It is assumed that rewardless flowers that are flowering early in the season attract naïve flower visitors or exploit the flower visitors' innate search image for flowers (Internicola & Harder, 2012). The presence of innate search images for flowers and floral traits has been tested with naïve and nontrained individuals of bumblebees (Heuschen, Gumbert, & Lunau, 2005; Lunau, Wacht, & Chittka, 1996), orchid bees and hummingbirds (Lunau, Papiorek, Eltz, & Sazima, 2011), diurnal hawkmoths (Kelber, 1997), papilionid butterflies (Weiss, 1997), hoverflies (Lunau & Wacht, 1994) and beetles (Döring & Cook, 2012) in regard to visual colour signals in laboratory settings. A similar approach is possible to demonstrate odour preference (Suchet et al., 2011). Even experienced flower visitors, e.g., honeybees and bumblebees, exhibit spontaneous preferences for distinct colours or colour patterns in preference tests with artificial flowers (Giurfa, Núnez, Chittka, & Menzel, 1995; Ings, Raine, & Chittka, 2009; Rohde, Papiorek, & Lunau, 2013). Also field experiments allow reliable conclusions about innate preferences, e.g., in nocturnal hawkmoths (Riffell et al., 2008) or honeybees (Goulson et al., 2007).

In other cases, additional model protagonists are occurring in the so-called multiple-model mimicry systems. This is valid for the Globe Orchid *Traunsteinera globosa* (Fig. 1A), which looks very similar to its assumed model species, the sympatric rewarding species *Knautia sylvatica* and *Scabiosa columbaria* (Dipsacaceae) and *Valeriana montana* (Valerianaceae) (Juillet, Gonzalez, Page, & Gigord, 2007). In field studies, Jersáková et al. (2016) showed that the visual similarity between *Traunsteinera* flowers and co-occurring rewarding species has no impact on the visitation frequency of *Trausteinera* blossoms, but the density of rewarding co-occurring flowers does. This has been interpreted as the magnet effect used by the generalized food-mimicking flower (Juillet et al., 2007; Pellissier, Vittoz, Internicola, & Gigord, 2010). Another well-studied case of multiple-model mimicry are the food-deceptive Oncidiinae orchids, which have evolved independently several times and look very similar as the rewarding oil-producing model species of Malpighiaceae; both are visited by oil-collecting bees (Papadopulos et al., 2013).

Gumbert and Kunze (2001) found that the perceptual similarity of mimic flowers to their models predicts choice in bumblebees. They stated that on the one hand the combined effect of colour and odour contributes to increased flower choice frequency of bumblebees in flower mimicry systems, but on the other hand odourless mimics are more successful than visual mimic flowers emitting a deviant odour. This also holds for deceptive flowers displaying a complex colour pattern referring to multiple-model species (Ma et al., 2015).

A particular aspect of flower mimicry is the frequency of model species in relation to the frequency of mimic species (Anderson & Johnson, 2006; Bierzychudek, 1981; Dafni & Ivri, 1981; Johnson, 1994, 2000). Many food-deceptive plants show discrete or continuous floral polymorphism within their populations. The deceptive orchid *Disa ferruginea* has two discrete allopatric red and orange floral morphs. The red floral morph mimics the visual appearance of the red-flowered *Tritoniopsis triticea* (Iridaceae) and the orange floral morph mimics *Kniphofia uvaria* (Asphodelaceae). All models and morphs of the mimic are pollinated by the nymphalid butterfly *Aeropetes tulbaghia* that relies strictly on nectar from the model species (Newman, Anderson, & Johnson, 2012). The rewardless orchid *Dactylorhiza sambucina* has two sympatric floral colour morphs, red and yellow. It has been suggested that negative frequency-dependent selection caused by the learning behaviour of pollinators was responsible for the maintenance of floral polymorphism (Gigord, Macnair, & Smithson, 2001; Gigord, Macnair, Stritesky, & Smithson, 2002). There are two aspects of this finding: The larger reproductive success of the rare morph indicates that indeed the relative rareness of the mimic as compared to the model plays a role for pollination

Figure 1 Flower mimicry. (A) The nectarless orchid *Traunsteinera globosa* (right) with one of its presumed models (*Scabiosa* sp.); (B) Similar size, form and colour in the inflorescences of *Orchis mascula*, *Trifolium pratense*, *Pedicularis verticillata* and *Trifolium alpinum*; (C) Dark central flower extending the umbel of *Daucus carota* deters gall midges; (D) Mushroom-mimicking flowers of *Aristolochia arborea*; (E) Model *Ophrys speculum* and its mimic, the female *Campsoscolia ciliata*; (F) Nectar mimicry by glossy structures in *Cleome monophylla*; (G) All yellow *Potentilla* flower with UV-absorbing androecium and central floral guide; (H) UV photo of the same flower; (I) Yellow disc florets in *Aster alpinus*; (J) Yellow flower of *Viola biflora* displays a UV-absorbing floral

success, and over the years another morph will become rare. In addition, flower polymorphism in a deceptive species leads to the inclusion of more models into that mimicry system thus shifting the relation in the frequency between model and mimic. However, in the deceptive colour polymorphic Crimean Iris, *Iris lutescens*, Imbert et al. (2014) observed a positive frequency-dependent selection and discuss various possible causes. Computer simulations suggest inaccurate discrimination of pollinators may cause the evolution of discrete trait polymorphism (Kagawa & Takimoto, 2016).

The deceptive orchid *Anacamptis morio* has quantitative petal colour variation. Pollinators mediated strong selection for increased petal brightness facilitating its detectability due to increased achromatic contrast against background. However, no indication of a rare-phenotype advantage maintaining the existing colour variation was found (Sletvold, Trunschke, Smit, Verbeek, & Ågren, 2016).

Mimicry represents a form of dishonest signalling that exploits the conditioned behaviour of animal 'operators'. It is expected that the fitness of mimics should decline as they become more numerous relative to their models, because operators would show weaker conditioned responses or learn to distinguish between mimics and models (Bierzychudek, 1981; Dafni & Ivri, 1981; Joron & Mallet, 1998). The relative frequency of model and mimic seems to play no role in mimicry systems in which innate responses towards the mimicked floral signals are exploited (Dafni, 1984). In fact, in the no-model mimicry system the nonrewarding flower mimics flowers without any coflowering model flower (Internicola & Harder, 2012; Tuomi et al., 2015). And many of the sexually deceptive orchids of the genus *Ophrys* (Fig. 1E) flower at that time when the pollinating bee and wasp males have already emerged, but the mimicked females not yet, and thus are unrivalled (Mant et al., 2005).

The fact that flower-visiting insects are attracted by artificial flowers is used for pan trapping. Coloured pan traps are in use to monitor abundance

guide; (K) UV photo of the same flower; (L) Heteranthery in *Commelina benghalensis* with three staminodes, one conspicuous feeding stamen and two inconspicuous pollination stamens; (M) Stamen-mimicking stigmatic lobes in *Crocus* flowers; (N) *Verbascum phoeniceum* flower with heteranthery comprising three stamens with conspicuous filamental hairs and two less conspicuous pollination stamens; (O) *Linaria alpina* lower lips mimic an orange anther; (P) The beard of the *Iris germanica* flower is mimicking a bunch of stamens; (Q) Floral colour change in *Myosotis arvensis* flowers and stamen-mimicking floral guides including yellow rewarding flowers.

of flower-visiting insects, mostly dipters and bees (Lunau, 2014; Vrdoljak & Samways, 2012; Westphal et al., 2008). The question whether pan traps that mimic the colour hue of food plant flowers are more attractive to flower visitors is plausible, but evidence is rare (Hicks et al., 2012).

4. DIVERSITY OF SIGNAL RECEIVERS IN FLOWER MIMICRY

The number of models and mimics in flower mimicry systems contribute to the diversity of floral mimicry. The diversity of flower mimicry systems is further increased by the number of signal receivers involved. Mimicry in flowers is a multifaceted phenomenon and comprises intraspecific as well as interspecific nutritive deception, sexual deception and some other forms of deception (Dafni, 1984). The most common floral mimicry systems (Fig. 1A and B) include nonrewarding flowers that mimic rewarding ones, rewarding as well as nonrewarding flowers that mimic reward, i.e., pollen, nectar (Lunau, 2006; Vogel, 1993) or flowers that mimic brood sites, prey, mates (Dafni, 1984; Lunau, 2011, p. 93; Schiestl & Johnson, 2013), galls (Fig. 1C; Polte & Reinhold, 2013) or shelter (Vereecken et al., 2013).

The mimic item of flowers may be restricted to only a part of a flower such as the stamens (Lunau, 2000, 2006), nectar spur (Dafni & Ivri, 1981; Jersáková et al., 2006) or nectary (Aldasoro, Aedo, & Navarro, 2000; Sandvik & Totland, 2003). The presence of empty nectar spurs might be interpreted as a deceptive trait in some rewardless orchids, but also as a relic of nectariferous ancestors (Johnson, Hobbhahn, & Bytebier, 2013). Mimic nectaries have been described only rarely (Fig. 1F; Vogel, 1993), whereas mimic stamens are a common attribute of flowers (Fig. 1G–Q) displaying conspicuous staminodes as in *Commelina* flowers (Hrycan & Davis, 2005), stamen-like protuberances as in *Iris germanica* or floral guides that resemble shape and colour of anthers (Lunau, 2000, 2006). Moreover, even the reward itself is mimicked by some flowers offering pseudopollen (Davies, Stpiczyńska, & Kamińska, 2013; Davies & Turner, 2004; Lunau, 2000, 2006) or pseudonectar (Vogel, 1993). Although traits mimicking pollen or pollen-producing organs are very common floral traits, the production of pseudopollen and its collection by bees is a rare phenomenon (Davies et al., 2013; Davies & Turner, 2004; Lunau, 2000, 2006).

Progress in the analysis of floral scents provided evidence for a number of spectacular flower mimicry systems. Sexual deception is mostly specific of one plant and one insect species due to the fact that sex pheromones of insects are species specific (Ayasse, 2007; Ayasse, Schiestl, Paulus, Ibarra, & Francke, 2003). However, this does not exclude that sympatric sexually deceptive orchids share the same pollinator (Gögler et al., 2009; Phillips, Xu, Hutchinson, Dixon, & Peakall, 2013). Sexual deception spans from attraction of mates via rendezvous sites (Keasar et al., 2010; Leys & Hogendoorn, 2008) over pseudocopulation (Ayasse, 2007; Ayasse et al., 2003) to ejaculation (Gaskett, Winnick, & Herberstein, 2008). Mimicry of rendezvous sites for the attraction of pollinators has evolved in unrelated flowering plants attracting Monkey beetles (Hopliini) in South Africa (Van Kleunen, Nänni, Donaldson, & Manning, 2007), in a guild of red and black-centred flowers in the Mediterranean to attract glaphyrid beetles (Dafni et al., 1990), and in the Asteraceae *Gorteria diffusa* to attract bombylid flies (Johnson & Midgley, 1997). Some sexually deceptive orchids such as *Ophrys heldreichii* obviously mimic not only the female bee, but also the flower the female is sitting on (Späthe, Moser, & Paulus, 2007; Späthe, Streinzer, & Paulus, 2010).

By contrast, brood site mimicry can be more or less specific (Jürgens, Dötterl, & Meve, 2006; van der Niet, Hansen, & Johnson, 2011; Urru, Stensmyr, & Hansson, 2011) due to the fact that the real breeding sites of some insects are very specific and some natural breeding sites are shared by many species attracted towards different stages and by different cues (Barton et al., 2013). Sexual deception seems to be mainly restricted to orchids (Gaskett, 2011) with only few exceptions and comprises extreme specialization to only males of a single pollinator species (Cozzolino & Widmer, 2005). Since the pollinators of orchids carry the entire pollen mass of each flower as pollinaria (Harder & Johnson, 2008), orchids might be constrained to very reliable pollen carriers and benefit from excluding occasional visitors. Some noteworthy exceptions are as follows: The flowers of *Gorteria diffusa* attract both sexes of the bombylid fly *Megapalpus nitidus*. Both sexes feed on nectar, but only males attempt to copulate with the fly marks displayed on some of the petals (de Jager & Ellis, 2012). The flowers of the red-bowl-shaped guild in the East Mediterranean, including *Papaver subpiriforme* and *Anemone coronaria*, attract both sexes of glaphyrid beetles that feed and mate on the flowers, but despite the presence of visual beetle marks on the flowers, no copulation attempts of male beetles with the flowers have been observed (Keasar et al., 2010). Contrary, brood site mimicry is rare among orchids (but see van der Niet et al., 2011; Policha et al., 2016), but common in many other plant families, at least in 10 Angiosperm families, i.e., Annonaceae, Araceae, Aristolochiaceae, Apocynaceae, Burmanniaceae, Hydnoraceae, Orchidaceae, Rafflesiaceae, Sterculiaceae, Taccaceae and Hyacinthaceae (Urru et al., 2011; Wiens, 1978). It occurs in dipteran as well as coleopteran insects of several families (Urru et al., 2011); the classification refers to the model of the brood site: saprophagy, coprophagy and necrophagy (Jürgens, Suk-Ling, Shuttleworth, & Johnson, 2013). Specialization also occurs in regard to mimic dung of herbivores and of carnivores (Jürgens et al., 2006). A good overview of mimicked insect oviposition sites based on volatile organic compounds emitted by angiosperms is given by Jürgens et al. (2013).

The enormous diversity of brood-site mimicry is illustrated by some examples: The deceptive *Aristolochia rotunda* (Aristolochiaceae) is pollinated by female Chloropidae flies which are food thieves that feed on secretions of true bugs (Miridae), while these are eaten by arthropod predators. *Aristolochia* flowers release the same scent components as freshly killed prey items of mirid bugs that attract the chloropid flies, mostly aliphatic esters with hexyl butyrate and (E)-2-hexenyl butyrate being the most abundant volatiles (Oelschlägel et al., 2015). The emission of green leaf volatiles is induced by the damage of leaves by herbivores and used to attract wasps that are predators or parasites of the herbivores (Turlings et al., 1995; Whitman & Eller, 1990; see also chapter: Plant Communication With Herbivores by Blande, 2017). Mimic green leaf volatiles are emitted by orchid flowers of *Epipactis helleborine* and *Epipactis purpurata* to attract several prey-hunting wasp species for pollination (Brodmann et al., 2008). Interestingly, flowers of other plant families, *Scrophularia nodosa* and *Scrophularia umbrosa*, that emit green leaf volatiles, are pollinated by various wasps, and share visual cues of the mentioned *Epipactis* flowers, i.e., small size and reddish colour (Brodmann, Emer, & Ayasse, 2012). A congeneric orchid, *Epipactis veratrifolia* is pollinated by aphidophagous hoverflies; the ovipositing females are attracted by multiple visual and chemical cues including an aphid-derived kairomone (Jin et al., 2014; Stökl, Brodmann, Dafni, Ayasse, & Hansson, 2010). Another orchid species, *Dendrobium sinense*, attracts hornets by mimicking a major component, (Z)-11-eicosen-1-ol, of a specific alarm pheromone of both Asian (*Apis cerana*) and European (*Apis mellifera*) Honeybees (Brodmann et al., 2009).

The Wild Carrot *Daucus carota* represents an outstanding example, since the black central floret of the white umbels (Fig. 1C) mimics a gall and thereby deters gall-producing midges which would otherwise harm the plant by laying eggs from which the herbivorous gall midges emerge (Polte & Reinhold, 2013).

5. THE COSTS OF DECEPTION

One of the major differences between sexual deception and brood site mimicry is the selective pressure on the deceived pollinators to avoid deception, the costs of being deceived (de Jager & Ellis, 2014). In sexually deceptive flowers, the mate-seeking males of bees, wasps, flies or fungus gnats waste time for searching conspecific females, but normally they do not ejaculate sperm during pseudocopulation. The costs for the deceived males are thus manageable, since the male copulation attempts typically are very short. In addition, the fact that many sexually deceptive orchids flower before the females of the pollinating bees or wasps have emerged reduces costs as well (Robbirt, Roberts, Hutchings, & Davy, 2014). There is, however, a notable exception with *Cryptostylis* orchids that deceive their male pollinators, *Lissopimpla excelsa* wasps, to make it ejaculate copious amount of sperm (Gaskett et al., 2008). Furthermore, sex determination in bees and wasps is based on fertilization of eggs with males developing from unfertilized eggs (Heimpel & de Boer, 2008); thus, orchids that inhibit exhaust sperm supply in their haplodiploid pollinators could cause an increase of unfertilized eggs and as a consequence generate an increase of males to act as pollinators during future flowering seasons (Gaskett et al., 2008).

With large costs of being deceived, an evolutionary arms race will start to reduce the costs. The sexually deceptive orchid *Chiloglottis trapeziformis* has a negative impact not only on its wasp pollinator, the males of *Neozeleboria cryptoides*, but also on the females. Male wasps cannot discriminate between the chemical cues of orchids and female wasps, and thus have to leave areas containing signalling orchids to avoid being deceived too often. As a consequence, the wingless females have less mating opportunities in areas containing signalling orchids (Wong & Schiestl, 2002).

Many flies, beetles, or fungus gnats that are deceived by flowers mimicking breeding sites, do not only waste time for finding real breeding sites, e.g., carcasses, feces or mushrooms, but also deposit eggs on the mimic brood site. Since the emerged larvae are unable to develop in the fake brood site, they die, which results in an immense fitness cost (Urru et al., 2011). This results in an evolutionary arms race between the mimics and the signal receivers to deceive or to avoid the deadly deception, respectively. One example is *Aristolochia arborea* (Fig. 1D) that mimics many details of a *Marasmius* mushroom (Barthlott, 1992).

Previously it had been assumed that the exploitation of innate preferences as in sexually deceptive orchids bears the advantage that the response of the deceived male bees or wasps does not habituate. Novel research indicates that some male bees learn very fast and avoid being duped after only few landings and pseudocopulations on conspecific females mimicking orchids. However, the orchids trick the males to impede their learning success. The intraspecific variability of the visual pattern on the lower lip of *Ophrys heldreichii* causes the males to land on more flowers (Stejskal, Streinzer, Dyer, Paulus, & Späthe, 2015). Receptive flowers of *Ophrys sphecodes* smell like a virgin female and pollinated flowers change their smell and smell like a mated female. What appears superfluous at first glance turns out to attract more males due to more scent-emitting flowers and direct the males at close range towards the receptive flowers (Schiestl, Ayasse, Paulus, Erdmann, & Francke, 1997).

ACKNOWLEDGEMENT

We thank our colleagues, Vinicius Brito, Amots Dafni, Thomas Eltz, Andre Rech, Zing-Xon Ren and Marlies Sazima, for stimulating discussions about flower mimicry.

REFERENCES

Aldasoro, J. J., Aedo, C., & Navarro, C. (2000). Insect attracting structure on *Erodium* petals (Geraniaceae). *Plant Biology, 2*, 471–481.

Anderson, B., & Johnson, S. D. (2006). The effects of floral mimics and models on each others' fitness. *Proceedings of the Royal Society B: Biological Sciences, 273*(1589), 969–974. http://dx.doi.org/10.1098/rspb.2005.3401.

Anderson, B., Johnson, S. D., & Carbutt, C. (2005). Exploitation of a specialized mutualism by a deceptive orchid. *American Journal of Botany, 92*, 1342–1349.

Armbruster, W. S., Antonsen, L., & Pélabon, C. (2005). Phenotypic selection on *Dalechampia* blossoms: honest signaling affects pollination success. *Ecology, 86*, 3323–3333. http://dx.doi.org/10.1890/04-1873.

von Arx, M. (2013). Floral humidity and other indicators of energy rewards in pollination biology. *Communicative & Integrative Biology, 6*(1), e22750. http://dx.doi.org/10.4161/cib.22750.

Ayasse, M. (2007). Chemical mimicry in sexually deceptive orchids of the genus *Ophrys*. *Phyton, 46*, 221–223.

Ayasse, M., & Dötterl, S. (2014). The role of preadaptations or evolutionary novelties for the evolution of sexually deceptive orchids. *New Phytologist, 203*(3), 710–712.

Ayasse, M., Schiestl, F., Paulus, H. F., Ibarra, F., & Francke, W. (2003). Pollinator attraction in a sexually deceptive orchid by means of unconventional chemicals. *Proceedings of the Royal Society B: Biological Sciences, 270*, 517–522.

Bailes, E. J., Ollerton, J., Pattrick, J. G., & Glover, B. J. (2015). How can an understanding of plant–pollinator interactions contribute to global food security? *Current Opinion in Plant Biology, 26*, 72–79.

Barrett, S. C. H. (1992). *Evolution and function of heterostyly*. Heidelberg, Berlin: Springer.

Barrett, S. C. H., & Hough, J. (2013). Sexual dimorphism in flowering plants. *Journal of Experimental Botany, 64*, 67–82. http://dx.doi.org/10.1093/jxb/ers308.

Barthlott, W. (1992). Mimikry – Nachahmung und Täuschung im Pflanzenreich. *Bonner Universitätsblätter, 1992*, 49–61.

Barton, P. S., Cunningham, S. A., Macdonald, B. C. T., McIntyre, S., Lindenmayer, D. B., & Manning, A. D. (2013). Species traits predict assemblage dynamics at ephemeral resource patches created by carrion. *PLoS One, 8*(1), e53961. http://dx.doi.org/10.1371/journal.pone.0053961.

Benitez-Vieyra, S., Hempel de Ibarra, N., Wertlen, A. M., & Cocucci, A. A. (2007). How to look like a mallow: evidence of floral mimicry between Turneraceae and Malvaceae. *Proceedings of the Royal Society B: Biological Sciences, 274*, 2239–2248.

Bierzychudek, P. (1981). *Asclepias, Lantana*, and *Epidendrum* – a floral mimicry complex. *Biotropica, 13*(2), 54–58.

Blande, J. D. (2017). Plant communication with herbivores. In G. Becard (Ed.). *How plants communicate with their biotic environment* (Vol. 82, 281–304).

Brito, V. L. G., Weynans, K., Sazima, M., & Lunau, K. (2015). Trees as huge flowers and flowers as oversized floral guides: the role of floral color change and retention of old flowers in *Tibouchina pulchra. Frontiers in Plant Science, 6*(362). http://dx.doi.org/10.3389/fpls.2015.00362.

Brodmann, J., Emer, D., & Ayasse, M. (2012). Pollinator attraction of the wasp-flower *Scrophularia umbrosa* (Scrophulariaceae). *Plant Biology, 14*(3), 500–505.

Brodmann, J., Twele, R., Francke, W., Hölzler, G., Zhang, Q.-H., & Ayasse, M. (2008). Orchid mimics green leaf volatiles to attract prey-hunting wasps for pollination. *Current Biology, 18*, 740–744.

Brodmann, J., Twele, R., Francke, W., Yi-Bo, L., Xi-Qiang, S., & Ayasse, M. (2009). Orchid mimics honeybee alarm pheromone in order to attract hornets for pollination. *Current Biology, 19*.

Cawoy, V., Deblauwe, V., Halbrecq, B., Ledent, J.-F., Kinet, J.-M., & Jacquemart, A.-L. (2006). Morph differences and honeybee morph preference in the distylous species *Fagopyrum esculentum* Moench. *International Journal of Plant Sciences, 167*(4), 853–861. http://dx.doi.org/10.1086/504924.

Costich, D. E., & Meagher, T. R. (2001). Impacts of floral gender and whole-plant gender on floral evolution in *Ecballium elaterium* (Cucurbitaceae). *Biological Journal of the Linnean Society, 74*, 475–487. http://dx.doi.org/10.1111/j.1095-8312.2001.tb01406.x.

Cozzolino, S., & Widmer, A. (2005). Orchid diversity: an evolutionary consequence of deception? *Trends in Ecology and Evolution, 20*(9), 487–494.

Dafni, A. (1984). Mimicry and deception in pollination. *Annual Review of Ecology and Systematics, 15*(1), 259–278.

Dafni, A., Bernhardt, P., Schmida, A., Ivri, Y., Greenbaum, S., O'Toole, C., & Losito, L. (1990). Red bowl-shaped flowers: convergence for beetle pollination in the Mediterranean region. *Israel Journal of Botany, 39*, 81–92.

Dafni, A., & Ivri, Y. (1981). Floral mimicry between *Orchis israelitica* Baumann and Dafni (Orchidaceae) and *Bellavalia flexuosa* Boiss (Liliaceae). *Oecologia, 49*(2), 229–232. http://dx.doi.org/10.1007/BF00349193.

Dai, C., & Galloway, L. F. (2011). Do dichogamy and herkogamy reduce sexual interference in a self-incompatible species? *Functional Ecology, 25*, 271–278. http://dx.doi.org/10.1111/j.1365–2435.2010.01795.x.

Davies, K. L., Stpiczyńska, M., & Kamińska, M. (2013). Dual deceit in pseudopollen-producing *Maxillaria* s.s. (Orchidaceae: Maxillariinae). *Botanical Journal of the Linnean Society, 173*, 744–763. http://dx.doi.org/10.1111/boj.12094.

Davies, K. L., & Turner, M. P. (2004). Pseudopollen in *Eria* Lindl. Section *Mycaranthes* Rchb.f. (Orchidaceae). *Annals of Botany, 94*, 707–715.

De Jong, T. J., Klinkhamer, P. G. L., Shmida, A., & Thuijsman, F. (2011). On the evolution of protandry and the distinction between preference and rank order in pollinator visitation. *Evolutionary Ecology Research, 13*, 307–314.

Delph, L. F., Galloway, L. F., & Stanton, M. L. (1996). Sexual dimorphism in flower size. *The American Naturalist, 148*(2), 299–320.

Dobson, H. E. M., Danielson, E. M., & van Wesep, I. D. (1999). Pollen odor chemicals as modulators of bumble bee foraging on *Rosa rugosa* Thunb. (Rosaceae). *Plant Species Biology, 14*(2), 153–166.

Döring, T. F., & Cook, S. (2012). Colour choice behaviour in the pollen beetle, *Meligethes aeneus* (Coleoptera: Nitidulidae). *Physiological Entomology, 37*, 360–368.

Ellis, A. G., Johnson, S. D., & Benkman, C. W. (2010). Floral mimicry enhances pollen export: the evolution of pollination by sexual deceit outside of the Orchidaceae. *The American Naturalist, 176*(5), E143–E151.

Endler, J. A. (1992). Signals, signal conditions, and the direction of evolution. *The American Naturalist, 139*, 125–153.

Endler, J. A., & Basolo, A. L. (1998). Sensory ecology, receiver biases and sexual selection. *Trends in Ecology and Evolution, 13*, 415–420.

Fahn, A., & Shimony, C. (2001). Nectary structure and ultrastructure of unisexual flowers of *Ecballium elaterium* (L.) A. Rich. (Cucurbitaceae) and their presumptive pollinators. *Annals of Botany, 87*(1), 27–33. http://dx.doi.org/10.1006/anbo.2000.1287.

Forero-Montaña, J., & Zimmerman, J. K. (2010). Sexual dimorphism in the timing of flowering in two dioecious trees in a subtropical wet forest, Puerto Rico. *Caribbean Journal of Science, 46*(1), 88–95.

Gaskett, A. C. (2011). Orchid pollination by sexual deception: pollinator perspectives. *Biological Reviews, 86*, 33–75. http://dx.doi.org/10.1111/j.1469-185X.2010.00134.x.

Gaskett, A. C., Winnick, C. G., & Herberstein, M. E. (2008). Orchid sexual deceit provokes ejaculation. *The American Naturalist, 171*, E206–E212.

Gigord, L. D. B., Macnair, M. R., & Smithson, A. (2001). Negative frequency-dependent selection maintains a dramatic flower color polymorphism in the rewardless orchid *Dactylorhiza sambucina* (L.) Soò. *Proceedings of the National Academy of Sciences of the United States of America, 98*(11), 6253–6255.

Gigord, L. D. B., Macnair, M. R., Stritesky, M., & Smithson, A. (2002). Experimental evidence for floral mimicry in a rewardless orchid. *Proceedings of the Royal Society B, Biological Sciences, 269*, 1389–1395. http://dx.doi.org/10.1098/rspb.2002.2018.

Giurfa, M., Núnez, J., Chittka, L., & Menzel, R. (1995). Colour preferences of flower-naive honeybees. *Journal of Comparative Physiology A, 177*(3), 247–259.

Gögler, J., Stökl, J., Sramkova, A., Twele, R., Francke, W., Cozzolino, S., … Ayasse, M. (2009). Menage a trois – two endemic species of deceptive orchids and one pollinator species. *Evolution, 63*, 2222–2234.

Gori, D. F. (1989). Floral color change in *Lupinus argenteus* (Fabaceae): why should plants advertise the location of unrewarding flowers to pollinators? *Evolution, 43*, 870–881. http://dx.doi.org/10.2307/2409314.

Goulson, D., Cruise, J. L., Sparrow, K. R., Harris, A. J., Park, K. J., Tinsley, M. C., & Gilburn, A. S. (2007). Choosing rewarding flowers; perceptual limitations and innate preferences influence decision making in bumblebees and honeybees. *Behavioral Ecology and Sociobiology, 61*(10), 1523–1529. http://dx.doi.org/10.1007/s00265-007-0384-4.

Gumbert, A., & Kunze, J. (2001). Colour similarity to rewarding model plants affects pollination in a food deceptive orchid, *Orchis boryi*. *Biological Journal of the Linnean Society, 72*(3), 419–433.

Harder, L. D. (1990). Behavioral responses by bumble bees to variation in pollen availability. *Oecologia, 85*, 41–47.

Harder, L. D., & Johnson, S. D. (2008). Function and evolution of aggregated pollen in angiosperms. *International Journal of Plant Sciences, 169*(1), 59–78.

Heimpel, G. E., & de Boer, J. G. (2008). Sex determination in the hymenoptera. *Annual Review of Entomology, 53*, 209–230.

Heuschen, B., Gumbert, A., & Lunau, K. (2005). A generalised mimicry system involving angiosperm flower colour, pollen and bumblebees' innate colour preferences. *Plant Systematics and Evolution, 252*, 121–137.

Hicks, B. J., Bartlett, T., Bishop, J., Cicciarella, S., Hiscock, B., & Woodford, R. (2012). Does trap colour influence sampling of bee pollinators on Blueberry fields? *The Osprey, 43*(4), 5–9.

Hrycan, W. C., & Davis, A. D. (2005). Comparative structure and pollen production of the stamens and pollinator-deceptive staminodes of *Commelina coelestis* and *C. dianthifolia* (Commelinaceae). *Annals of Botany, 95*(7), 1113–1130.

Imbert, E., Wang, H., Conchou, L., Vincent, H., Talavera, M., & Schatz, B. (2014). Positive effect of the yellow morph on female reproductive success in the flower colour polymorphic *Iris lutescens* (Iridaceae), a deceptive species. *Journal of Evolutionary Biology, 27*, 1965–1974. http://dx.doi.org/10.1111/jeb.12451.

Ings, T. C., Raine, N. E., & Chittka, L. (2009). A population comparison of the strength and persistence of innate colour preference and learning speed in the bumblebee *Bombus terrestris*. *Behavioral Ecology and Sociobiology, 63*, 1207–1218.

Internicola, A. I., & Harder, L. D. (2012). Bumble-bee learning selects for both early and long flowering in food-deceptive plants. *Proceedings of the Royal Society B, Biological Sciences, 279*, 1538–1543.

Ish-Am, G. (2008). Daily patterns of Avocado bloom and Honeybee activity. *California Avocado Society 2008 Yearbook, 91*, 105–127.

de Jager, M. L., & Ellis, A. G. (2012). Gender-specific pollinator preference for floral traits. *Functional Ecology, 26*, 1197–1204. http://dx.doi.org/10.1111/j.1365-2435.2012.02028.x.

de Jager, M. L., & Ellis, A. G. (2014). Costs of deception and learned resistance in deceptive interactions. *Proceedings of the Royal Society B: Biological Sciences, 281*(1779), 20132861.

Jersáková, J., Johnson, S. D., & Kindlmann, P. (2006). Mechanisms and evolution of deceptive pollination in orchids. *Biological Reviews, 81*(2), 219–235. http://dx.doi.org/10.1017/S1464793105006986.

Jersáková, J., Späthe, J., Streinzer, M., Neumayer, J., Paulus, H., Dötterl, S., & Johnson, S. D. (2016). Does *Traunsteinera globosa* (the globe orchid) dupe its pollinators through generalized food deception or mimicry? *Botanical Journal of the Linnean Society, 180*, 269–294. http://dx.doi.org/10.1111/boj.12364.

Jin, X.-H., Ren, Z.-X., Xu, S.-Z., Wang, H., Li, D.-Z., & Li, Z.-Y. (2014). The evolution of floral deception in *Epipactis veratrifolia* (Orchidaceae): from indirect defense to pollination. *BioMed Central Plant Biology, 14*, 63. http://dx.doi.org/10.1186/1471-2229-14-63.

Johnson, S. D. (1994). Evidence for Batesian mimicry in a butterfly-pollinated orchid. *Biological Journal of the Linnean Society, 53*, 91–104. http://dx.doi.org/10.1006/bijl.1994.1062.

Johnson, S. D. (2000). Batesian mimicry in the nonrewarding orchid *Disa pulchra*, and its consequences for pollinator behaviour. *Biological Journal of the Linnean Society, 71*, 119–132.

Johnson, S. D., Alexandersson, R., & Peter Linder, H. (2003). Experimental and phylogenetic evidence for floral mimicry in a guild of fly-pollinated plants. *Biological Journal of the Linnean Society, 80*, 289–304. http://dx.doi.org/10.1046/j.1095-8312. 2003.00236.x.

Johnson, S. D., Hobbhahn, N., & Bytebier, B. (2013). Ancestral deceit and labile evolution of nectar production in the African orchid genus *Disa*. *Biology Letters, 9*(5), 20130500. http://dx.doi.org/10.1098/rsbl.2013.0500.

Johnson, S. D., & Midgley, J. J. (1997). Fly pollination of *Gorteria diffusa* (Asteraceae), and a possible mimetic function for dark spots on the capitulum. *American Journal of Botany, 84*(4), 429–436. http://www.jstor.org/stable/2446018.

Joron, M., & Mallet, J. (1998). Diversity in mimicry: paradox or paradigm? *Trends in Ecology and Evolution, 13*, 461–466. http://dx.doi.org/10.1016/S0169-5347(98)01483-9.

Juillet, N., Gonzalez, M. A., Page, P. A., & Gigord, L. D. B. (2007). Pollination of the European food-deceptive *Traunsteinera globosa* (Orchidaceae): the importance of nectar-producing neighbouring plants. *Plant Systematics and Evolution, 265*, 123–129.

Jürgens, A., Dötterl, S., & Meve, U. (2006). The chemical nature of fetid floral odours in stapeliads (Apocynaceae – Asclepiadoideae – Ceropegieae). *New Phytologist, 172*(3), 452–468.

Jürgens, A., Suk-Ling, W., Shuttleworth, A., & Johnson, S. D. (2013). Chemical mimicry of insect oviposition sites: a global analysis of convergence in angiosperms. *Ecology Letters, 16*, 1157–1167.

Kagawa, K., & Takimoto, G. (2016). Inaccurate color discrimination by pollinators promotes evolution of discrete color polymorphism in food-deceptive flowers. *The American Naturalist, 187*(2), 194–204.

Keasar, T., Harari, A. R., Sabatinelli, G., Keith, D., Dafni, A., Shavit, O., … Shmida, A. (2010). Red anemone guild flowers as focal places for mating and feeding by Levant glaphyrid beetles. *Biological Journal of the Linnean Society, 99*(4), 808–817.

Kelber, A. (1997). Innate preferences for flower features in the hawk moth *Macroglossum stellatarum*. *Journal of Experimental Biology, 200*(4), 827–836.

Knauer, A. C., & Schiestl, F. P. (2015). Bees use honest floral signals as indicators of reward when visiting flowers. *Ecology Letters, 18*(2), 135–143. http://dx.doi.org/10.1111/ele.12386.

Le Corff, J., Agren, J., & Schemske, D. W. (1998). Foral display, pollinator discrimination, and female reproductive success in two monoecious *Begonia* species. *Ecology, 79*, 1610–1619. http://dx.doi.org/10.2307/176781.

Leys, R., & Hogendoorn, K. (2008). Correlated evolution of mating behaviour and morphology in large carpenter bees (*Xylocopa*). *Apidologie, 39*(1), 119–132.

Lunau, K. (1995). Notes on the colour of pollen. *Plant Systematics and Evolution, 198*, 235–252.

Lunau, K. (1996). Unidirectionality of floral colour changes. *Plant Systematics and Evolution, 200*, 125–140.

Lunau, K. (2000). The ecology and evolution of visual pollen signals. *Plant Systematics and Evolution, 222*(1), 89–111.

Lunau, K. (2006). Stamens and mimic stamens as components of floral colour patterns. *Botanische Jahrbücher für Systematik, Pflanzengeschichte und Pflanzengeographie, 127*, 13–41.

Lunau, K. (2011). *Warnen, Tarnen, Täuschen. Mimikry und Nachahmung bei Pflanze, Tier und Mensch*. Darmstadt: Wissenschaftliche Buchgesellschaft.

Lunau, K. (2014). Visual ecology of flies with particular reference to colour vision and colour preferences. *Journal of Comparative Physiology A, 200*(6), 497–512.

Lunau, K., Papiorek, S., Eltz, T., & Sazima, M. (2011). Avoidance of achromatic colours by bees provides a private niche for hummingbirds. *Journal of Experimental Biology, 214*(9), 1607–1612.

Lunau, K., & Wacht, S. (1994). Optical releasers of the innate proboscis extension in the hoverfly *Eristalis tenax L. (Syrphidae, Diptera)*. *Journal of Comparative Physiology A, 174*(5), 575–579.

Lunau, K., Wacht, S., & Chittka, L. (1996). Colour choices of naive bumble bees and their implications for colour perception. *Journal of Comparative Physiology A, 178*, 477–489.

Ma, X., Shi, J., Bänziger, H., Sun, Y., Guo, Y., Liu, Z., … Luo, Y. (2015). The functional significance of complex floral colour pattern in a food-deceptive orchid. *Functional Ecology*. http://dx.doi.org/10.1111/1365-2435.12571.

Mant, J., Brändli, C., Vereecken, N. J., Schulz, C. M., Francke, W., & Schiestl, F. P. (2005). Cuticular hydrocarbons as sex pheromone of the bee *Colletes cunicularius* and the key to its mimicry by the sexually deceptive orchid, *Ophrys exaltata*. *Journal of Chemical Ecology, 31*(8), 1765–1787.

Matsumura, C., & Washitani, I. (2000). Effects of population size and pollinator limitation on seed-set of *Primula sieboldii* populations in a fragmented landscape. *Ecological Research, 15*(3), 307–322.

Newman, E., Anderson, B., & Johnson, S. D. (2012). Flower colour adaptation in a mimetic orchid. *Proceedings of the Royal Society B, Biological Sciences, 279*, 2309–2313.

Nicholls, E., & Hempel de Ibarra, N. (2014). Bees associate colour cues with differences in pollen rewards. *Journal of Experimental Biology, 217*, 2783–2788. http://dx.doi.org/10.1242/jeb.106120.

van der Niet, T., Hansen, D. M., & Johnson, S. D. (2011). Carrion mimicry in a South African orchid: flowers attract a narrow subset of the fly assemblage on animal carcasses. *Annals of Botany, 107*(6), 981–992. http://dx.doi.org/10.1093/aob/mcr048.

Nilsson, L. A. (1992). Orchid pollination biology. *Trends in Ecology and Evolution, 7*, 255–259. http://dx.doi.org/10.1016/0169-5347(92)90170-G.

Oelschlägel, B., Nuss, M., von Tschirnhaus, M., Pätzold, C., Neinhuis, C., Dötterl, S., & Wanke, S. (2015). The betrayed thief – the extraordinary strategy of *Aristolochia rotunda* to deceive its pollinators. *New Phytologist, 206*, 342–351. http://dx.doi.org/10.1111/nph.13210.

Ohashi, K., Makino, T. T., & Arikawa, K. (2015). Floral colour change in the eyes of pollinators: testing possible constraints and correlated evolution. *Functional Ecology, 29*, 1144–1155. http://dx.doi.org/10.1111/1365-2435.12420.

Ohashi, K., & Thomson, J. D. (2005). Efficient harvesting of renewing resources. *Behavioral Ecology, 16*, 592–605.

Papadopulos, A. S., Powell, M. P., Pupulin, F., Warner, J., Hawkins, J. A., Salamin, N., … Savolainen, V. (2013). Convergent evolution of floral signals underlies the success of Neotropical orchids. *Proceedings of the Royal Society B: Biological Sciences, 280*(1765), 20130960. http://dx.doi.org/10.1098/rspb.2013.0960.

Pellissier, L., Vittoz, P., Internicola, A. I., & Gigord, L. D. B. (2010). Generalized food-deceptive orchid species flower earlier and occur at lower altitudes than rewarding ones. *Journal of Plant Ecology, 3*(4), 243–250.

Phillips, R. D., Xu, T., Hutchinson, M. F., Dixon, K. W., & Peakall, R. (2013). Convergent specialization – the sharing of pollinators by sympatric genera of sexually deceptive orchids. *Journal of Ecology, 101*(3), 826–835.

Pohl, M., Watolla, T., & Lunau, K. (2008). Anther-mimicking floral guides exploit a conflict between innate preference and learning in bumblebees (*Bombus terrestris*). *Behavioral Ecology and Sociobiology, 63*, 295–302.

Policha, T., Davis, A., Barnadas, M., Dentinger, B. T. M., Raguso, R. A., & Roy, B. A. (2016). Disentangling visual and olfactory signals in mushroom-mimicking *Dracula* orchids using realistic three-dimensional printed flowers. *New Phytologist, 210*(3), 1058–1071. http://dx.doi.org/10.1111/nph.13855.

Polte, S., & Reinhold, K. (2013). The function of the wild carrot's dark central floret: attract, guide or deter? *Plant Species Biology, 28*, 81–86. http://dx.doi.org/10.1111/j.1442-1984.2012.00368.x.

Raguso, R. A. (2004). Flowers as sensory billboards: progress towards an integrated understanding of floral advertisement. *Current Opinion in Plant Biology, 7*(4), 434–440.

Raguso, R. A. (2008). Wake up and smell the roses: the ecology and evolution of floral scent. *Annual Review of Ecology Evolution, and Systematics, 39*(1), 549–569.

Raine, N. E., & Chittka, L. (2007). The adaptive significance of sensory bias in a foraging context: floral colour preferences in the bumblebee *Bombus terrestris*. *PLoS One, 2*(6), e556. http://dx.doi.org/10.1371/journal.pone.0000556.

Riffell, J. A., Alarcón, R., Abrell, L., Davidowitz, G., Bronstein, J. L., & Hildebrand, J. G. (2008). Behavioral consequences of innate preferences and olfactory learning in hawkmoth–flower interactions. *Proceedings of the National Academy of Sciences, 105*(9), 3404–3409. http://dx.doi.org/10.1073/pnas.0709811105.

Robbirt, K. M., Roberts, D. L., Hutchings, M. J., & Davy, A. J. (2014). Potential disruption of pollination in a sexually deceptive orchid by climatic change. *Current Biology, 24*(23), 2845–2849.

Rohde, K., Papiorek, S., & Lunau, K. (2013). Bumblebees (*Bombus terrestris*) and honeybees (*Apis mellifera*) prefer similar colours of higher spectral purity over trained colours. *Journal of Comparative Physiology A, 199*, 197–210.

Roy, B. A., & Widmer, A. (1999). Floral mimicry: a fascinating yet poorly understood phenomenon. *Trends in Plant Science, 4*(8), 325–330.

Rust, R. W., Vaissiere, B. E., & Westrich, P. (2003). Pollinator biodiversity and floral resource use in *Ecballium elaterium* (Cucurbitaceae), a Mediterranean endemic. *Apidologie, 34*, 29–42.

Sandvik, S. M., & Totland, Ø. (2003). Quantitative importance of staminodes for female reproductive success in *Parnassia palustris* under contrasting environmental conditions. *Canadian Journal of Botany, 81*(1), 49–56.

Schäfer, H. M., & Ruxton, G. D. (2009). Deception in plants: mimicry or perceptual exploitation? *Trends in Ecology and Evolution, 24*(12), 676–685.

Schäfer, H. M., Schäfer, V., & Levey, D. J. (2004). Plant-animal interactions signal new insights in communication. *Trends in Ecology and Evolution, 19*, 577–584.

Schemske, D. W., & Agren, J. (1995). Deceit pollination and selection on female flower size in *Begonia involucrata*: an experimental approach. *Evolution, 49*, 207–214.

Schiestl, F. P. (2005). On the success of a swindle: pollination by deception in orchids. *Die Naturwissenschaften, 92*(6), 255–264.

Schiestl, F. P., Ayasse, M., Paulus, H. F., Erdmann, D., & Francke, W. (1997). Variation of floral scent emission and post pollination changes in individual flowers of *Ophrys sphegodes* subsp. *sphegodes* (Miller) (Orchidaceae). *Journal of Chemical Ecology, 23*(12), 2881–2896.

Schiestl, F. P., Ayasse, M., Paulus, H. F., Löfstedt, C., Hansson, B. S., Ibarra, F., & Francke, W. (1999). Orchid pollination by sexual swindle. *Nature, 399*, 421–422.

Schiestl, F. P., & Johnson, S. D. (2013). Pollinator-mediated evolution of floral signals. *Trends in Ecology and Evolution, 28*(5), 307–315. http://dx.doi.org/10.1016/j.tree.2013.01.019.

Sletvold, N., Trunschke, J., Smit, M., Verbeek, J., & Ågren, J. (2016). Strong pollinator-mediated selection for increased flower brightness and contrast in a deceptive orchid. *Evolution, 70*, 716–724. http://dx.doi.org/10.1111/evo.12881.

Späthe, J., Moser, W., & Paulus, H. F. (2007). Increase of pollinator attraction by means of a visual signal in the sexually deceptive orchid, *Ophrys heldreichii* (Orchidaceae). *Plant Systematics and Evolution, 264*, 31–40. http://dx.doi.org/10.1007/s00606-006-0503-0.

Späthe, J., Streinzer, M., & Paulus, H. F. (2010). Why sexually deceptive orchids have colored flowers. *Communicative & Integrative Biology, 3*, 139–141. http://dx.doi.org/10.4161/cib.3.2.10333.

Stejskal, K., Streinzer, M., Dyer, A., Paulus, H. F., & Späthe, J. (2015). Functional significance of labellum pattern variation in a sexually deceptive orchid (*Ophrys heldreichii*): evidence of individual signature learning effects. *PLoS One, 10*(11), e0142971. http://dx.doi.org/10.1371/journal.pone.0142971.

Stökl, J., Brodmann, J., Dafni, A., Ayasse, M., & Hansson, B. S. (2010). Smells like aphids: orchid flowers mimic aphid alarm pheromones to attract hoverflies for pollination. *Proceedings of the Royal Society B: Biological Sciences, 278*(1079), 1612–1622. http://dx.doi.org/10.1098/rspb.2010.1770.21.

Suchet, C., Dormont, L., Schatz, B., Giurfa, M., Simon, V., Raynaud, C., & Chave, J. (2011). Floral scent variation in two *Antirrhinum majus* subspecies influences the choice of naïve bumblebees. *Behavioral Ecology and Sociobiology, 65*(5), 1015–1027.

Theis, N. B., & Raguso, R. A. (2005). The effect of pollination on floral fragrance in thistles (*Cirsium*, Asteraceae). *Journal of Chemical Ecology, 31*, 2581–2600.
Tuomi, J., Lamsa, J., Wannas, L., Abeli, T., & Jakalaniem, A. (2015). Pollinator behaviour on a food-deceptive orchid *Calypso bulbosa* and coflowering species. *The Scientific World Journal, 482161*. http://dx.doi.org/10.1155/2015/482161.
Turlings, T. C. J., Loughrin, J. H., McCall, P. J., Rose, U. S. R., Lewis, W. J., & Tumlinson, J. H. (1995). How caterpillar-damaged plants protect themselves by attracting parasitic wasps. *Proceedings of the National Academy of Sciences, 92*, 4169–4174.
Urru, I., Stensmyr, M. C., & Hansson, B. S. (2011). Pollination by brood-site deception. *Phytochemistry, 72*(13), 1655–1666. http://dx.doi.org/10.1016/j.phytochem.2011.02.014.
Van Kleunen, M., Nänni, I., Donaldson, J. S., & Manning, J. C. (2007). The role of beetle marks and flower colour on visitation by monkey beetles (Hopliini) in the Greater Cape Floral Region, South Africa. *Annals of Botany, 100*(7), 1483–1489. http://dx.doi.org/10.1093/aob/mcm256.
Vereecken, N. J., Dorchin, A., Dafni, A., Hötling, S., Schulz, S., & Watts, S. (2013). A pollinators' eye view of a shelter mimicry system. *Annals of Botany, 111*(6), 1155–1165.
Vogel, S. (1993). Betrug bei Pflanzen: Die Täuschblumen. *Abhandlungen der Mathematisch-Naturwissenschaftlichen Klasse der Akademie der Wisseenschaften und Literatur Mainz, 1*, 1–48.
Vrdoljak, S. M., & Samways, M. J. (2012). Optimising coloured pan traps to survey flower visiting insects. *Journal of Insect Conservation, 16*(3), 345–354.
Weiss, M. R. (1991). Floral colour change as cues for pollinators. *Nature, 354*(6350), 227–229.
Weiss, M. R. (1997). Innate colour preferences and flexible colour learning in the pipevine swallowtail. *Animal Behaviour, 53*(5), 1043–1052.
Wester, P., & Lunau, K. (2017). Plant–pollinator communication. In G. Becard (Ed.). *How plants communicate with their biotic environment* (Vol. 82, 225–257).
Westphal, C., Bommarco, R., Carre, G., Lamborn, E., Morison, N., Petanidou, T., ... Steffan-Dewenter, I. (2008). Measuring bee diversity in different European habitats and biogeographical regions. *Ecological Monographs, 78*, 653–671.
Whitehead, M. R., Linde, C. L., & Peakall, R. (2015). Pollination by sexual deception promotes outcrossing and mate diversity in self-compatible clonal orchids. *Journal of Evolutionary Biology, 28*, 1526–1541.
Whitman, D. W., & Eller, F. J. (1990). Parasitic wasps orient to green leaf volatiles. *Chemoecology, 1*, 69–76.
Wickler, W. (1968). *Mimicry in plants and animals*. New York: McGraw-Hill.
Wiens, D. (1978). Mimicry in plants. *Evolutionary Biology, 11*, 365–403.
Willson, M. F. (1982). Sexual selection and Dicliny in angiosperms. *The American Naturalist, 119*(4), 579–583.
Wilms, J., & Eltz, T. (2008). Foraging scent marks of bumblebees: footprint cues rather than pheromone signals. *Die Naturwissenschaften, 95*(2), 149–153.
Wolfe, L. M., & Barrett, S. C. H. (1987). Pollinator foraging behaviour and pollen collection on the floral morphs of tristylous *Pontederia cordata* L. *Oecologia, 74*, 347–351.
Wong, B. B., & Schiestl, F. P. (2002). How an orchid harms its pollinator. *Proceedings of the Royal Society B, Biological Sciences, 269*(1500), 1529–1532.
Wright, G. A., & Schiestl, F. P. (2009). The evolution of floral scent: the influence of olfactory learning by insect pollinators on the honest signalling of floral rewards. *Functional Ecology, 23*, 841–851. http://dx.doi.org/10.1111/j.1365-2435.2009.01627.x.
Wyatt, G. E., & Sazima, M. (2011). Pollination and reproductive biology of thirteen species of *Begonia* in the Serra do Mar State Park, São Paulo, Brazil. *Journal of Pollination Ecology, 6*, 95–107.
Zhang, F.-P., Larson-Rabin, Z., Li, D.-Z., & Wang, H. (2012). Colored nectar as an honest signal in plant-animal interactions. *Plant Signaling & Behavior, 7*(7), 811–812.

CHAPTER ELEVEN

Plant Communication With Herbivores

J.D. Blande
University of Eastern Finland, Kuopio, Finland
E-mail: james.blande@uef.fi

Contents

Abstract

Plants and herbivores both release volatile organic compounds that have important roles in mediating important biological functions related to defence and reproduction. Plants emit complex blends of chemicals that are involved in multitrophic interactions, coordination of systemic defence responses and pollination, whereas herbivorous insects release pheromones that play important roles in attracting mates, instigating defence responses and initiating aggregation. Interactions between plants and herbivores have been subject to a wealth of studies and knowledge on their biology, biochemistry, ecology and evolution is constantly expanding. In this chapter the idea of communication between plants and herbivores will be explored. Communication between organisms of consecutive trophic levels is somewhat controversial due to unidirectional reliance and competition precluding some of the requirements of a conventional communication process, but there are growing examples of where chemically mediated interactions between plants and herbivores can be viewed as eavesdropping by a signal recipient, or even as true communication where both chemical emitter and

Advances in Botanical Research, Volume 82
ISSN 0065-2296
http://dx.doi.org/10.1016/bs.abr.2016.09.004

receiver gain a benefit from communication. Examples of herbivores responding to plant-emitted cues and plants responding to herbivore-emitted cues are both explored, and suggestions for future directions in this field are provided.

1. INTRODUCTION

1.1 Plant Communication With Herbivores – Communication or Arms Race?

Plants and herbivores coexist in an evolutionary arms race, where plants evolve new ways to defend themselves from attack and herbivores evolve means to circumvent, tolerate or even utilize those defences (Ehrlich & Raven, 1964; Howe & Jander, 2008; Kant et al., 2015; Petschenka & Agrawal, 2016). Plants produce an enormously diverse range of secondary metabolites that mediate numerous ecological interactions and constitute a central focus of ecological and evolutionary studies into plant defence theory (Kessler, 2015). Some metabolites involved in plant defence also serve roles in plant communication. For example, plants constitutively emit volatile organic compounds into the atmosphere, but when they are damaged they alter the composition and quantity of the chemical blend; compounds within this induced blend may defend plants by intoxicating herbivores (Veyrat, Robert, Turlings, & Erb, 2016) or signal to a multitude of other organisms in the plant's community (e.g., Dicke, 2009; Dicke & Baldwin, 2010; Heil & Karban, 2010). As a consequence of this multifunctionality, the concept of plant communication with herbivores – the focus of this chapter – is difficult to disentangle from other functional hypotheses for the evolution of plant secondary metabolites and requires careful defining. Plant sensing, communication and the idea of plant intelligence have been the subjects of several recent books and definitions of plant behaviour, communication, cues and signals abound (Chamovitz, 2012; Karban, 2015; Mancuso & Viola, 2013; Trewavas, 2014).

In his book, *Plant Sensing and Communication*, Richard Karban (2015) provided a synthesis of the various definitions put forward by ecologists for the whole, or parts, of the communication process. Karban (2015) emphasized that there are no universally agreed-upon definitions of communication for either animals or plants, but that there are many advantages to viewing a signal to have evolved because of the effects it causes and that both the sender and the receiver of the signal should therefore experience a benefit of communicating. This definition will be utilized as something of a holy grail, whereby interactions between plants and herbivores that satisfy these requirements can truly be considered to be a process of communication. However, there are simpler processes that can also come under the broader umbrella of signalling and communication. In this chapter the simplest process, signalling, whereby a receiving organism responds to an informative cue, an example of which would be a herbivore locating a host plant by using constitutively emitted volatiles to orientate, will be given some consideration. However, greater attention will be given to the process of eavesdropping, where emitters release an informative plastic cue that is received by a receiver organism without the emitter or receiver necessarily benefitting. A form of communication will also be considered to occur when receivers respond to a plastic cue that results in a benefit to the signal sender. Together, these ascendingly complex definitions of eavesdropping, communication and true communication will be used to satisfy, to different extents, the phenomenon of plant communication with herbivores.

Herbivores can utilize olfactory and visual cues to locate their host plants. While visual cues play essential roles in a plant's interactions with insects, including herbivores and pollinators (see chapters: Plant—Pollinator Communication by Wester & Lunau, 2017 and Mimicry and Deception in Pollination by Lunau & Wester, 2017 for an overview of plant communication with pollinators), this chapter will focus on communication based on secondary metabolites, and particularly volatile organic compounds. Communication between plants and herbivores based on herbivores responding to volatiles emitted by plants will constitute the first part of the chapter, we will then explore the phenomenon of plants detecting and responding to volatile pheromones and cues produced by herbivorous insects and then briefly consider the effects of herbivore-induced plant volatiles (HIPVs) on other organisms that impact on herbivore behaviour or survival. The chapter ends with a short summary and suggestions for future research directions.

2. HERBIVORES USE PLANT VOLATILE SIGNALS TO LOCATE THEIR HOSTS

Foraging by herbivorous insects can include oriented flight and selection of a host plant by winged herbivores that feed in the adult stage, oriented flight and selection of a host plant for oviposition by winged gravid female insects, and short distance on land searching for host plants by nonwinged insects. Plants constitutively emit a blend of volatile chemicals, which provides cues that herbivores can utilize to recognize their hosts at a distance (Bruce & Pickett, 2011; Bruce, Wadhams, & Woodcock, 2005).

The constitutive emissions of plants vary considerably throughout their lifecycle with changes reflecting the ontogenetic stage of the plant and its phenology (Hare, 2010). Effective processing of information encoded in the constitutive emissions of plants allows herbivores to colonize host plants when they are at the most suitable phenological stage for their exploitation (Brilli et al., 2009; Magalhães et al., 2016). For a wide range of herbivorous insects, there is substantial evidence demonstrating that peripheral receptors in the insect antennae are tuned to detect ubiquitous plant volatiles (Bruce et al., 2005 and references therein). Bruce et al. (2005) argued that if the majority of peripheral receptors of herbivorous insects are tuned to detect chemical compounds that are not unique to their host, the ratios of compounds in a blend emitted by their host plant becomes a vital component of the olfactory signal. Indeed, there are abundant examples of herbivores displaying behavioural responses to host plant volatile blends at levels that far exceed those of blend components tested individually (reviewed by Bruce & Pickett, 2011). These studies support an alternative to the token stimulus theory put forward in a seminal paper by Fraenkel (1959), which is based on host–plant recognition relying on volatiles that are highly specific to the host plant and absent from unrelated species. However, it should be noted that there are a few cases where herbivorous insects utilize taxonomically specific compounds as kairomones, chemicals that mediate interspecific interactions that benefit the receiver and harm the emitter. The most studied examples are the isothiocyanates, volatile catabolites of glucosinolates that are characteristic of the Brassicaceae and can be utilized by a range of insects among the Lepidoptera, Hemiptera, Diptera and Coleoptera (Bruce et al., 2005; Kostal, 1992).

There are numerous documented cases of herbivorous insects having a preference for a particular plant phenological stage, many of which relate to flowering or fruiting stages (Magalhães et al., 2016), that also involve the active recruitment of pollinating and seed dispersing animals by plants. Constitutive volatile cues from clover *Trifolium pratense* L. are attractive to the herbivorous root-boring beetle *Hylastinus obscurus*, but the beetle has particular preferences for plants of a certain age range, preferring plants aged 1.5–2.5 years over older and younger plants (Quiroz, Ortego, Ramirez, Wadhams, & Pinilla, 2005). These preferences are based on volatile cues, which reflect plants that are sufficiently advanced in their development to act as a host and yet young enough to be less likely recipients of earlier insect infestation (Quiroz et al., 2005). During the vegetative stages of a plant's development, it is unclear why it would advertize its presence to insects, especially to herbivores that feed on the plant and may result in extensive damage and even death. Therefore volatile chemicals that make a plant more apparent to herbivores without serving a function that improves plant fitness would appear nonadaptive.

In addition to having roles in plant defence against biotic stress and communication with other organisms, plant volatiles also protect plants from some abiotic stress factors and the dynamics of plant emissions are strongly affected by abiotic conditions, such as temperature, humidity, light intensity, drought, ozone, CO_2 and nutrient availability (Blande, Holopainen, & Niinemets, 2014; Gouinguené & Turlings, 2002; Pinto, Blande, Souza, Nerg, & Holopainen, 2010; Staudt & Lhoutellier, 2011). Plant volatiles — especially isoprene — have been linked to thermotolerance (Behnke et al., 2007; Sharkey & Singsaas, 1995) and many compounds are emitted in larger quantities as temperatures increase, although floral volatile emissions do not always follow ambient temperature (Theis, Lerdau, & Raguso, 2007). De novo synthesized compounds, emissions of which are photosynthesis reliant, increase with temperature up to a threshold over which emissions can be decreased (Kleist et al., 2012). High temperatures that quantitatively increase volatile emissions may make plants more apparent to foraging herbivores, but this increased apparency would only be relevant if the high temperature also correlates with foraging activity of herbivores. Interestingly, isoprene has been shown to be avoided by foraging *Manduca sexta* larvae (Laothawornkitkul et al., 2008), which may reduce the attractiveness of some plants as hosts under high temperature conditions. However, the poplar leaf beetle *Chrysomela populi* is not affected by isoprene and electroantennography showed that although *C. populi* detects a range of monoterpenes and sesquiterpenes, it does not detect isoprene (Müller et al., 2015). According to the definitions outlined in the introduction, the process of herbivores locating undamaged host plants by volatile chemicals represents signalling, whereby a constitutively emitted signal is utilized in the process of host finding by a herbivore. It does not equate to a communication process in the common sense, but is a vital step in the ecology and evolution of plant—insect interactions.

3. INDUCTION OF VOLATILES BY HERBIVORES

Once a herbivore has located a host plant, accepts it as a suitable host, and starts to feed, it induces substantial changes in the volatiles emitted by the plant (Heil, 2014; Holopainen & Blande, 2013; Mithöfer & Boland, 2012). This induction of plant volatiles has been the subject of a large body of work and a number of reviews have been published on the underlying induction mechanisms and the ecological significance of induced volatiles (e.g., Dicke & Hilker, 2003; Dicke & van Loon, 2000; Dicke, van Loon, & Soler, 2009; Kant, Bleeker, Van Wijk, Schuurink, Haring, 2009; Maffei, Mithöfer, & Boland, 2007; Paré et al., 2005). In addition to the feeding damage caused by herbivores, oviposition on or in a plant can also induce changes in plant defences and volatile emissions (Büchel et al., 2011; Fatouros et al., 2005; Hilker & Meiners, 2002; Meiners & Hilker, 2000; Mumm, Schrank, Wegener, Schulz, & Hilker, 2003; Pinto-Zevallos, Hellen, Hakola, van Nouhuys, & Holopainen, 2013; Wegener, Schulz, Meiners, Hadwich, & Hilker, 2001). The changes in plants induced by oviposition can have a significant effect on the plant responses to subsequent feeding by the larvae that emerge from the eggs (Pashalidou et al., 2015; and see Hilker & Fatouros, 2016). Even insects walking on a plant have been shown to stimulate early defence signalling events (Hall, MacGregor, Nijsse, & Bown, 2004). These changes are central to plant defence and communication between the plant and other organisms. Defence may be direct, through the repulsion of pests, and indirect, through the attraction of predatory and parasitic insects that prey on the herbivores. Communication based on HIPVs are known to exist between plants and a number of other community members including other herbivores, predatory and parasitic insects, other plants (see chapter: From the Lab Bench to the Forest: Ecology and Defence Mechanisms of Volatile-Mediated 'Talking Trees' by Arimura & Pearse, 2017), and hyperparasitoids. The emission of volatiles by plants has great plasticity, with the emitted blends often specific to the herbivore feeding mode, and even to the herbivore species. The herbivore pressure may also affect the quantity of emitted volatiles (e.g., Blande, Korjus, & Holopainen, 2010); thus, herbivore-damaged plants emit volatiles that can convey information on the type of stress the plant is experiencing, and the extent of that stress. However, it should be noted that there can be substantial variation in both constitutive and induced plant emissions and it has been shown with tomato plants that the degree of uniformity of induced volatile emissions varies with herbivore species (Bautista-Lozada & Espinosa-Garcia, 2013). Herbivores can use volatile signals as indicators of host presence, competition and potential high density of natural enemies that can also utilize the same signals. These induced volatiles have been hypothesized to have first evolved as a means of direct defence serving a primary benefit and to have subsequently become a source of information to other organisms (Veyrat et al., 2016).

3.1 Herbivore Oral Secretions as Signal Providers or Plant Manipulators

It is well documented that different insect feeding modes affect the volatile emissions of plants, with substantial variation between herbivores with chewing mouthparts and those with stealthier feeding modes such as hemipterans that feed from the phloem via stylets. In addition to the extent and temporal dynamics of mechanical feeding damage, both of which affect the blends of volatiles (Mithöfer, Wanner, & Boland, 2005), other feeding-related cues derived from the saliva, regurgitant and faeces of herbivores, can trigger herbivore-specific responses in plants (Acevedo, Rivera-Vega, Chung, Ray, & Felton, 2015). There is an expanding body of work on the identification of the elicitors and effectors derived from herbivores, but our knowledge on plant receptors that perceive these herbivore-specific cues is still limited (Acevedo et al., 2015). It is essential to better understand how plants can perceive specific herbivores and activate responses that are tuned to a specific threat.

Factors in insect saliva are now recognized as having a critical role to play in herbivore-induced blends of volatiles. Elicitors in the oral secretions of Lepidoptera larvae were found to play a critical role in the differences observed between volatiles induced by a mechanical wound and volatiles induced by insect feeding. A β-glucosidase isolated from *Pieris brassicae* oral secretions (Mattiacci, Dicke, & Posthumus, 1995) and the compound N-(17-hydroxylinolenoyl)-L-glutamine, known as volicitin, isolated from the oral secretion of beet armyworm (*Spodoptera exigua*) are two well-documented elicitors (Alborn et al., 1997; Turlings, Alborn, Loughrin, & Tumlinson, 2000). However, factors in oral secretions can also suppress plant responses to damage, an example being glucose oxidase which was the first insect salivary enzyme shown to suppress wound-inducible plant defences (Musser et al., 2005, 2002). These examples of salivary constituents are among a number of factors derived from herbivores — often gathered under the term herbivore-associated molecular patterns — that fine-tune the information content of a volatile blend.

4. HERBIVORES EAVESDROPPING ON INFORMATIVE CHEMICAL CUES

HIPVs constitute a signal that can be utilized by herbivores in their foraging for and selection of host plants. Classifying this induced volatile-mediated communication into a communication type is not as simple as for the constitutive emissions of plants, which are signals used by herbivores to find undamaged host plants. If the induced signal is viewed to have evolved either as a 'cry for help' to beneficial insects (predators and parasitoids that feed on the herbivores) (Dicke, 2009; Dicke, Sabelis, & Takabayashi, 1990), or a within plant signal to coordinate systemic defences within a damaged plant (Frost et al., 2007; Heil & Silva Bueno, 2007), utilization of those signals by herbivores could be classified as eavesdropping. That is the use of a plastic cue in an interaction that does not necessarily benefit the sender or the receiver. Transfer of information in the form of HIPVs has been shown to play a role in herbivore foraging in a range of empirical studies. HIPVs may make a plant more apparent to foraging herbivores and thus lead to increased colonization and feeding pressure on the plant. Conversely, herbivores may utilize HIPVs to determine that a potential host plant is already colonized by herbivores, thus indicating competition and a potentially poor host plant option (see Fig. 1).

Spodoptera frugiperda females have been shown to avoid plants infested with conspecifics as hosts for oviposition. Using the technique of gas chromatography coupled to electroantennography, which enables the

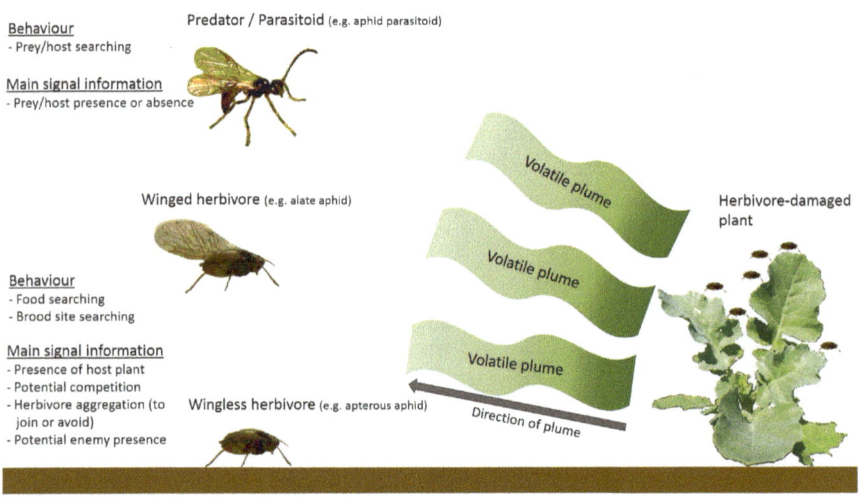

Figure 1 Summary of the information content in a plume of herbivore-induced plant volatiles (HIPVs). HIPVs convey information to predatory or parasitic insects foraging for prey or hosts, and winged or wingless herbivores searching for food or a host for offspring. The responses of the various HIPV recipients depend on interpretation of the information in the volatile plume and can result in attraction, repulsion or no behavioural response.

identification of compounds inducing an antennal response in insects, a range of the volatiles induced by conspecific larvae were shown to be detected by *S. frugiperda* adult females (Pinto-Zevallos, Strapasson, & Zarbin, 2016). *S. frugiperda* larvae are cannibalistic under field conditions, with cannibalism accounting for 40% of mortality in field experiments (Chapman et al., 2000). Greater larval density also correlates with a greater abundance of predatory and parasitic insects (Chapman et al., 2000). Therefore avoidance of plants infested with conspecific larvae has clear benefits for gravid *S. frugiperda* females.

Herbivores may also benefit from avoiding competition from heterospecific competitors. In behavioural tests in the laboratory, the whitefly *Bemisia tabaci* was shown to preferentially orientate towards undamaged plants over plants that had been previously infested with the aphid *Myzus persicae* (Saad, Roff, Hallett, & Idris, 2015). In this case, HIPVs may indicate competition, but may also indicate the presence of common natural enemies. Interestingly, in a study assessing the selection of host plants for oviposition by *Pieris rapae* butterflies, plants infested with *M. persicae* were selected as hosts at a level equal to noninfested plants (Layman & Lundgren, 2015). However, plants hosting a single predatory pink ladybird larva (*Coleomegilla maculata*) were selected significantly less often than noninfested plants and plants hosting aphids and a ladybird larva were selected even less often, which may have been due to a combination of HIPVs and aphid alarm pheromone released in response to predation by the ladybird larva (Layman & Lundgren, 2015).

The studies given as examples above have mostly focussed on choices made by adult insects foraging for host plants to feed on or host their young. Larval herbivore stages also make host choices based on perceived competition, which was demonstrated with *M. sexta* larvae, which preferentially select undamaged *Solanum carolinense* plants over plants that had been previously damaged by herbivores (Kariyat et al., 2014).

Relatively few studies have provided information on the chemical and sensory mechanisms underlying the avoidance of herbivore-infested plants, but a recent study has shed some new light on a mechanism underlying the phenomenon with cotton and the herbivore *Spodoptera littoralis*. As with the examples described earlier, herbivore-induced cotton volatiles suppressed the orientation of *S. littoralis* to host plants, but also orientation towards mates (Hatano et al., 2015). The homoterpene (*E*)-4,8-dimethyl-1,3,7-nonatriene (DMNT) was emitted in large amounts by herbivore-damaged plants and its role in suppressing the orientation of *S. littoralis* was tested by adding synthetically produced compound to attractive plant volatile blends. Addition of DMNT to the blends strongly inhibited take off and upwind flight orientation of female *S. littoralis*. Males are less attracted to plant volatiles than females and the addition of DMNT did not have a significant effect on male orientation to cotton, but it did have a significant effect on male orientation toward a main component of female pheromones, which suggests that DMNT significantly interferes with odour perception in both males and females (Hatano et al., 2015). Further studies are needed to determine whether DMNT interferes with the olfactory systems of other herbivores and if other volatiles emitted by different plants have similar effects.

Examples of herbivores avoiding plants damaged by either conspecific or heterospecific herbivores are abundant, but there are also examples of herbivore-induced volatiles being attractive to herbivores. The cotton boll weevil, *Anthonomus grandis*, is attracted by volatiles induced by conspecific herbivores over undamaged control plants, but not by volatiles induced by other herbivore species. The cotton boll weevils also showed a significant preference for reproductive plants over those at the vegetative stage (Magalhães et al., 2012). The responses of the insects to aggregation pheromone were also enhanced by the presence of conspecific HIPVs. It follows that herbivores that aggregate for either defence or reproductive reasons would utilize conspecific HIPVs as attractants, whereas herbivores that do not aggregate or have smaller aggregations would perceive similar cues as repellent.

In a recent study, a model was created that included a number of ecological parameters related to induced plant resistance with the aim of gaining information on how induced resistance affects aggregation patterns of herbivory (Rubin, Ellner, Kessler, & Morrell, 2015). Interestingly, the model showed that both increased aggregation and distribution evenness of herbivores can result from induced plant resistance based on volatile emissions and communication. The model supports the range of contrasting empirical observations for the responses of different herbivores.

5. TRUE COMMUNICATION BETWEEN PLANTS AND HERBIVORES

Instances of true communication between plants and herbivores are contingent on communication providing a benefit to both the plant and the herbivore. Due to the inherent necessity for herbivores to gain sustenance from their host plant, the benefit relayed by the herbivore will not be without cost. However, there are conceivably circumstances where plants will gain a fitness benefit from the presence of certain herbivores, either through defence against other more damaging herbivores or services that are essential to a plant's lifecycle, such as pollination. If HIPVs are avoided by herbivores, ultimately preventing overcolonization of a plant and resulting in improved survival of both the plant and herbivores, there is an argument to say that the signal process is genuine communication between the organisms. Indeed, honest signalling of toxicity through induced repellent compounds is conceptually the simplest form of plant—herbivore communication. In this scenario a plant that is toxic could provide an honest signal of toxicity and thereby reduce damage caused by herbivores, and the herbivores could also benefit through reduced effort on lower quality plants. This form of communication would effectively prevent feeding on a plant by a herbivore, but there are several examples in the literature of true communication between plants and herbivorous insects that do feed on the plant.

One example of where a herbivorous insect may provide a fitness benefit to a plant is if that herbivore can deter or eliminate other herbivorous insects. Omnivores fit squarely into that category as they feed on plants in the absence of animal prey and may provide a beneficial role for plants as long as they do not themselves become pests and if their presence results in plants receiving less damage from herbivores and ultimately having improved reproductive success. One omnivorous predator, *Anthocoris nemorum*, responds to both constitutive and systemic HIPVs (Lehrman, Boddum, Stenberg, Orians, & Bjorkman, 2013). This insect is attracted to the volatiles of undamaged plants, but plants damaged by the herbivorous leaf beetle *Phratora vulgatissima* are more attractive. The effectiveness of this communication channel for the plant relies on the omnivore relieving the plant of more damage than it causes, while the omnivore can gain sustenance from the plant and locate a site for location of potential prey. Differential use of constitutive and induced volatiles by *A. nemorum* is consistent with animal prey being the preferred and more difficult to locate food source and the plant being an easier to locate resource of lower food quality.

In general, plant investment into defence and reproduction trade-off against each other. Plants under attack from herbivores tend to invest more resources into defence, whereas plants that are less under threat invest in reproductive success. In a recent study, the effects of *S. littoralis* feeding on *Silene latifolia* floral volatiles and pollination success were investigated (Cozzolino et al., 2015). The authors found that *S. littoralis*—infested plants emitted higher amounts of two floral volatiles, (*Z*)-3-hexenyl acetate and β-ocimene, which appeared to attract more nocturnal pollinators and resulted in higher fruit production. These results suggest that the presence of a herbivore can potentially make a plant more attractive to pollinators and increase reproductive fitness. This is an intriguing observation as it implies that attraction of herbivores at a certain time in a plant's lifecycle can provide reproductive fitness benefits. The use of plant volatiles by such herbivores could thus fit the requirements of a true communication event between plants and herbivores in which the signal benefits both sender and receiver. Another interesting example of herbivory seemingly providing a fitness benefit to plants was found for field grown potato in the Colombian Andes. Herbivory by low numbers of potato moth larvae induced a 2.5-fold increase in potato yield over undamaged plants in a response that was consistent with a herbivore-elicited increase in primary productivity (Poveda, Gómez Jiménez, Halitschke, & Kessler, 2012; Poveda, Gómez Jiménez, & Kessler, 2010).

It is possible that some true herbivores could also be beneficial if they prevent more damaging herbivores from being present or if they effectively perform pollination tasks in their adult phases. Evaluating the beneficial versus negative effects of herbivore presence can be challenging, but to fully understand where communication channels exist in parallel to the plant–herbivore arms race, careful experiments need to be performed.

A prime example of where herbivores may be considered to have an indisputable beneficial role for plants is in brood site pollination mutualisms, where a plant provides a breeding ground and food resources for insects that perform pollination services (Borges, 2016; Hossaert-Mckey, Soler, Schatz, & Proffit, 2010). One of the best described brood site pollination mutualisms is between figs and fig wasps, which are coevolved and mutually dependent upon each other for successful reproduction. Scent plays an essential role in mediating fig–fig wasp mutualism. During foraging fig wasps are required to make decisions that, if wrong, could result in reproductive failure (Borges, 2016). The fig is an enclosed globular inflorescence known as a syconium, which has a restricted opening that often damages the fig wasp wings and antennae during entrance making a decision to enter the syconium one of utmost importance. Adult fig wasps live for 24–48 h (Ghara & Borges, 2010), consequently, female fig wasps must find and select a syconium in the pollen-receptive stage within that time period. Coordination of the floral scent messages, the ability to receive those messages by the fig wasp, and the ability to gall fig flowers within the selected syconium are critical steps in successful reproduction of fig and fig wasp alike (Borges, 2016). While the focus of this chapter is not to give exhaustive details on the mechanisms of these interactions, especially for interactions that are pollination based and will be covered elsewhere in this book (chapters: Plant—Pollinator Communication by Wester & Lunau, 2017 and Mimicry and Deception in Pollination by Lunau & Wester, 2017), it is important to consider that offering reward to other organisms can allow plants to accomplish functions that their sedentary lifestyle makes otherwise impossible.

6. PLANT EAVESDROPPING ON HERBIVORE-EMITTED CHEMICAL CUES

We earlier looked at the phenomenon of eavesdropping, whereby herbivores are able to utilize chemical cues that make their hosts more apparent or provide information on their physiological condition. In this section we will take a look at eavesdropping in the opposite direction, whereby plants are able to detect chemical cues emitted by insects, utilize them as a signal of a potential threat and modify their defences in response. Insects emit pheromones to communicate important information that is often necessary for their survival. Alarm pheromones and aggregation pheromones can play roles in processes related to defence and enemy avoidance. Sex pheromones are used to signal to potential mates and are essential for successful reproduction. From the perspective of the plant, detecting these pheromones may provide information about a potential threat and the form and immediacy of that threat. Alarm pheromones may indicate that herbivores on nearby plants are in an agitated state and could be about to disperse, this could indicate an immediate threat by either a small or large number of herbivores. Aggregation pheromones may indicate that large numbers of herbivores are present on neighbouring plants and could represent a future threat of large numbers of herbivores. Sex pheromones indicate that mating herbivores are in the vicinity and that plants may be subject to oviposition and future herbivory. If plants can utilize these cues to optimize their defences, they may gain a tangible fitness benefit (see Fig. 2).

The phenomenon of plants eavesdropping on insect-released chemicals is an emerging research area. Consequently, there is scant literature documenting these interactions. However, recent work with tall goldenrod, *Solidago altissima* L., and the specialist gall-inducing tephritid fruit fly, *Eurosta solidaginis*, has yielded evidence that exposure to a putative sex pheromone of the fly leads to tall goldenrod exhibiting increased defences and a reduced susceptibility to feeding damage by insects (Helms, De Moraes, Tooker, & Mescher, 2013). *S. altissima* and *E. solidaginis* are thought to be closely

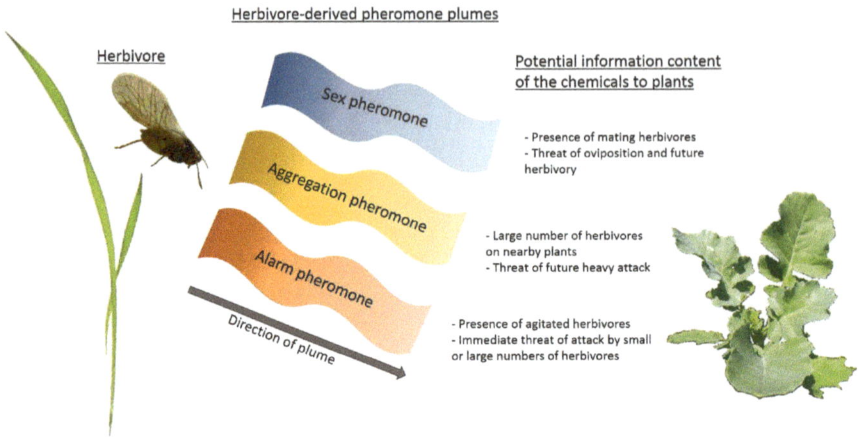

Figure 2 Summary of potential chemically encoded information that can be transmitted from insects to plants. Insect pheromones could function as kairomones to plants, providing information of benefit to the receiving plant and detriment to the fitness of the herbivore.

coevolved, with larvae of *E. solidaginis* inducing spherical galls in the stems of *S. altissima*, which substantially reduce plant growth and fitness (Helms et al., 2013 and references therein). Therefore preempting attack and optimizing defence allocation can potentially provide a substantial boost to plant fitness. The male *E. solidaginis* flies emerge before females and are found perching on leaves towards the top of tall goldenrod ramets. They release a volatile blend dominated by spiroacetal compounds, which were found to be attractive to female *E. solidaginis* flies in behavioural tests conducted in Y-tube olfactometers (Helms et al., 2013). The effect of exposure to male *E. solidaginis* volatiles on the susceptibility of *S. altissima* to oviposition and herbivore-feeding damage was tested in a field study (Helms et al., 2013). The tops of tall goldenrod ramets were enclosed in mesh nets and divided into four treatment groups. Three of the treatment groups contained an individual adult fly, either a male *E. solidaginis* fly, a female *E. solidaginis* fly or a single common housefly (*Musca domestica*), and a fourth group consisted of empty mesh nets as a control. The nets were removed after 3 days and the plants were observed weekly over a 4-week period to score for herbivore-damage and wounds caused by ovipuncture into the terminal buds. Plants exposed to *E. solidaginis* males had significantly lower incidence of ovipuncture and were subject to significantly lower levels of herbivore-feeding damage than each of the other treatments. Treated plants were also observed to have a stronger induction of jasmonic acid, a defence-related phytohormone, than the control plants (Helms et al., 2013) and a stronger induction of HIPVs (Helms, De Moraes, Mescher, & Tooker, 2014). These observations provide evidence that the plant is responding to the insect-released chemicals, but the phenomenon appears to be rather specific to this particular plant–herbivore association. When similar tests were conducted with the same herbivore but with plants that have not coevolved, HIPV emissions were not altered and the plants did not have enhanced levels of defence against their herbivores (Helms et al., 2014).

Interestingly, a couple of recent studies have been conducted to assess the potential for the application of volatile 3-pentanol, a bacterial volatile derivative and a component of several insect pheromones, for the priming and induction of resistance to bacterial pathogens (Choi, Song, Yi, & Ryu, 2014; Song & Ryu, 2013) and aphids (Song & Ryu, 2013). These studies were conducted with the intention of elucidating the role of bacterial volatile organic compounds in defence against pathogenic infection. However, in a twist of fate, a field trial conducted by Song and Ryu (2013) involving the application of selected volatiles (3-pentanol and 2-butanone) to cucumber plants coincided with an unexpected outbreak of the peach–potato aphid, *M. persicae*. The authors took advantage of the situation to assess the effects of the volatile treatment on the susceptibility of plants to aphid infestation and found that 3-pentanol and 2-butanone treated plants had substantially reduced aphid infestation rates. This coincided with increased presence of ladybirds, a key natural enemy of aphids. In terms of herbivore to plant communication, these studies are rather preliminary and circumstantial, but they do provide evidence that a component of several insect pheromones has the potential to increase resistance of plants to herbivore attack.

The examples outlined above are based on volatile pheromone components, but herbivores also produce nonvolatile chemicals that come into contact with plants. A study of the snail, *Helix aspersa*, feeding on *Brassica nigra* found that seedlings regularly exposed to the mucus and faeces of the snail experienced reduced rates of attack (Orrock, 2013). The presence of mucus alone was not found to be a feeding deterrent, and kairomones were suspected of playing a key role in inducing the increased resistance of plants. However, the identity and source of those kairomones has not yet been determined. Interestingly, a screen for phytohormones in mucus from 13 different snail species showed that one species, *Deroceras reticulatum*, contained significant amounts of salicylic acid (Kästner et al., 2014). Application of

D. reticulatum mucus to leaves of *Arabidopsis thaliana* activated the promotor of the pathogenesis related 1 (*PR1*) gene, which demonstrated the potential for mucus to regulate plant defences. *H. aspersa* was not one of the snails screened by Kästner et al. (2014), which leaves the kairomonal mechanism observed by Orrock (2013) open for further experimentation.

Together, these studies highlight that a range of different chemical cues can be released by herbivores and received by plants. Although a relatively new research direction, it could be common for plants to eavesdrop on herbivore-derived chemical cues. This line of research could lead to enhanced understanding of coevolution between plants and herbivores and could also open up new strategies for manipulating plant defences in an applied agricultural context.

7. COMMUNICATION BETWEEN PLANTS AND HIGHER TROPHIC LEVELS

One of the most widely investigated volatile-mediated interactions is between herbivore-damaged plants and the predatory and parasitic natural enemies of the herbivores. The process of recruiting natural enemies of herbivores by releasing volatile attractants is referred to as indirect defence and has been observed in a range of different plant–insect combinations. As described above for plant–herbivore interactions, HIPV provide information that can be specific to the plant and the damaging agent, which provides information for predatory and parasitic insects to utilize in their pursuit of food prey or hosts (see Fig. 1). Blends of volatile organic compounds provide important foraging signals, but some specific individual compounds have also been shown to play important roles. For example, emission of 3-butenyl isothiocyanate, a product of glucosinolate degradation and characteristic of the Brassicaceae, was found to be induced by aphid feeding and attractive to the parasitoid *Diaeretiella rapae* (Blande, Pickett, & Poppy, 2007). A vast body of literature has been accumulated on tritrophic interactions involving plants, arthropod herbivores and arthropod natural enemies (see reviews by Dicke, 2009; Mumm & Dicke, 2010; Pierik, Ballaré, & Dicke, 2014; de Rijk, Dicke, & Poelman, 2013; Turlings & Benrey, 1998; Vet & Dicke, 1992). In recent times it has also been found that birds can also utilize HIPVs in their search for prey (Amo, Jansen, van Dam, Dicke, & Visser, 2013; Mäntylä et al., 2008). Evidence suggests that birds learn to utilize HIPVs and are not innately attracted to prey-infested trees by exposure to HIPVs alone (Amo, Dicke, & Visser, 2016), utilizing predominantly visual cues in early foraging forays. Recent studies have also extended our knowledge beyond the responses of third trophic level organisms and elucidated the responses of hyperparasitoids – parasitoids that parasitize primary parasitoids – to HIPVs. Elegant experiments determined that plants respond to feeding by parasitized herbivores by releasing volatile blends that differ from those induced by nonparasitized individuals and that hyperparasitoids are able to exploit these differences in searching for hosts (Poelman et al., 2012). It was further shown that the hyperparasitoid *Lysibia nana* can utilize HIPV cues under field conditions and that they are equally attracted by *Brassica oleracea* plants damaged by two different herbivore species (*P. brassicae* and *P. rapae*) that are parasitized by the gregarious parasitoid *Cotesia glomerata* (Zhu et al., 2015). Parasitoid species affects the magnitude of the response of *B. oleracea* to herbivory by *Pieris* species, which can also be detected by gravid *Plutella xylostella* searching for a host for oviposition (Poelman et al., 2011). While signalling from plants to predators and parasitoids is a relatively clear cut example of interkingdom communication, it appears fantastical to view interactions with hyperparasitoids as anything more than eavesdropping on infochemicals. However, this eavesdropping has the potential to dramatically influence the interaction dynamics in complex ecological systems.

8. SUMMARY AND FUTURE DIRECTIONS

Plants communicate their physiological condition to a diverse range of organisms in their community through the emission of volatile organic compounds. Herbivores can utilize the chemical signals emitted by plants as foraging cues, and there are examples of plants gaining a benefit either through the repulsion of herbivores or the attraction of herbivores that can perform beneficial functions, such as roles in defence or pollination, either directly or indirectly. Recent research has shown that plants can also detect and process chemical information derived from herbivores (Helms et al., 2014, 2013). Pheromones emitted to coordinate essential parts of a herbivore's lifecycle can impart specific information to plants regarding a type of threat and its immediacy. Therefore detecting such information and utilizing it for the effective deployment of defences is potentially of great benefit to the plant. Where the arms race of plants evolving new defences and herbivores evolving ways to overcome them ends and communication begins is demarcated by an extremely blurry line. Even so, there are clear examples of insects that feed on plants having positive (or even essential) effects on plants in interactions that are mediated by volatile organic compounds.

Future research on plant—herbivore communication should focus on discovering further examples of where herbivore presence can have a fitness benefit on plants. Determination of such a benefit opens doors to explore the functioning of plant herbivore interactions from the perspective of communication. Where the biology of herbivorous insects leads them to release large quantities of pheromone in the vicinity of plants, the effects of those pheromones on plant defences should be explored. If plant responses to insect pheromones are commonplace, the potential for utilizing pheromones not only in insect traps, but also as a mechanism to manipulate plant defences should be explored.

ACKNOWLEDGEMENTS

I thank Martín Pareja for providing insightful comments on the draft of this chapter. My work on plant—herbivore communication was funded by the Academy of Finland decision number 251898.

REFERENCES

Acevedo, F. E., Rivera-Vega, L. J., Chung, S. H., Ray, S., & Felton, G. W. (2015). Cues from chewing insects — the intersection of DAMPs, HAMPs, MAMPs and effectors. *Current Opinion in Plant Biology, 26*, 80—86. http://dx.doi.org/10.1016/j.pbi.2015.05.029.

Alborn, H. T., Turlings, T. C. J., Jones, T. H., Stenhagen, G., Loughrin, J. H., & Tumlinson, J. H. (1997). An elicitor of plant volatiles from beet armyworm oral secretion. *Science, 276*, 945—949.

Amo, L., Dicke, M., & Visser, M. E. (2016). Are naive birds attracted to herbivore-induced plant defences? *Behaviour, 153*, 353—366. http://dx.doi.org/10.1163/1568539X-00003345.

Amo, L., Jansen, J. J., van Dam, N. M., Dicke, M., & Visser, M. E. (2013). Birds exploit herbivore-induced plant volatiles to locate herbivorous prey. *Ecology Letters, 16*, 1348—1355. http://dx.doi.org/10.1111/ele.12177.

Arimura, G., & Pearse, I. S. (2017). From the lab bench to the forest: ecology and defence mechanisms of volatile-mediated 'talking trees'. In G. Becard (Ed.). *How plants communicate with their biotic environment* (Vol. 82, 3—17).

Bautista-Lozada, A., & Espinosa-Garcia, J. F. (2013). Odor uniformity among tomato individuals in response to herbivore depends on insect species. *PLoS One, 8*, e77199. http://dx.doi.org/10.1371/journal.pone.0077199.

Behnke, K., Ehlting, B., Teuber, M., Bauerfeind, M., Louis, S., Hänsch, R., ... Schnitzler, J. (2007). Transgenic, non-isoprene emitting poplars don't like it hot. *Plant Journal, 51*, 485—499. http://dx.doi.org/10.1111/j.1365-313X.2007.03157.x.

Blande, J. D., Holopainen, J. K., & Niinemets, Ü. (2014). Plant volatiles in polluted atmospheres: stress responses and signal degradation. *Plant, Cell and Environment, 37*, 1892–1904. http://dx.doi.org/10.1111/pce.12352.

Blande, J. D., Korjus, M., & Holopainen, J. K. (2010). Foliar methyl salicylate emissions indicate prolonged aphid infestation on silver birch and black alder. *Tree Physiology, 30*, 404–416. http://dx.doi.org/10.1093/treephys/tpp124.

Blande, J. D., Pickett, J. A., & Poppy, G. M. (2007). A comparison of semiochemically mediated interactions involving specialist and generalist *Brassica*-feeding aphids and the braconid parasitoid *Diaeretiella rapae*. *Journal of Chemical Ecology, 33*, 767–779. http://dx.doi.org/10.1007/s10886-007-9264-7.

Borges, R. M. (2016). On the air: broadcasting and reception of volatile messages in brood-site pollination mutualisms. In J. D. Blande, & R. Glinwood (Eds.), (1st ed.). Deciphering the Chemical Language of Plant Communication. Springer International Publishing. http://dx.doi.org/10.1007/978-3-319-33498-1_10

Brilli, F., Ciccioli, P., Frattoni, M., Prestininzi, M., Spanedda, A. F., & Loreto, F. (2009). Constitutive and herbivore-induced monoterpenes emitted by *Populus × euroamericana* leaves are key volatiles that orient *Chrysomela populi* beetles. *Plant, Cell and Environment, 32*, 542–552. http://dx.doi.org/10.1111/j.1365-3040.2009.01948.x.

Bruce, T. J. A., & Pickett, J. A. (2011). Perception of plant volatile blends by herbivorous insects – finding the right mix. *Phytochemistry, 72*, 1605–1611. http://dx.doi.org/10.1016/j.phytochem.2011.04.011.

Bruce, T. J. A., Wadhams, L. J., & Woodcock, C. M. (2005). Insect host location: a volatile situation. *Trends in Plant Science, 10*, 269–274. http://dx.doi.org/10.1016/j.tplants.2005.04.003.

Büchel, K., Malskies, S., Mayer, M., Fenning, T. M., Gershenzon, J., Hilker, M., & Meiners, T. (2011). How plants give early herbivore alert: volatile terpenoids attract parasitoids to egg-infested elms. *Basic and Applied Ecology, 12*, 403–412. http://dx.doi.org/10.1016/j.baae.2011.06.002.

Chamovitz, D. (2012). *What a plant knows* (1st ed.). London: Oneworld Publications.

Chapman, J. W., Williams, T., Martinez, A. M., Cisneros, J., Caballero, P., Cave, R. D., & Goulson, D. (2000). Does cannibalism in *Spodoptera frugiperda* (Lepidoptera: Noctuidae) reduce the risk of predation? *Behavioral Ecology and Sociobiology, 48*, 321–327. http://dx.doi.org/10.1007/s002650000237.

Choi, H. K., Song, G. C., Yi, H., & Ryu, C. (2014). Field evaluation of the bacterial volatile derivative 3-pentanol in priming for induced resistance in pepper. *Journal of Chemical Ecology, 40*, 882–892. http://dx.doi.org/10.1007/s10886-014-0488-z.

Cozzolino, S., Fineschi, S., Litto, M., Scopece, G., Trunschke, J., & Schiestl, F. P. (2015). Herbivory increases fruit set in *Silene latifolia*: a consequence of induced pollinator-attracting floral volatiles? *Journal of Chemical Ecology, 41*, 622–630. http://dx.doi.org/10.1007/s10886-015-0597-3.

Dicke, M. (2009). Behavioural and community ecology of plants that cry for help. *Plant, Cell and Environment, 32*, 654–665. http://dx.doi.org/10.1111/j.1365-3040.2008.01913.x.

Dicke, M., & Baldwin, I. T. (2010). The evolutionary context for herbivore-induced plant volatiles: beyond the 'cry for help'. *Trends in Plant Science, 15*, 167–175. http://dx.doi.org/10.1016/j.tplants.2009.12.002.

Dicke, M., & Hilker, M. (2003). Induced plant defences: from molecular biology to evolutionary ecology. *Basic and Applied Ecology, 4*, 3–14. http://dx.doi.org/10.1078/1439-1791-00129.

Dicke, M., & van Loon, J. J. A. (2000). Multitrophic effects of herbivore-induced plant volatiles in an evolutionary context. *Entomologia Experimentalis et Applicata, 97*, 237–249. http://dx.doi.org/10.1046/j.1570-7458.2000.00736.x.

Dicke, M., van Loon, J. J. A., & Soler, R. (2009). Chemical complexity of volatiles from plants induced by multiple attack. *Nature Chemical Biology, 5*, 317–324. http://dx.doi.org/10.1038/nchembio.169.

Dicke, M., Sabelis, M. W., & Takabayashi, J. (1990). Do plants cry for help? Evidence related to a tritrophic system of predatory mites, spider mites and their host plants. *Symposia Biologica Hungarica, 39*, 127–134.

Ehrlich, P. R., & Raven, P. H. (1964). Butterflies and plants: a study in coevolution. *Evolution, 18*, 586–608. http://dx.doi.org/10.2307/2406212.

Fatouros, N. E., Bukovinszkine'Kiss, G., Kalkers, L. A., Gamborena, R. S., Dicke, M., & Hilker, M. (2005). Oviposition-induced plant cues: do they arrest *Trichogramma* wasps during host location? *Entomologia Experimentalis et Applicata, 115*, 207–215. http://dx.doi.org/10.1111/j.1570-7458.2005.00245.x.

Fraenkel, G. S. (1959). The raison d'etre of secondary plant substances. *Science, 129*, 1466–1470. http://dx.doi.org/10.1126/science.129.3361.1466.

Frost, C. J., Appel, M., Carlson, J. E., De Moraes, C. M., Mescher, M. C., & Schultz, J. C. (2007). Within-plant signalling via volatiles overcomes vascular constraints on systemic signalling and primes responses against herbivores. *Ecology Letters, 10*, 490–498.

Ghara, M., & Borges, R. M. (2010). Comparative life-history traits in a fig wasp community: implications for community structure. *Ecological Entomology, 35*, 139–148. http://dx.doi.org/10.1111/j.1365-2311.2010.01176.x.

Gouinguené, S. P., & Turlings, T. C. J. (2002). The effects of abiotic factors on induced volatile emissions in corn plants. *Plant Physiology, 129*, 1296–1307. http://dx.doi.org/10.1104/pp.001941.

Hall, D., MacGregor, K., Nijsse, J., & Bown, A. (2004). Footsteps from insect larvae damage leaf surfaces and initiate rapid responses. *European Journal of Plant Pathology, 110*, 441–447. http://dx.doi.org/10.1023/B:EJPP.0000021072.89968.de.

Hare, J. D. (2010). Ontogeny and season constrain the production of herbivore-inducible plant volatiles in the field. *Journal of Chemical Ecology, 36*, 1363–1374. http://dx.doi.org/10.1007/s10886-010-9878-z.

Hatano, E., Saveer, A. M., Borrero-Echeverry, F., Strauch, M., Zakir, A., Bengtsson, M., … Dekker, T. (2015). A herbivore-induced plant volatile interferes with host plant and mate location in moths through suppression of olfactory signalling pathways. *BMC Biology, 13*, 75. http://dx.doi.org/10.1186/s12915-015-0188-3.

Heil, M. (2014). Herbivore-induced plant volatiles: targets, perception and unanswered questions. *New Phytologist, 204*, 297–306. http://dx.doi.org/10.1111/nph.12977.

Heil, M., & Karban, R. (2010). Explaining evolution of plant communication by airborne signals. *Trends in Ecology and Evolution, 25*, 137–144. http://dx.doi.org/10.1016/j.tree.2009.09.010.

Heil, M., & Silva Bueno, J. C. (2007). Within-plant signaling by volatiles leads to induction and priming of an indirect plant defense in nature. *Proceedings of the National Academy of Sciences of the United States of America, 104*, 5467–5472. http://dx.doi.org/10.1073/pnas.0610266104.

Helms, A. M., De Moraes, C. M., Mescher, M. C., & Tooker, J. F. (2014). The volatile emission of *Eurosta solidaginis* primes herbivore-induced volatile production in *Solidago altissima* and does not directly deter insect feeding. *BMC Plant Biology, 14*, 173. http://dx.doi.org/10.1186/1471-2229-14-173.

Helms, A. M., De Moraes, C. M., Tooker, J. F., & Mescher, M. C. (2013). Exposure of *Solidago altissima* plants to volatile emissions of an insect antagonist *(Eurosta solidaginis)* deters subsequent herbivory. *Proceedings of the National Academy of Sciences of the United States of America, 110*, 199–204. http://dx.doi.org/10.1073/pnas.1218606110.

Hilker, M., & Fatouros, N. E. (2016). Resisting the onset of herbivore attack: plants perceive and respond to insect eggs. *Current Opinion in Plant Biology, 32*, 9–16. http://dx.doi.org/10.1016/j.pbi.2016.05.003.

Hilker, M., & Meiners, T. (2002). Induction of plant responses to oviposition and feeding by herbivorous arthropods: a comparison. *Entomologia Experimentalis et Applicata, 104*, 181–192. http://dx.doi.org/10.1046/j.1570-7458.2002.01005.x.

Holopainen, J. K., & Blande, J. D. (2013). Where do herbivore-induced plant volatiles go? *Frontiers in Plant Science, 4*. http://dx.doi.org/10.3389/fpls.2013.00185.

Hossaert-McKey, M., Soler, C., Schatz, B., & Proffit, M. (2010). Floral scents: their roles in nursery pollination mutualisms. *Chemoecology, 20*, 75–88. http://dx.doi.org/10.1007/s00049-010-0043-5.

Howe, G. A., & Jander, G. (2008). Plant immunity to insect herbivores. *Annual Review of Plant Biology, 59*, 41–66. http://dx.doi.org/10.1146/annurev.arplant.59.032607.092825.

Kant, M. R., Bleeker, P. M., Van Wijk, M., Schuurink, R. C., & Haring, M. A. (2009). Plant volatiles in defence. *Advances in Botanical Research, 51*, 613–666. http://dx.doi.org/10.1016/S0065-2296(09)51014-2.

Kant, M. R., Jonckheere, W., Knegt, B., Lemos, F., Liu, J., Schimmel, B. C. J., … Alba, J. M. (2015). Mechanisms and ecological consequences of plant defence induction and suppression in herbivore communities. *Annals of Botany, 115*, 1015–1051. http://dx.doi.org/10.1093/aob/mcv054.

Karban, R. (2015). *Plant sensing and communication* (1st ed.). Chicago and London: The University of Chicago Press.

Kariyat, R. R., Scanlon, S. R., Moraski, R. P., Stephenson, A. G., Mescher, M. C., & De Moraes, C. M. (2014). Plant inbreeding and prior herbivory influence the attraction of caterpillars (*Manduca sexta*) to odors of the host plant *Solanum carolinense* (Solanaceae). *American Journal of Botany, 101*, 376–380. http://dx.doi.org/10.3732/ajb.1300295.

Kästner, J., von Knorre, D., Himanshu, H., Erb, M., Baldwin, I. T., & Meldau, S. (2014). Salicylic acid, a plant defense hormone, is specifically secreted by a Molluscan herbivore. *PLoS One, 9*, e86500. http://dx.doi.org/10.1371/journal.pone.0086500.

Kessler, A. (2015). The information landscape of plant constitutive and induced secondary metabolite production. *Current Opinion in Insect Science, 8*, 47–53. http://dx.doi.org/10.1016/j.cois.2015.02.002.

Kleist, E., Mentel, T. F., Andres, S., Bohne, A., Folkers, A., Kiendler-Scharr, A., … Wildt, J. (2012). Irreversible impacts of heat on the emissions of monoterpenes, sesquiterpenes, phenolic BVOC and green leaf volatiles from several tree species. *Biogeosciences, 9*, 5111–5123. http://dx.doi.org/10.5194/bg-9-5111-2012.

Kostal, V. (1992). Orientation behavior of newly hatched larvae of the cabbage maggot, *Delia radicum* (L) (Diptera, Anthomyiidae), to volatile plant metabolites. *Journal of Insect Behavior, 5*, 61–70. http://dx.doi.org/10.1007/BF01049158.

Laothawornkitkul, J., Paul, N. D., Vickers, C. E., Possell, M., Taylor, J. E., Mullineaux, P. M., & Hewitt, C. N. (2008). Isoprene emissions influence herbivore feeding decisions. *Plant, Cell and Environment, 31*, 1410–1415. http://dx.doi.org/10.1111/j.1365-3040.2008.01849.x.

Layman, M. L., & Lundgren, J. G. (2015). The influence of aphids (*Myzus persicae*) and pink lady beetle larvae (*Coleomegilla maculata*) on host plant preference of imported cabbageworm (*Pieris rapae*). *Arthropod-Plant Interactions, 9*, 507–514. http://dx.doi.org/10.1007/s11829-015-9392-x.

Lehrman, A., Boddum, T., Stenberg, J. A., Orians, C. M., & Bjorkman, C. (2013). Constitutive and herbivore-induced systemic volatiles differentially attract an omnivorous biocontrol agent to contrasting Salix clones. *AoB Plants, 5*, plt005. http://dx.doi.org/10.1093/aobpla/plt005.

Lunau, K., & Wester, P. (2017). Mimicry and deception in pollination. In G. Becard (Ed.). *How plants communicate with their biotic environment* (Vol. 82, 259–279).

Maffei, M. E., Mithöfer, A., & Boland, W. (2007). Insects feeding on plants: rapid signals and responses preceding the induction of phytochemical release. *Phytochemistry, 68*, 2946–2959. http://dx.doi.org/10.1016/j.phytochem.2007.07.016.

Magalhães, D. M., Borges, M., Laumann, R. A., Sujii, E. R., Mayon, P., Caulfield, J. C., … Blassioli-Moraes, M. C. (2012). Semiochemicals from herbivory induced cotton plants enhance the foraging behavior of the cotton boll weevil, *Anthonomus grandis*. *Journal of Chemical Ecology, 38*, 1528–1538. http://dx.doi.org/10.1007/s10886-012-0216-5.

Magalhães, D. M., Borges, M., Laumann, R. A., Woodcock, C. M., Pickett, J. A., Birkett, M. A., & Blassioli-Moraes, M. C. (2016). Influence of two acyclic homoterpenes (tetranorterpenes) on the foraging behavior of *Anthonomus grandis* Boh. *Journal of Chemical Ecology, 42*, 305–313. http://dx.doi.org/10.1007/s10886-016-0691-1.

Mancuso, S., & Viola, A. (2013). *Brilliant green: The surprising history and science of plant intelligence* (1st ed.). Washington, DC: Island Press.

Mäntylä, E., Alessio, G. A., Blande, J. D., Heijari, J., Holopainen, J. K., Laaksonen, T., … Klemola, T. (2008). From plants to birds: higher avian predation rates in trees responding to insect herbivory. *PLoS One, 3*, e2832. http://dx.doi.org/10.1371/journal.pone.0002832.

Mattiacci, L., Dicke, M., & Posthumus, M. (1995). Beta-glucosidase – an elicitor of herbivore-induced plant odor that attracts host-searching parasitic wasps. *Proceedings of the National Academy of Sciences of the United States of America, 92*, 2036–2040. http://dx.doi.org/10.1073/pnas.92.6.2036.

Meiners, T., & Hilker, M. (2000). Induction of plant synomones by oviposition of a phytophagous insect. *Journal of Chemical Ecology, 26*, 221–232. http://dx.doi.org/10.1023/A:1005453830961.

Mithöfer, A., & Boland, W. (2012). Plant defense against herbivores: chemical aspects. *Annual Review of Plant Biology, 63*, 431–450.

Mithöfer, A., Wanner, G., & Boland, W. (2005). Effects of feeding *Spodoptera littoralis* on lima bean leaves. II. Continuous mechanical wounding resembling insect feeding is sufficient to elicit herbivory-related volatile emission. *Plant Physiology, 137*, 1160–1168. http://dx.doi.org/10.1104/pp.104.054460.

Müller, A., Kaling, M., Faubert, P., Gort, G., Smid, H. M., Van Loon, J. J. A., … Rosenkranz, M. (2015). Isoprene emission by poplar is not important for the feeding behaviour of poplar leaf beetles. *BMC Plant Biology, 15*, 165. http://dx.doi.org/10.1186/s12870-015-0542-1.

Mumm, R., & Dicke, M. (2010). Variation in natural plant products and the attraction of bodyguards involved in indirect plant defense. *Canadian Journal of Zoology, 88*, 628–667. http://dx.doi.org/10.1139/Z10-032.

Mumm, R., Schrank, K., Wegener, R., Schulz, S., & Hilker, M. (2003). Chemical analysis of volatiles emitted by *Pinus sylvestris* after induction by insect oviposition. *Journal of Chemical Ecology, 29*, 1235–1252. http://dx.doi.org/10.1023/A:1023841909199.

Musser, R. O., Cipollini, D. F., Hum-Musser, S. M., Williams, S. A., Brown, J. K., & Felton, G. W. (2005). Evidence that the caterpillar salivary enzyme glucose oxidase provides herbivore offense in Solanaceous plants. *Archives of Insect Biochemistry and Physiology, 58*, 128–137. http://dx.doi.org/10.1002/arch.20039.

Musser, R. O., Hum-Musser, S. M., Eichenseer, H., Peiffer, M., Ervin, G., Murphy, J. B., & Felton, G. W. (2002). Herbivory: caterpillar saliva beats plant defences - a new weapon emerges in the evolutionary arms race between plants and herbivores. *Nature, 416*, 599–600. http://dx.doi.org/10.1038/416599a.

Orrock, J. L. (2013). Exposure of unwounded plants to chemical cues associated with herbivores leads to exposure-dependent changes in subsequent herbivore attack. *PLoS One, 8*, e79900. http://dx.doi.org/10.1371/journal.pone.0079900.

Paré, P. W., Farag, M. A., Krishnamachari, V., Zhang, H. M., Ryu, C. M., & Kloepper, J. W. (2005). Elicitors and priming agents initiate plant defense responses. *Photosynthesis Research, 85*, 149–159. http://dx.doi.org/10.1007/s11120-005-1001-x.

Pashalidou, F. G., Gols, R., Berkhout, B. W., Weldegergis, B. T., van Loon, J. J. A., Dicke, M., & Fatouros, N. E. (2015). To be in time: egg deposition enhances plant-mediated detection of young caterpillars by parasitoids. *Oecologia, 177*, 477–486. http://dx.doi.org/10.1007/s00442-014-3098-0.

Petschenka, G., & Agrawal, A. A. (2016). How herbivores coopt plant defenses: natural selection, specialization, and sequestration. *Current Opinion in Insect Science, 14*, 17–24. http://dx.doi.org/10.1016/j.cois.2015.12.004.

Pierik, R., Ballaré, C. L., & Dicke, M. (2014). Ecology of plant volatiles: taking a plant community perspective. *Plant, Cell and Environment, 37*, 1845–1853. http://dx.doi.org/10.1111/pce.12330.

Pinto, D. M., Blande, J. D., Souza, S. R., Nerg, A. M., & Holopainen, J. K. (2010). Plant volatile organic compounds (VOCs) in ozone (O(3)) polluted atmospheres: the ecological effects. *Journal of Chemical Ecology, 36*, 22–34. http://dx.doi.org/10.1007/s10886-009-9732-3.

Pinto-Zevallos, D. M., Hellen, H., Hakola, H., van Nouhuys, S., & Holopainen, J. K. (2013). Induced defenses of Veronica spicata: variability in herbivore- induced volatile organic compounds. *Phytochemistry Letters, 6*, 653–656. http://dx.doi.org/10.1016/j.phytol.2013.08.015.

Pinto-Zevallos, D. M., Strapasson, P., & Zarbin, P. H. G. (2016). Herbivore-induced volatile organic compounds emitted by maize: electrophysiological responses in *Spodoptera frugiperda* females. *Phytochemistry Letters, 16*, 70–74. http://dx.doi.org/10.1016/j.phytol.2016.03.005.

Poelman, E. H., Bruinsma, M., Zhu, F., Weldegergis, B. T., Boursault, A. E., Jongema, Y., … Dicke, M. (2012). Hyperparasitoids use herbivore-induced plant volatiles to locate their parasitoid host. *PLoS Biology, 10*, e1001435. http://dx.doi.org/10.1371/journal.pbio.1001435.

Poelman, E. H., Zheng, S., Zhang, Z., Heemskerk, N. M., Cortesero, A., & Dicke, M. (2011). Parasitoid-specific induction of plant responses to parasitized herbivores affects colonization by subsequent herbivores. *Proceedings of the National Academy of Sciences of the United States of America, 108*, 19647–19652. http://dx.doi.org/10.1073/pnas.1110748108.

Poveda, K., Gómez Jiménez, M. I., Halitschke, R., & Kessler, A. (2012). Overcompensating plants: their expression of resistance traits and effects on herbivore preference and performance. *Entomologia Experimentalis et Applicata, 143*, 245–253. http://dx.doi.org/10.1111/j.1570-7458.2012.01256.x.

Poveda, K., Gómez Jiménez, M. I., & Kessler, A. (2010). The enemy as ally: herbivore-induced increase in crop yield. *Ecological Applications, 20*, 1787–1793. http://dx.doi.org/10.1890/09-1726.1.

Quiroz, A., Ortega, F., Ramirez, C. C., Wadhams, L. J., & Pinilla, K. (2005). Response of the beetle *Hylastinus obscurus* Marsham (Coleoptera: Scolytidae) to red clover (*Trifolium pratense* L.) volatiles in a laboratory olfactometer. *Environmental Entomology, 34*, 690–695.

de Rijk, M., Dicke, M., & Poelman, E. H. (2013). Foraging behaviour by parasitoids in multiherbivore communities. *Animal Behaviour, 85*, 1517–1528. http://dx.doi.org/10.1016/j.anbehav.2013.03.034.

Rubin, I. N., Ellner, S. P., Kessler, A., & Morrell, K. A. (2015). Informed herbivore movement and interplant communication determine the effects of induced resistance in an

individual-based model. *Journal of Animal Ecology, 84*, 1273–1285. http://dx.doi.org/10.1111/1365-2656.12369.
Saad, K. A., Roff, M. N. M., Hallett, R. H., & Idris, A. B. (2015). Aphid-induced defences in chilli affect preferences of the whitefly, *Bemisia tabaci* (Hemiptera: Aleyrodidae). *Scientific Reports, 5*, 13697. http://dx.doi.org/10.1038/srep13697.
Sharkey, T. D., & Singsaas, E. L. (1995). Why plants emit isoprene. *Nature, 374*, 769. http://dx.doi.org/10.1038/374769a0.
Song, G. C., & Ryu, C. (2013). Two volatile organic compounds trigger plant self-defense against a bacterial pathogen and a sucking insect in cucumber under open field conditions. *International Journal of Molecular Sciences, 14*, 9803–9819. http://dx.doi.org/10.3390/ijms14059803.
Staudt, M., & Lhoutellier, L. (2011). Monoterpene and sesquiterpene emissions from *Quercus coccifera* exhibit interacting responses to light and temperature. *Biogeosciences, 8*, 2757–2771. http://dx.doi.org/10.5194/bg-8-2757-2011.
Theis, N., Lerdau, M., & Raguso, R. A. (2007). The challenge of attracting pollinators while evading floral herbivores: patterns of fragrance emission in *Cirsium arvense* and *Cirsium repandum* (Asteraceae). *International Journal of Plant Sciences, 168*, 587–601. http://dx.doi.org/10.1086/513481.
Trewavas, A. (2014). *Plant behaviour and intelligence* (1st ed.). Oxford: Oxford university Press.
Turlings, T. C. J., Alborn, H. T., Loughrin, J. H., & Tumlinson, J. H. (2000). Volicitin, an elicitor of maize volatiles in oral secretion of *Spodoptera exigua*: isolation and bioactivity. *Journal of Chemical Ecology, 26*, 189–202. http://dx.doi.org/10.1023/A:1005449730052.
Turlings, T. C. J., & Benrey, B. (1998). Effects of plant metabolites on the behavior and development of parasitic wasps. *Ecoscience, 5*, 321–333.
Vet, L. E. M., & Dicke, M. (1992). Ecology of infochemical use by natural enemies in a tritrophic context. *Annual Review of Entomology, 37*, 141–172. http://dx.doi.org/10.1146/annurev.en.37.010192.001041.
Veyrat, N., Robert, C. A. M., Turlings, T. C. J., & Erb, M. (2016). Herbivore intoxication as a potential primary function of an inducible volatile plant signal. *Journal of Ecology, 104*, 591–600. http://dx.doi.org/10.1111/1365-2745.12526.
Wegener, R., Schulz, S., Meiners, T., Hadwich, K., & Hilker, M. (2001). Analysis of volatiles induced by oviposition of elm leaf beetle *Xanthogaleruca luteola* on *Ulmus minor*. *Journal of Chemical Ecology, 27*, 499–515. http://dx.doi.org/10.1023/A:1010397107740.
Wester, P., & Lunau, K. (2017). Plant–pollinator communication. In G. Becard (Ed.). *How plants communicate with their biotic environment* (Vol. 82, 225–257).
Zhu, F., Broekgaarden, C., Weldegergis, B. T., Harvey, J. A., Vosman, B., Dicke, M., & Poelman, E. H. (2015). Parasitism overrides herbivore identity allowing hyperparasitoids to locate their parasitoid host using herbivore-induced plant volatiles. *Molecular Ecology, 24*, 2886–2899. http://dx.doi.org/10.1111/mec.13164.

CHAPTER TWELVE

Communication of Sedentary Plant-Parasitic Nematodes With Their Host Plants

T. Hewezi*[,1], T.J. Baum[§]
*University of Tennessee, Knoxville, TN, United States
[§]Iowa State University, Ames, IA, United States
[1]Corresponding author: E-mail: thewezi@utk.edu

Contents

Abstract

Parasitism by sedentary plant-parasitic cyst and root-knot nematodes exhibits intriguing and fascinating examples of plant modifications induced by animals. This is reflected by the ability of these parasites to locate host plants, to trigger plant responses when approaching host roots and to induce the redifferentiation of plant root cells into specialized feeding structures. This chapter reviews current knowledge of the many layers of communication and signal exchange between plant-parasitic nematodes and host plants. In particular, it highlights host- and nematode-derived chemical signals and bioactive compounds that may facilitate plant–nematode recognition. In addition, this chapter emphasizes how this recognition results in the formation of nematode feeding sites and describes the underlying molecular mechanisms, which involve both nematode secretions and host factors.

1. INTRODUCTION

Nematodes are roundworms that inhabit virtually every ecological niche on Earth. It therefore comes as no surprise that the phylum Nematoda also contains independent lineages that harbour plant-parasitic nematodes.

Advances in Botanical Research, Volume 82
ISSN 0065-2296
http://dx.doi.org/10.1016/bs.abr.2016.11.004

Among all plant-parasitic nematodes, the so-called sedentary parasites are considered to be most damaging worldwide. In addition to their economic importance, sedentary nematodes also exhibit most intriguing examples of plant modifications induced by animals. These modifications are the direct result of the nematodes' sedentary life styles. In this context, the term sedentary refers to the fact that, while these nematodes are migratory during infection and penetration phases, they feed only after they have committed to a single feeding site and have lost their ability to move to different sites within their host plant, i.e., they have become sedentary. While this sedentary life style undoubtedly provides evolutionary advantages resulting in profound parasitic success and increased reproduction, it also requires impressive host–parasite interactions to govern an intricate network of mutual manipulations. In short, sedentary plant-parasitic nematodes are completely reliant on their ability to induce the redifferentiation of plant cells into specialized feeding cells that are not found elsewhere in the plant. In other words, these nematodes induce the formation of completely novel plant organs from which they are deriving their nourishment. This most intriguing biological feat is accomplished by many layers of communication and signal exchange.

2. PLANT–NEMATODE RECOGNITION

One critical aspect of the interaction between host plants and parasitic nematodes is for hatched second-stage juveniles (J2) to find their host plants. This is because unlike the free-living nematode *Caenorhabditis elegans*, plant-parasitic nematodes cannot enter into a survival state known as *dauer* stage, and to survive they must find and recognize their host plants in a very short period of time. Efforts in the last several years indicated that chemotaxis, a movement towards a gradient of increasing concentration of chemical signals or bioactive compounds, may guide the nematode to distant hosts (Curtis, Robinson, & Perry, 2009; Perry, 2005; Reynolds et al., 2011). Host-derived chemical signals and bioactive compounds can function as long- and short-distance attractants to direct the nematode into the host plant's root zone. Also, these so-called semiochemicals can act as local attractants enabling nematodes to invade roots at a specific area. For example, it has been suggested that a fine gradient of carbon dioxide in the rhizosphere may attract parasitic nematodes to host plants (Robinson, 1995). In addition, root-derived volatile compounds were found to attract potato cyst nematodes (Farnier et al., 2012). Combinations of volatile, nonvolatile and chemotactic compounds originating from host roots may constitute long- and short-distance attractants, which would guide the J2s to host roots. In another study, root tip and border cell exudates were found to modulate the attraction of the root-knot nematode *Meloidogyne incognita* to host plants in species- and cultivar-dependent manners (Zhao, Schmitt, & Hawes, 2000). In an elegant study, the movement patterns of two root-knot nematode species, *M. incognita* and *Meloidogyne graminicola*, were studied using Y-chamber olfactometers, which provide two-route options for the nematodes to find their hosts (Reynolds et al., 2011). Both nematode species were attracted to their preferred hosts by taking the short route, whereas they took the long route towards the nonpreferred hosts. In addition to these examples of plants emitting signals to nematodes, long-distance communication appears to also flow from nematode to plant. For example, *M. incognita* J2s were able to trigger morphological changes in host plant border cells that resembled those induced by exogenous application of phytohormones (Zhao et al., 2000). In control experiments, the free-living nematode *C. elegans* did not elicit any noticeable changes in border cells in cocultivation assays, suggesting that such chemical nematode signals are unique to parasitic nematodes and are responsible for plant–nematode recognition.

While such compounds released from parasitic nematode have not yet been identified, experimental evidence suggested that root-knot nematodes may produce molecules functionally equivalent to the *Rhizobium* signalling nodulation factors (Nod factors) (Weerasinghe, Bird, & Allen, 2005). In this study, root-knot nematodes were shown to induce cytoskeleton rearrangement and nuclear relocations leading to root-hair deformation, a phenotype mimicking what was triggered by Nod factors in legume root hairs. This response was eliminated in the NF receptor mutants *nrf1*, *nrf5* and *symRK*, suggesting that these receptors are involved in the perception of nematode-derived signalling molecules that are functionally resembling Nod factors. The wavy root-hair phenotype observed in *Lotus japonicus* was also induced in tomato, but not in *Arabidopsis* (Weerasinghe et al., 2005), implying that the response to root-knot nematodes is not a specific characteristic of legumes.

Taken together, it is tempting to speculate that the chemical properties of root-derived chemotactic compounds and nematode responsiveness as well as nematode-generated chemicals mediate plant–nematode recognition before direct contact. In this context, parasitic nematodes with a broad host range such as *M. incognita* may respond to general semiochemicals, while cyst nematodes, which have narrower host ranges, may respond to or produce more specific chemicals. Identification and isolation of such host-derived compounds that mediate nematode attraction to host plants will provide opportunities for genetic manipulation to reduce nematode attraction and infectivity.

3. NEMATODE INVASION OF HOST PLANTS AND FORMATION OF FEEDING STRUCTURES

Root invasion: Cyst and root-knot nematodes differ significantly in their modes of penetration and migration inside plant roots. Once nematodes have reached the root systems, the J2s of cyst nematodes penetrate roots specifically in the elongation, lateral root and root-hair formation zones, and then migrate into the vascular cylinder. Cyst nematodes migrate intracellularly with the aid of the stylet and a collection of secreted cell-wall degrading enzymes produced in the nematode subventral gland cells. This type of migration significantly damages root cells causing cell necrosis that marks the nematodes' migration paths (Hussey & Grundler, 1998). Infective J2s of root-knot nematodes, on the other hand, once they have reached the roots, aggregate at the surface of the elongation zone, a phenomenon that suggests communication may occur between the juveniles prior to root invasion (Bird et al., 2009). Alternatively, the presence of local root attractants may be responsible for this aggregation, but no such attractants were reported so far. As has been observed in *Arabidopsis*, the J2s of root-knot nematodes migrate intercellularly towards the root tips using both mechanical force and enzymatic stylet secretions. Once they reach the meristematic region, these nematodes reversed their direction of travel and migrated upwards within the vascular tissues. As soon as this migration path was established, several J2 nematodes used the same track of vascular migration (Wyss, Grundler, & Munch, 1992). This less-destructive intercellular movement may be a tactic employed by the nematodes to escape host recognition and activation of host defences.

Syncytium formation: In the vascular tissues, cyst and root-knot nematodes induce structurally different feeding sites. Cyst nematodes induce the formation of a feeding site termed 'syncytium' by targeting single cambial or procambial cells as the initial feeding cell (IFC). If the selected cells trigger rapid defence responses such as callose deposition and cytoplasm collapse, the infective J2s retract their stylets and target other cells. If the targeted cell does not degenerate and its responses are recognized by the nematode, the IFC will develop into a young syncytium, which is mediated and accompanied by notable cellular modifications. These modifications include nuclear polyploidization resulting from continuous DNA replication, fragmentation of large vacuoles and proliferation of organelles (Fig. 1) (Golinowski, Grundler, & Sobczak, 1996; Wyss, 1992). Changes in cell-wall structure and an increase in cytoplasm density coupled with an enrichment of ribosomes, plastids, mitochondria and endoplasmic reticulum have also been observed (Fig. 1). These changes are also triggered in hundreds of neighbouring cells, which progressively fuse with each other to form one large, fully developed, multinucleated syncytium as a permanent feeding structure. During this initial syncytium formation period, the J2s become sedentary, i.e., completely committed to a single location. After a few hours of a feeding preparation period, the sedentary J2 nematodes start repeated feeding periods from their syncytia for an extended time and moult into third-stage juvenile (J3), fourth-stage juvenile (J4) and finally adult male or female nematodes. While female nematodes withdraw food from the syncytium during all stages of development, male nematodes feed from the syncytia only during J2 and J3 stages. It is estimated that a female nematode

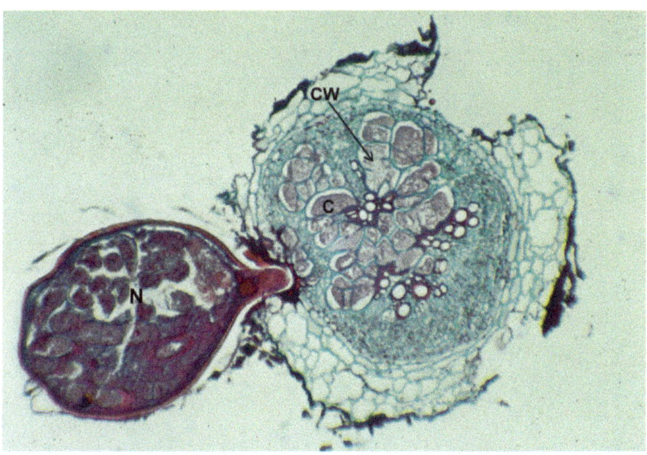

Figure 1 Cross section of a fully developed syncytium formed by a cyst nematode. Syncytium cells, formed adjacent to nematode (N) head, are marked by dense cytoplasm (C) and disrupted cell walls (CW). *Used with permission by Dr. Richard Hussey, University of Georgia.*

withdraws about 30 times more food than a male (Müller, Rehbock, & Wyss, 1981), a finding that is consistent with the observation that syncytia associated with females are frequently much larger than those of males.

Giant-cell formation: Within 24 h of infection, the J2s of root-knot nematodes induce at a single feeding site the formation of between four and nine so-called giant-cells, each originating from a separate precursor cell in the parenchyma (Williamson & Hussey, 1996). Concomitant to giant-cell induction, the surrounding cortical cells proliferate quickly causing root swelling and the formation of gall-like organs. In many cases, a single gall can contain a number of feeding sites and developing nematodes. Root galling is the obvious characteristic of root-knot nematode—plant interactions that led to the name root-knot nematode. As the term giant-cells suggests, these feeding cells dramatically expand to an at least 100 times larger size than neighbouring parenchyma cells and undergo severe morphological and metabolic changes to meet the new function as a nutrient sink for the developing nematodes (Fig. 2). The reprogramming of infected root cells into giant-cells involves significant increases in cytoplasm density and protein content (Fig. 2). Also, substantial increases in cell-wall thickening, numbers of organelles and DNA content are evident. In contrast to the endoreduplication observed in nuclei of syncytia, giant-cell nuclei undergo karyokinesis without cytokinesis, i.e., while nuclei divide, the harbouring cells do not. As a consequence,

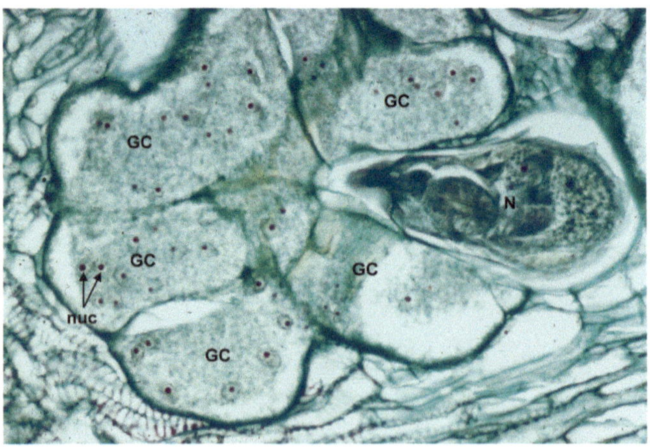

Figure 2 Cross section of fully developed giant-cells formed by a root-knot nematode. The individual giant-cells (GC) are multinucleated (nuc) and formed adjacent to the nematode (N) head. *Used with permission by the Society of Nematologists.*

giant-cells are highly multinuclear (Escobar, Barcala, Cabrera, & Fenoll, 2015; Williamson & Hussey, 1996). Inside the gall, the now sedentary and immobile nematodes feed from their giant-cells in an alternative fashion and continue their development into males or females.

Feeding tube formation: During feeding, both cyst and root-knot nematodes induce the formation of feeding tubes inside the cytoplasm of their respective feeding cells. These tubes appear to be vital structures that mediate nutrient uptake from the feeding sites (Wyss et al., 1992). The feeding tube connects on one end to the stylet orifice and extends into the cytoplasm of the syncytium or giant-cell. Cytological research using light and electron microscopy indicated that feeding tubes induced by various nematode species are structurally different (Berg, Fester, & Taylor, 2008; Hussey & Mims, 1990, 1991; Sobczak, Golinowski, & Grundler, 1999), despite their apparent common function. However, feeding tubes induced by the same nematode species in different host plants are structurally similar (Hussey & Mims, 1991), suggesting that they may be of nematode origin, most likely as a result of nematode secretory activity. Although the structural components of the feeding tube are currently unknown, it is convincing that nematode secretions stimulate the formation of the feeding tube whether it is composed of plant proteins, nematode secretions or both. The feeding tubes may function as molecular sieves to differentiate host cell components and control the size of assimilates passing through the stylet (Escobar et al., 2015; Eves-van den Akker et al., 2014). Several studies aiming at determining the size exclusion limit of feeding tubes formed by cyst and root-knot nematodes using fluorescent proteins yielded conflicting results (Bockenhoff & Grundler, 1994; Goverse et al., 1998; Urwin, Moller, Lilley, McPherson, & Atkinson, 1997; Valentine et al., 2007). However, it is largely acknowledged that the size exclusion limit of feeding tubes formed by root-knot nematodes is larger than of those formed by cyst nematodes (Lilley, Bakhetia, Charlton, & Urwin, 2007; Urwin et al., 1997). This may be attributed to an apparent uniform pore size of the root-knot nematode feeding tubes versus the irregular pore size of the cyst nematodes feeding tubes (Eves-van den Akker et al., 2014; Hussey & Mims, 1991).

4. NEMATODE ADAPTATIONS TO PARASITISM

Plant-parasitic nematode evolution has resulted in the development of a number of adaptations for parasitism of host plants. These adaptations include the development of a stylet and large oesophageal secretory gland cells (Davis, Hussey, & Baum, 2004; Hussey, 1989). The stylet is a hollow, needle-like structure in the nematode head with several functions. It provides the mechanical force to breach the plant cell walls as it can be actively and forcefully protruded by the contraction of nematode muscles. Furthermore, due to its hollow centre, the stylet is used as a feeding structure enabling the nematode to uptake assimilates and nutrients from the feeding sites. On the other hand, the stylet allows the nematode to accurately deliver secretions into the plant apoplast as well as directly into feeding cell cytoplasm. These secretions usually are referred to as effector proteins and are produced in the oesophageal secretory gland cells, which are believed to have enlarged significantly during the evolution of plant-parasitic nematodes from microbe-feeding nematodes (Davis et al., 2000; Hussey, 1989). These gland cells are comprised of a single dorsal cell and two subventral cells, which are connected to the oesophagus through valves. The size and morphology of the dorsal and subventral gland cells differ among nematode species and even among parasitic stages within the same species (Hussey, Davis, & Baum, 2002; Hussey & Mims, 1990). The two subventral gland cells are specifically active in the infective J2 during migration into and through host plant tissues and apparently become inactive after nematodes have become sedentary (Hewezi & Baum, 2013; Hussey & Mims, 1990). In contrast, the dorsal gland cell then becomes predominantly active in the sedentary parasitic stages, when nematodes maintain their feeding sites and excessively feed to develop into adult males or females (Hussey & Mims, 1990).

5. FUNCTIONS OF NEMATODE EFFECTOR PROTEINS

Nematode effectors, as the main determinants of nematode pathogenicity, share common criteria. These criteria include exclusive expression in the nematode oesophageal gland cells during infective stages, the existence of an N-terminal secretion signal peptide in the effector protein sequence and the absence of any transmembrane motifs. These criteria along with the technical advances in the isolation of gland cells and sequencing technology resulted in the identification of significant numbers of putative nematode effector candidates in various phytonematode species (Haegeman, Mantelin, Jones, & Gheysen, 2012; Hewezi, 2015; Hewezi & Baum, 2013; Maier, Hewezi, Peng, & Baum, 2013; Noon et al., 2015;

Rutter et al., 2014). Additionally, the sequencing of the genomes of the root-knot nematodes *M. incognita* (Abad et al., 2008) and *Meloidogyne hapla* (Opperman et al., 2008) and the potato cyst nematodes *Globodera pallida* (Cotton et al., 2014) and *Globodera rostochiensis* (Eves-van den Akker et al., 2016) provided unprecedented insights into the composition and diversity of nematode effector repertoires. Nevertheless, only a very limited number of these effectors has been functionally characterized.

Cell wall modification: One striking feature of the nematode effector repertoires is the presence of high numbers of cell-wall modifying enzymes that can degrade different structural polysaccharide components of the plant cell wall (Bohlmann & Sobczak, 2014; Hewezi, 2015; Wieczorek, 2015), a finding that is consistent with the ability of parasitic nematode to penetrate and migrate through plant cells and tissues. For example, secreted cellulases (endo-β 1,4-glucanases) function in degrading cellulose by hydrolysing the β (1,4) bonds. The majority of these cellulases belong to the glycosyl hydrolase family 5 (GHF5), which are believed to have been acquired from bacterial plant pathogens through horizontal gene transfer (Danchin et al., 2010; Haegeman, Jones, & Danchin, 2011). In addition to cellulases, cyst and root-knot nematodes secrete pectate lyases (Davis, Haegeman, & Kikuchi, 2011; Wieczorek, 2015). Pectate lyases catalyse cleavage of the internal α-1,4-linkages of pectate by β-elimination, a critical step in pectin degradation. Effector genes encoding other cell wall–degrading enzymes, including expansin, polygalacturonase, xylanase, arabinogalactan endo-β-1,4-galactosidase and arabinase, have recently been identified in the genome of cyst and root-knot nematodes (Abad et al., 2008; Cotton et al., 2014; Eves-van den Akker et al., 2016; Opperman et al., 2008).

While the synthesis of these extracellular effectors in the subventral gland cells suggested functions restricted to early stages of infection (i.e., penetration and migration), characterization studies of a cellulose-binding protein (CBP) from the cyst nematode *Heterodera schachtii*, indicated a function during the sedentary stage of syncytium initiation and formation (Hewezi Hussey et al., 2008). Overexpression of *CBP* with or without the signal peptide sequences proved to be functional *in planta*, increasing root length and enhancing plant susceptibility to *H. schachtii* (Hewezi Hussey et al., 2008). Detailed functional assays indicated that these effects are mediated through the interaction of CBP with plant pectin methylesterase protein3 (PME3). This direct interaction resulted in a significant increase in the activity of PME3, which is most likely required to facilitate syncytium formation and development via pectin demethylesterification (Hewezi Hussey et al., 2008).

Suppression of host defence responses: Sedentary endoparasitic nematodes spend an extended period of time in direct contact with host plants. Thus, suppression and/or avoidance of host defence responses is critical for successful parasitism. For example, expression of the cyst nematode *10A06* effector in *Arabidopsis* was found to inhibit salicylic acid–induced defence signalling and to activate the plant antioxidant machinery to protect the nematode from reactive oxygen species (ROS) generated by the host during infection. These functions are most likely mediated through changes in polyamine-signalling as a result of the 10A06 interaction with the *Arabidopsis* host plant spermidine synthase 2, an essential enzyme for polyamine biosynthesis (Hewezi et al., 2010). Another cyst nematode effector that inhibits host defences through direct interaction with host proteins is the 30C02 protein. 30C02 physically associates with β-1,3-endoglucanase, a pathogenesis-related (PR) protein of the PR-2 group (Hamamouch et al., 2012). Analysis of β-1,3-endoglucanase overexpression and T-DNA mutant lines in *Arabidopsis* confirmed the function of this enzymes as a positive regulator of host defence responses, which negatively impacts nematode parasitism of *Arabidopsis* plants (Hamamouch et al., 2012). Recently, a plasmodium-like effector protein from the soybean cyst nematode (SCN, *H. glycines*) (HgGLAND18) was identified and found to inhibit both basal immune responses and hypersensitive cell death (Noon et al., 2015, 2016). Transgenic soybean hairy roots expressing an RNAi construct of HgGLAND18 were generated in both SCN-susceptible and SCN-resistant cultivars and then inoculated with avirulent and virulent isolates of SCN (Noon et al., 2016). Interestingly, the effector mRNA abundance was decreased and nematode susceptibility was significantly reduced in all cases, suggesting the HgGLAND18 functions as a general suppressor of the immune responses in the susceptible interactions.

The secreted *M. incognita* calreticulin (Mi-CRT) effector was also found to suppress host innate immune responses (Jaouannet et al., 2013). However, the underlying mechanism remains to be determined. Very recently, the subventral glad effector MjTTL5 from the root-knot nematode *Meloidogyne javanica* was found to interact with the catalytic subunit of *Arabidopsis* ferredoxin-thioredoxin reductase, an important enzyme of the plant antioxidant machinery (Lin et al., 2016). This interaction seems to augment plant ROS scavenging activity, a process that may result in suppression of plant basal defence responses to promote nematode parasitism. In addition, a secreted translationally controlled tumour protein (TCTP) from *Meloidogyne enterolobii* was found to subvert programmed cell death induced by the mammalian proapoptotic BAX protein in agroinfiltration assays (Zhuo et al., 2016). Similarly, a number of cyst and root-knot nematodes have been shown to suppress cell death mediated by nucleotide-binding leucine-rich repeat (NB-LRR) resistance proteins in agroinfiltration assays (reviewed in Goverse & Smant, 2014; Hewezi, 2015; Hewezi & Baum, 2013). Taken together, suppression of host defences clearly is a common mechanism of various parasitic nematodes to promote infection.

Alteration of Auxin signalling: Molecular and genetic data indicate that auxin signalling is involved in the early stage of differentiating infected root cells into syncytia (Absmanner, Stadler, & Hammes, 2013; Goverse et al., 2000; Grunewald, Cannoot, Friml, & Gheysen, 2009). A recent study also pointed to a role of auxin signalling in shaping the maximum size and sustaining the functional phenotype of fully developed syncytia (Hewezi, Piya, Richard, & Rice, 2014). Two functionally characterized effector proteins from the cyst nematode *H. schachtii* (19C07 and 10A07) have provided novel insights into the mechanisms underlying modulation of auxin signalling during infection. The 19C07 effector has been shown to physically associate with auxin influx transporter LAX3 both in yeast and in planta (Lee et al., 2011). This interaction may be of biological significance because constitutive overexpression of 19C07 in *Arabidopsis* accelerated lateral root emergence, a phenotype related to the LAX3 function of increasing auxin influx and upregulating cell wall–modifying genes (Péret et al., 2013; Swarup et al., 2008). Thus, 19C07 may function in increasing the activity of LAX3 in developing syncytia and surrounding root cells to facilitate the development of mature syncytia. Recently, we found that the 10A07 effector protein interacts specifically with an *Arabidopsis* serine/threonine protein kinase in the cytoplasm. This plant kinase phosphorylates 10A07 and, hence facilitates its translocation to the plant nucleus (Hewezi et al., 2015). In the nucleus, 10A07 targets the indole-3-acetic acid inducible16 (IAA16) transcription factor. IAA16 belongs to the 29 AUXIN/IAA family members that, in the absence of auxin, repress the activity of auxin response factors (ARFs) via direct interaction (Piya, Shrestha, Binder, Stewart, & Hewezi, 2014). Detailed functional analyses indicated that 10A07 precludes IAA16 from being able to suppress the activity of several ARFs, leading to upregulation of ARFs in the syncytium and subsequently triggering auxin-specific transcriptional responses required for syncytium development (Hewezi et al., 2015).

In contrast to the key roles that auxin plays in syncytium initiation and development, the importance of auxin signalling in giant-cell formation is still poorly understood, despite the fact that a number of gene expression studies using auxin-responsive reporters have suggested a role of auxin in giant-cell formation (Absmanner et al., 2013; Karczmarek, Overmars, Helder, & Goverse, 2004; Mazarei, Lennon, Puthoff, Rodermel, & Baum, 2003). Recently, Kyndt et al. (2016) documented that *M. incognita* affects auxin distribution patterns controlled by influx and efflux proteins during the formation of giant-cells in *Arabidopsis* roots. Analysis of mutant lines of these proteins further supported the involvement of AUX1, LAX3 and PIN4 in feeding site formation and expansion (Kyndt et al., 2016). The mechanism through which *M. incognita* modulates auxin distribution remains to be determined, particularly when considering that no homologue of the cyst nematode 19C07 effector was identified in *M. incognita*. Thus, it seems most likely that both cyst and root-knot nematodes employ different tactics to manipulate auxin distribution during infection. This hypothesis is consistent with the fact that both nematode species are evolutionarily divergent and induce feeding sites whose aetiology is different (Bert, Karssen, & Helder, 2011; Quist, Smant, & Helder, 2015).

6. microRNAs AND EPIGENETIC REGULATION

During the last 10 years, small noncoding RNAs of the microRNA (miRNA) class have emerged as new regulators of gene expression. The mature miRNA molecules, generally 21-nt, negatively regulate the expression of their targets, which contain perfect or near-perfect complementary sequences. Binding miRNA to the target gene typically results in mRNA degradation or translation inhibition, causing posttranscriptional silencing. The involvement of miRNA in different aspects of plant–nematode interactions is only recently being revealed. Isolation of small RNAs from root tissues post infection with cyst or root-knot nematodes resulted in the identification of a significant number of miRNAs that change expression in response to nematode infection (Cabrera et al., 2016; Hewezi, Howe, Maier, & Baum, 2008; Li et al., 2012; Xu et al., 2014; Zhao et al., 2015). Nevertheless, only very few of these miRNA genes have been functionally characterized. miR396 is the first miRNA gene shown to be involved in plant–nematode interactions (Hewezi, Maier, Nettleton, & Baum, 2012). miR396 and its growth regulation factor (GRF) 1 and 3 target genes exhibited two distinct phases of temporal expression patterns, which delineate the two developmental stages of syncytium formation versus syncytium maintenance. During the syncytium initiation and formation stage, miR396 is downregulated and *GRF1* and *3* are upregulated. Once this stage is completed, miR396 is again upregulated in the syncytium and posttranscriptionally silences *GRF1* and *3* (Hewezi et al., 2012). GRF1 and 3 regulate a significant number of syncytium-differentially expressed genes, specifically those involved in developmental processes and defence signalling (Hewezi et al., 2012; Liu, Rice, Chen, Baum, & Hewezi, 2014). miR396 and GRF1/3 are tightly connected through a feedback loop (Hewezi & Baum, 2012), and modulation of the equilibrium of this regulatory system negatively impacts syncytium size and nematode development (Hewezi et al., 2012).

Another recent study provided novel insights into the role of jasmonic acid (JA)–mediated regulation of miR319 and its target *TCP4* transcription factor in response to infection by *M. incognita* in tomato (Zhao et al., 2015). The opposite expression patterns of miR319 and its target gene observed both in leaves and roots in response to nematode infection suggested a role of miR319/TCP4 regulatory unit in modulating systemic defence responses. Overexpression of miR319 significantly enhanced plant susceptibly to nematode infection. This phenotype is most likely the result of reduced endogenous JA levels because of the downregulation of *TCP4*, which is involved in JA biosynthesis (Zhao et al., 2015). Also in this study, miR396 and GRF1, 2 and 4 were found to respond to *M. incognita* parasitism. Because *Arabidopsis* TCP4 has been shown to regulate the expression of miR396 as well as the activity of GRFs independently of miR396 (Rodriguez et al., 2010; Schommer et al., 2014), a possible interference between the miR319/TCP4 and miR396/GRFs regulatory modules has been suggested to occur in the roots during the course of *M. incognita* infection (Zhao et al., 2015).

Sequencing of small RNA libraries from galls formed by the root-rot nematode *M. javanica* in *Arabidopsis* identified miR390 as one of few upregulated miRNAs (Cabrera et al., 2016). miR390 controls the biogenesis of trans-acting short-interfering RNAs (tasiRNAs) generated from the primary transcripts of TAS3, which contain two sites complementary to miR390 mature sequences (Xia, Meyers, Liu, Beers, & Ye, 2013). TAS3–derived tasiRNAs target ARF2, 3 and 4 (Cho, Coruh, & Axtell, 2012; Marin et al., 2010). Promoter activity of miR390s, TAS3a and ARF3 showed similar expression patterns of upregulation both in the giant-cells and surrounding cells (Cabrera et al., 2016). In addition, mutant lines of miR390a and TAS3a showed significant reduction in the number of galls after *M. javanica* infection, suggesting a role of miR390a-TAS3a-ARF3 pathway in gall formation and successful nematode parasitism.

A role of miR827 in suppressing immune responses in the syncytium has been recently described (Hewezi et al., 2016). In *Arabidopsis*, miR827 is encoded by a single gene and negatively regulates the expression of the *nitrogen limitation adaptation* (*NLA*) target gene, which codes for an ubiquitin E3 ligase enzyme. Using reporter constructs designed to discriminate between transcriptional and posttranscriptional regulation of *NLA*, it has been shown that *NLA* is posttranscriptionally silenced by miR827 during all stages of syncytium formation and development. Agroinfiltration and PCR quantification of several pathogenesis-related genes indicated that NLA activates basal defences (Hewezi et al., 2016). Thus, maintaining a silenced state of *NLA* in the syncytium is crucial for successful infection. This conclusion was further supported by the finding that suppression of miR827 expression significantly reduced plant susceptibility to *H. schachtii*, whereas overexpression resulted in a hypersusceptible phenotype (Hewezi et al., 2016). Because miR827 has been shown to regulate plant responses to phosphate and nitrate deficiency (Kant, Peng, & Rothstein, 2011; Lin, Huang, & Chiou, 2013), a putative function in homoeostatic regulation of macronutrient contents in the feeding sites cannot be ruled out and deserves further analysis.

Nematode-induced genome-wide epigenetic modifications have recently emerged as a mechanism that may shape the interactions between plant-parasitic nematodes and host plants. In a recent study, it has been shown that SCN changes DNA methylation patterns genome-wide during the susceptible interaction with soybean (Rambani et al., 2015). Thousands of differentially methylated regions have been identified, and many of them are overlapping with protein-coding genes. The methylome changes induced by SCN are marked by increasing hypomethylation relative to hypermethylation specifically in the gene body in CG and CHG sequence contexts. The methylome changes were also found to occur predominantly in the recently duplicated genes relative to old copies. In addition, differential methylation patterns in gene body and promoter regions showed significant associations with gene expression changes, suggesting that DNA methylation changes may contribute to genome-wide transcriptome regulation during SCN infection. Interestingly, a significant number of differentially methylated genes overlapped with genes previously identified as differentially expressed in the soybean syncytium, specifically those involved in epigenetic regulation, transcriptional control, phytohormones, and carbohydrate degradation and glycolysis (Rambani et al., 2015). While these intriguing data indicate that DNA methylation contributes significantly to gene expression changes in the nematode feeding sites, the mechanisms through which plant-parasitic nematodes induce these epigenetic marks remain to be elucidated.

ACKNOWLEDGEMENTS

Some of the research described here was supported by funds from NSF (Award #: IOS-1145053), Tennessee Soybean Promotion Board, the United Soybean Board, the Iowa Soybean Association, USDA-AFRI, the University of Tennessee, Institute of Agriculture, and the Iowa Agriculture Experiment Station, Hatch Act and the State of Iowa.

REFERENCES

Abad, P., Gouzy, J., Aury, J. M., Castagnone-Sereno, P., Danchin, E. G. J., Deleury, E., ... Wincker, P. (2008). Genome sequence of the metazoan plant-parasitic nematode *Meloidogyne incognita. Nature Biotechnology, 26*, 909–915.

Absmanner, B., Stadler, R., & Hammes, U. Z. (2013). Phloem development in nematode-induced feeding sites: the implications of auxin and cytokinin. *Frontiers in Plant Science, 4*, 241.

Berg, R. H., Fester, T., & Taylor, C. G. (2008). Development of the root-knot nematode feeding cell. In R. H. Berg, & C. G. Taylor (Eds.), *Cell biology of plant nematode parasitism* (pp. 115–152). Berlin: Springer.

Bert, W., Karssen, G., & Helder, J. (2011). Phylogeny and evolution of nematodes. In J. Jones, G. Gheysen, & C. Fenoll (Eds.), *Genomics and molecular genetics of plant nematode interactions* (pp. 45–59). Dordrecht: Springer.

Bird, D., Opperman, C. H., & Williamson, V. M. (2009). Plant infection by root-knot nematode. In R. H. Berg, & C. G. Taylor (Eds.), *Cell biology of plant nematode parasitism* (pp. 1–14). Heidelberg: Springer.

Bockenhoff, A., & Grundler, F. M. W. (1994). Studies on the nutrient uptake by the beet cyst nematode *Heterodera schachtii* by in situ microinjection of fluorescent probes into the feeding structures in *Arabidopsis thaliana. Parasitology, 109*, 249–254.

Bohlmann, H., & Sobczak, M. (2014). The plant cell wall in the feeding sites of cyst nematodes. *Frontiers in Plant Science, 5*, 89.

Cabrera, J., Barcala, M., García, A., Rio-Machín, A., Medina, C., Jaubert-Possamai, S., ... Fenoll, C. (2016). Differentially expressed small RNAs in *Arabidopsis* galls formed by *Meloidogyne javanica*: a functional role for miR390 and its TAS3-derived tasiRNAs. *New Phytologist, 209*, 1625–1640.

Cho, S. H., Coruh, C., & Axtell, M. J. (2012). miR156 and miR390 regulate tasiRNA accumulation and developmental timing in *Physcomitrella patens. The Plant Cell, 24*, 4837–4849.

Cotton, J. A., Lilley, C. J., Jones, L. M., Kikuchi, T., Reid, A. J., Thorpe, P., ... Urwin, P. E. (2014). The genome and life-stage specific transcriptomes of *Globodera pallida* elucidate key aspects of plant parasitism by a cyst nematode. *Genome Biology, 15*, R43.

Curtis, R. H. C., Robinson, A. F., & Perry, R. N. (2009). Hatch and host location. In R. N. Perry, M. Moens, & J. L. Starr (Eds.), *Root knot nematodes* (pp. 139–162). Wallingford: CABI Publishing.

Danchin, E. G., Rosso, M. N., Vieira, P., de Almeida-Engler, J., Coutinho, P. M., Henrissat, B., & Abad, P. (2010). Multiple lateral gene transfers and duplications have promoted plant parasitism ability in nematodes. *Proceedings of the National Academy of Sciences of the United States of America, 107*, 17651–17656.

Davis, E. L., Haegeman, A., & Kikuchi, T. (2011). Degradation of the plant cell wall by nematodes. In J. Jones, G. Gheysen, & C. Fenoll (Eds.), *Genomics and molecular genetics of plant-nematode interactions* (pp. 255–272). Dordrecht: Springer.

Davis, E. L., Hussey, R. S., & Baum, T. J. (2004). Getting to the roots of parasitism by nematodes. *Trends in Parasitology, 20*, 134–141.

Davis, E. L., Hussey, R. S., Baum, T. J., Bakker, J., Schots, A., Rosso, M. N., & Abad, P. (2000). Nematode parasitism genes. *Annual Review of Phytopathology, 38*, 341–372.

Escobar, C., Barcala, M., Cabrera, J., & Fenoll, C. (2015). Overview of root-knot nematodes and giant cells. In C. Escobar, & C. Fenoll (Eds.), *Plant nematode interactions — a view on compatible interrelationships: Vol. 73. Advances in botanical research* (pp. 1–32). Oxford, UK: Elsevier.

Eves-van den Akker, S., Laetsch, D. R., Thorpe, P., Lilley, C. J., Danchin, E. G., Da Rocha, M., … Jones, J. T. (2016). The genome of the yellow potato cyst nematode, *Globodera rostochiensis*, reveals insights into the basis of parasitism and virulence. *Genome Biology, 17*, 124.

Eves-van den Akker, S., Lilley, C. J., Ault, J. R., Ashcroft, A. E., Jones, J. T., & Urwin, P. E. (2014). The feeding tube of cyst nematodes: characterization of protein exclusion. *PLoS One, 9*, e87289.

Farnier, K., Bengtsson, M., Becher, P. G., Witzell, J., Witzgall, P., & Manduric, S. (2012). Novel bioassay demonstrates attraction of the white potato cyst nematode *Globodera pallida* (Stone) to non-volatile and volatile host plant cues. *Journal of Chemical Ecology, 38*, 795–801.

Golinowski, W., Grundler, F. M. W., & Sobczak, M. (1996). Changes in the structure of *Arabidopsis thaliana* during female development of the plant-parasitic nematode *Heterodera schachtii*. *Protoplasma, 194*, 103–116.

Goverse, A., Biesheuvel, J., Wijers, G. J., Gommers, F. J., Bakker, J., Schots, A., & Helder, J. (1998). In planta monitoring of the activity of two constitutive promoters, CaMV 35S and TR20, in developing feeding cells induced by *Globodera rostochiensis* using green fluorescent protein in combination with confocal laser scanning microscopy. *Physiological and Molecular Plant Pathology, 52*, 275–284.

Goverse, A., Overmars, H., Engelbertink, J., Schots, A., Bakker, J., & Helder, J. (2000). Both induction and morphogenesis of cyst nematode feeding cells are mediated by auxin. *Molecular Plant-Microbe Interactions, 13*, 1121–1129.

Goverse, A., & Smant, G. (2014). The activation and suppression of plant innate immunity by parasitic nematodes. *Annual Review of Phytopathology, 52*, 243–265.

Grunewald, W., Cannoot, B., Friml, J., & Gheysen, G. (2009). Parasitic nematodes modulate PIN-mediated auxin transport to facilitate infection. *PLoS Pathogens, 5*, e1000266.

Haegeman, A., Jones, J. T., & Danchin, E. G. (2011). Horizontal gene transfer in nematodes: a catalyst for plant parasitism? *Molecular Plant-Microbe Interactions, 24*, 879–887.

Haegeman, A., Mantelin, S., Jones, J. T., & Gheysen, G. (2012). Functional roles of effectors of plant-parasitic nematodes. *Gene, 492*, 19–31.

Hamamouch, N., Li, C., Hewezi, T., Baum, T. J., Mitchum, M. G., Hussey, R. S., … Davis, E. L. (2012). The interaction of the novel 30C02 cyst nematode effector protein with a plant β-1,3-endoglucanase may suppress host defense to promote parasitism. *Journal of Experimental Botany, 63*, 3683–3695.

Hewezi, T. (2015). Cellular Signaling Pathways and posttranslational modifications mediated by nematode effector proteins. *Plant Physiology, 169*, 1018–1026.

Hewezi, T., & Baum, T. J. (2012). Complex feedback regulations govern the expression of miRNA396 and its GRF target genes. *Plant Signaling & Behavior, 7*, 749–751.

Hewezi, T., & Baum, T. J. (2013). Manipulation of plant cells by cyst and root-knot nematode effectors. *Molecular Plant-Microbe Interactions, 26*, 9–16.

Hewezi, T., Howe, P., Maier, T. R., & Baum, T. J. (2008). *Arabidopsis* small RNAs and their targets during cyst nematode parasitism. *Molecular Plant-Microbe Interactions, 21*, 1622–1634.

Hewezi, T., Howe, P., Maier, T. R., Hussey, R. S., Mitchum, M. G., Davis, E. L., & Baum, T. J. (2008). Cellulose binding protein from the parasitic nematode *Heterodera schachtii* interacts with *Arabidopsis* pectin methylesterase: cooperative cell wall modification during parasitism. *The Plant Cell, 20*, 3080–3093.

Hewezi, T., Howe, P. J., Maier, T. R., Hussey, R. S., Mitchum, M. G., Davis, E. L., & Baum, T. J. (2010). *Arabidopsis* spermidine synthase is targeted by an effector protein of the cyst nematode *Heterodera schachtii*. *Plant Physiology, 152*, 968–984.

Hewezi, T., Juvale, P. S., Piya, S., Maier, T. R., Rambani, A., Rice, J. H., … Baum, T. J. (2015). The cyst nematode effector protein 10A07 targets and recruits host posttranslational machinery to mediate its nuclear trafficking and to promote parasitism in *Arabidopsis*. *The Plant Cell, 27*, 891–907.

Hewezi, T., Maier, T. R., Nettleton, D., & Baum, T. J. (2012). The *Arabidopsis* microRNA396-GRF1/GRF3 regulatory module acts as a developmental regulator in the reprogramming of root cells during cyst nematode infection. *Plant Physiology, 159*, 321–335.

Hewezi, T., Piya, S., Qi, M., Balasubramaniam, M., Rice, J. H., & Baum, T. J. (2016). *Arabidopsis* miR827 mediates post-transcriptional gene silencing of its ubiquitin E3 ligase target gene in the syncytium of the cyst nematode *Heterodera schachtii* to enhance susceptibility. *Plant Journal, 88*, 179–192. http://dx.doi.org/10.1111/tpj.13238.

Hewezi, T., Piya, S., Richard, G., & Rice, J. H. (2014). Spatial and temporal expression patterns of auxin response transcription factors in the syncytium induced by the beet cyst nematode *Heterodera schachtii* in *Arabidopsis*. *Molecular Plant Pathology, 15*, 730–736.

Hussey, R. S. (1989). Disease-inducing secretions of plant parasitic nematodes. *Annual Review of Phytopathology, 27*, 123–141.

Hussey, R. S., Davis, E. L., & Baum, T. J. (2002). Secrets in secretions: genes that control nematode parasitism of plants. *Brazilian Journal of Plant Physiology, 14*, 183–194.

Hussey, R. S., & Grundler, F. M. W. (1998). Nematode parasitism of plants. In R. N. Perry, & D. J. Wright (Eds.), *Physiology and biochemistry of free-living and plant parasitic nematodes* (pp. 213–243). England: CAB International Press.

Hussey, R. S., & Mims, C. W. (1990). Ultrastructure of esophageal glands and their secretory granules in the root-knot nematode *Meloidogyne incognita*. *Protoplasma, 156*, 9–18.

Hussey, R. S., & Mims, C. W. (1991). Ultrastructure of feeding tubes formed in giant-cells induced in plants by the root-knot nematode *Meloidogyne incognita*. *Protoplasma, 162*, 99–107.

Jaouannet, M., Magliano, M., Arguel, M. J., Gourgues, M., Evangelisti, E., Abad, P., & Rosso, M. N. (2013). The root-knot nematode calreticulin Mi-CRT is a key effector in plant defense suppression. *Molecular Plant-Microbe Interaction, 26*, 97–105.

Kant, S., Peng, M., & Rothstein, S. J. (2011). Genetic regulation by NLA and microRNA827 for maintaining nitrate-dependent phosphate homeostasis in *Arabidopsis*. *PLoS Genetics*, 7, 1002021.

Karczmarek, A., Overmars, H., Helder, J., & Goverse, A. (2004). Feeding cell development by cyst and root-knot nematodes involves a similar early, local and transient activation of a specific auxin-inducible promoter element. *Molecular Plant Pathology, 5*, 343–346.

Kyndt, T., Goverse, A., Haegeman, A., Warmerdam, S., Wanjau, C., Jahani, M., … Gheysen, G. (2016). Redirection of auxin flow in *Arabidopsis thaliana* roots after infection by root-knot nematodes. *Journal of Experimental Botany*. http://dx.doi.org/10.1093/jxb/erw230.

Lee, C., Chronis, D., Kenning, C., Peret, B., Hewezi, T., Davis, E. L., ... Mitchum, M. G. (2011). The novel cyst nematode effector protein 19C07 interacts with the *Arabidopsis* auxin influx transporter LAX3 to control feeding site development. *Plant Physiology, 155*, 866–880.

Lilley, C. J., Bakhetia, M., Charlton, W. L., & Urwin, P. E. (2007). Recent progress in the development of RNA interference for plant parasitic nematodes. *Molecular Plant Pathology, 8*, 701–711.

Lin, W. Y., Huang, T. K., & Chiou, T. J. (2013). NITROGEN LIMITATION ADAPTATION, a target of microRNA827, mediates degradation of plasma membrane–localized phosphate transporters to maintain phosphate homeostasis in *Arabidopsis*. *The Plant Cell, 25*, 4061–4074.

Lin, B., Zhuo, K., Chen, S., Hu, L., Sun, L., Wang, X., ... Liao, J. (2016). A novel nematode effector suppresses plant immunity by activating host reactive oxygen species-scavenging system. *New Phytologist, 209*, 1159–1173.

Liu, J., Rice, J. H., Chen, N., Baum, T. J., & Hewezi, T. (2014). Synchronization of developmental processes and defense signaling by growth regulating transcription factors. *PLoS One*, e98477.

Li, X., Wang, X., Zhang, S., Liu, D., Duan, Y., & Dong, W. (2012). Identification of soybean microRNAs involved in soybean cyst nematode infection by deep sequencing. *PLoS One, 7*, e39650.

Maier, T. R., Hewezi, T., Peng, J., & Baum, T. J. (2013). Isolation of whole esophageal gland cells from plant-parasitic nematodes for transcriptome analyses and effector identification. *Molecular Plant-Microbe Interactions, 26*, 31–35.

Marin, E., Jouannet, V., Herz, A., Lokerse, A. S., Weijers, D., Vaucheret, H., & Maizel, A. (2010). miR390, *Arabidopsis* TAS3 tasiRNAs, and their AUXIN RESPONSE FACTOR targets define an autoregulatory network quantitatively regulating lateral root growth. *The Plant Cell, 22*, 1104–1117.

Mazarei, M., Lennon, K. A., Puthoff, D. P., Rodermel, S. R., & Baum, T. J. (2003). Expression of an *Arabidopsis* phosphoglycerate mutase homologue is localized to apical meristems, regulated by hormones, and induced by sedentary plant-parasitic nematodes. *Plant Molecular Biology, 53*, 513–530.

Müller, J., Rehbock, K., & Wyss, U. (1981). Growth of *Heterodera schachtii* with remarks on amounts of food consumed. *Revue de Nematologie, 4*, 227–234.

Noon, J. B., Hewezi, T., Maier, T. R., Simmons, C., Wei, J. Z., Wu, G., ... Baum, T. J. (2015). Eighteen new candidate effectors of the phytonematode *Heterodera glycines* produced specifically in the secretory esophageal gland cells during parasitism. *Phytopathology, 105*, 1362–1372.

Noon, J. B., Qi, M., Sill, D. N., Muppirala, U., Eves-van den Akker, S., Maier, T. R., ... Baum, T. J. (2016). A Plasmodium-like virulence effector of the soybean cyst nematode suppresses plant innate immunity. *New Phytologist*. http://dx.doi.org/10.1111/nph.14047.

Opperman, C. H., Bird, D. M., Williamson, V. M., Rokhsar, D. S., Burke, M., Cohn, J., ... Windham, E. (2008). Sequence and genetic map of *Meloidogyne hapla*: a compact nematode genome for plant parasitism. *Proceedings of the National Academy of Sciences of the United States of America, 105*, 14802–14807.

Péret, B., Middleton, A. M., French, A. P., Larrieu, A., Bishopp, A., Njo, M., ... Bennett, M. J. (2013). Sequential induction of auxin efflux and influx carriers regulates lateral root emergence. *Molecular Systems Biology, 9*, 699.

Perry, R. N. (2005). An evaluation of types of attractants enabling plant-parasitic nematodes to locate plant roots. *Russian Journal of Nematology, 13*, 83–88.

Piya, S., Shrestha, S. K., Binder, B., Stewart, C. N., Jr., & Hewezi, T. (2014). Protein-protein interaction and gene co-expression maps of ARFs and Aux/IAAs in *Arabidopsis*. *Frontiers in Plant Science, 5*, 744.

Quist, C. W., Smant, G., & Helder, J. (2015). Evolution of plant parasitism in the phylum Nematoda. *Annual Review of Phytopathology, 53*, 289–310.

Rambani, A., Rice, J. H., Liu, J., Lane, T., Ranjan, P., Mazarei, M., ... Hewezi, T. (2015). The methylome of soybean roots during the compatible interaction with the soybean cyst nematode. *Plant Physiology, 168*, 1364–1377.

Reynolds, A. M., Dutta, T. K., Curtis, R. H. C., Powers, S. J., Gaur, H. S., & Kerry, B. R. (2011). Chemotaxis can take plant-parasitic nematodes to the source of a chemo-attractant via the shortest possible routes. *Journal of the Royal Society Interface, 8*, 568–577.

Robinson, A. F. (1995). Optimal release rates for attracting *Meloidogyne incognita, Rotylenchulus reniformis*, and other nematodes to carbon dioxide in sand. *Journal of Nematology, 27*, 42–50.

Rodriguez, R. E., Mecchia, M. A., Debernardi, J. M., Schommer, C., Weigel, D., & Palatnik, J. F. (2010). Control of cell proliferation in *Arabidopsis thaliana* by microRNA miR396. *Development, 137*, 103–112.

Rutter, W. B., Hewezi, T., Abubucker, S., Maier, T., Huang, G., Mitreva, M., ... Baum, T. J. (2014). Mining novel effector proteins from the esophageal gland cells of *Meloidogyne incognita*. *Molecular Plant-Microbe Interactions, 27*, 965–974.

Schommer, C., Juan, M., Debernardi, J. M., Bresso1, E. G., Rodriguez, R. E., & Palatnik, J. F. (2014). Repression of cell proliferation by miR319-regulated TCP4. *Molecular Plant, 10*, 1533–1544.

Sobczak, M., Golinowski, W., & Grundler, F. (1999). Ultrastructure of feeding plugs and feeding tubes formed by *Heterodera schachtii*. *Nematology, 1*, 363–374.

Swarup, K., Benková, E., Swarup, R., Casimiro, I., Péret, B., Yang, Y., & Bennett, M. J. (2008). The auxin influx carrier LAX3 promotes lateral root emergence. *Nature Cell Biology, 10*, 946–954.

Urwin, P. E., Moller, S. G., Lilley, C. J., McPherson, M. J., & Atkinson, H. J. (1997). Continual green-fluorescent protein monitoring of cauliflower mosaic virus 35S promoter activity in nematode-induced feeding cells in *Arabidopsis thaliana*. *Molecular Plant-Microbe Interactions, 10*, 394–400.

Valentine, T. A., Randall, E., Wypijewski, K., Chapman, S., Jones, J., & Oparka, K. J. (2007). Delivery of macromolecules to plant parasitic nematodes using a tobacco rattle virus vector. *Plant Biotechnology Journal, 5*, 827–834.

Weerasinghe, R. R., Bird, D. McK., & Allen, N. S. (2005). Root-knot nematodes and bacterial Nod factors elicit common signal transduction events in *Lotus japonicus* root hair cells. *Proceedings of the National Academy of Sciences of the United States of America, 102*, 3147–3152.

Wieczorek, K. (2015). Cell wall alterations in nematode-infected roots. In C. Escobar, & C. Fenoll (Eds.), *Plant nematode interactions: A view on compatible interrelationships: Vol. 73. Advances in botanical research* (pp. 61–90). New York: Academic Press.

Williamson, V. M., & Hussey, R. S. (1996). Nematode pathogenesis and resistance in plants. *The Plant Cell, 8*, 1735–1745.

Wyss, U. (1992). Observations on the feeding behaviour of *Heterodera schachtii* throughout development, including events during moulting. *Fundamental and Applied Nematology, 15*, 75–89.

Wyss, U., Grundler, F. M. W., & Munch, A. (1992). The parasitic behavior of second stage juveniles in *Meloidogyne incognita* in roots of *Arabidopsis thaliana*. *Nematologica, 38*, 98–111.

Xia, R., Meyers, B. C., Liu, Z., Beers, E. P., & Ye, S. (2013). MicroRNA superfamilies descended from miR390 and their roles in secondary small interfering RNA biogenesis in Eudicots. *The Plant Cell, 25*, 1555–1572.
Xu, M., Li, Y., Zhang, Q., Xu, T., Qiu, L., Fan, Y., & Wang, L. (2014). Novel miRNA and phasiRNA biogenesis networks in soybean roots from two sister lines that are resistant and susceptible to SCN race 4. *PLoS One, 9*, e110051.
Zhao, W., Li, Z., Fan, J., Hu, C., Yang, R., Qi, X., … Wang, S. (2015). Identification of jasmonic acid-associated microRNAs and characterization of the regulatory roles of the miR319/TCP4 module under root-knot nematode stress in tomato. *Journal of Experimental Botany, 66*, 4653–4667.
Zhao, X., Schmitt, M., & Hawes, M. C. (2000). Species-dependent effects of border cell and root tip exudates on nematode behavior. *Phytopathology, 90*, 1239–1245.
Zhuo, K., Chen, J., Lin, B., Wang, J., Sun, F., Hu, L., & Liao, J. (2016). A novel *Meloidogyne enterolobii* effector MeTCTP promotes parasitism by suppressing programmed cell death in host plants. *Molecular Plant Pathology*. http://dx.doi.org/10.1111/mpp.12374.

SUBJECT INDEX

'*Note*: Page numbers followed by "f" indicate figures, "t" indicate tables.'

E

F

K

L

M

O

P

T

U

V

AUTHOR INDEX

'*Note:* Page numbers followed by "f" indicate figures, "t" indicate tables.'

A

B

D

E

F

H

K

M

N

O

P

Q

R

S

T

U

V

W

X

Y

Z

Kolibri 360 GmbH
Ferdinand-Jühlke-Straße 7
99095 Erfurt, Deutschland
produktsicherheit@kolibri360.de